T0293819

Power Electronics: Devices, Circuits and Applications

Power Electronics: Devices, Circuits and Applications

Edited by Blair Cole

CLANRYE INTERNATIONAL
www.clanryeinternational.com

Clanrye International,
750 Third Avenue, 9th Floor,
New York, NY 10017, USA

Copyright © 2023 Clanrye International

This book contains information obtained from authentic and highly regarded sources. Copyright for all individual chapters remain with the respective authors as indicated. All chapters are published with permission under the Creative Commons Attribution License or equivalent. A wide variety of references are listed. Permission and sources are indicated; for detailed attributions, please refer to the permissions page and list of contributors. Reasonable efforts have been made to publish reliable data and information, but the authors, editors and publisher cannot assume any responsibility for the validity of all materials or the consequences of their use.

Trademark Notice: Registered trademark of products or corporate names are used only for explanation and identification without intent to infringe.

ISBN: 978-1-64726-672-1

Cataloging-in-Publication Data

Power electronics : devices, circuits and applications / edited by Blair Cole.
    p. cm.
Includes bibliographical references and index.
ISBN 978-1-64726-672-1
1. Power electronics. 2. Electronics. 3. Electric power. I. Cole, Blair.
TK7881.15 .F86 2023
621.317--dc23

For information on all Clanrye International publications
visit our website at www.clanryeinternational.com

Contents

Preface...VII

Chapter 1   **A Novel Zero Dead-Time PWM Method to Improve the Current Distortion of a Three-Level NPC Inverter**...1
Jin-Wook Kang, Seung-Wook Hyun, Yong Kan, Hoon Lee and Jung-Hyo Lee

Chapter 2   **A Fast and Accurate Maximum Power Point Tracking Approach Based on Neural Network Assisted Fractional Open-Circuit Voltage**.................................27
Ahmad Alzahrani

Chapter 3   **Design, Simulation and Hardware Implementation of Shunt Hybrid Compensator using Synchronous Rotating Reference Frame (SRRF)-Based Control Technique**..43
R. Balasubramanian, K. Parkavikathirvelu, R. Sankaran and Rengarajan Amirtharajan

Chapter 4   **Line Frequency Instability of One-Cycle-Controlled Boost Power Factor Correction Converter**...63
Rui Zhang, Wei Ma, Lei Wang, Min Hu, Longhan Cao, Hongjun Zhou and Yihui Zhang

Chapter 5   **Extended Kalman Filter Based Sliding Mode Control of Parallel-Connected Two Five-Phase PMSM Drive System**..75
Tounsi Kamel, Djahbar Abdelkader, Barkat Said, Sanjeevikumar Padmanaban and Atif Iqbal

Chapter 6   **All SiC Grid-Connected PV Supply with HF Link MPPT Converter: System Design Methodology and Development of a 20 kHz, 25 kVA Prototype**..........................94
Serkan Öztürk, Mehmet Canver, Işik Çadırcı and Muammer Ermiş

Chapter 7   **Open Circuit Fault Diagnosis and Fault Tolerance of Three-Phase Bridgeless Rectifier**...124
Hong Cheng, Wenbo Chen, Cong Wang and Jiaqing Deng

Chapter 8   **Performance Evaluation of a Semi-Dual-Active-Bridge with PPWM Plus SPS Control**...139
Ming Lu and Xiaodong Li

Chapter 9   **Performance Improvement for PMSM DTC System through Composite Active Vectors Modulation**...152
Tianqing Yuan and Dazhi Wang

Chapter 10   **Series Active Filter Design Based on Asymmetric Hybrid Modular Multilevel Converter for Traction System**..168
Muhammad Ali, Muhammad Mansoor Khan, Jianming Xu, Muhammad Talib Faiz, Yaqoob Ali, Khurram Hashmi and Houjun Tang

Chapter 11   **New Fault-Tolerant Control Strategy of Five-Phase Induction Motor with Four-Phase and Three-Phase Modes of Operation**..189
Sonali Chetan Rangari, Hiralal Murlidhar Suryawanshi and Mohan Renge

Chapter 12   **SHIL and DHIL Simulations of Nonlinear Control Methods Applied for Power Converters using Embedded Systems**..205
Arthur H. R. Rosa, Matheus B. E. Silva, Marcos F. C. Campos, Renato A. S. Santana, Welbert A. Rodrigues, Lenin M. F. Morais and Seleme I. Seleme Jr.

Permissions

List of Contributors

Index

Preface

Power electronics refers to a subfield of electrical engineering concerned with processing of high voltages and currents in order to deliver power that supports a variety of needs. It encompasses a multidisciplinary study of various concepts in physics, including electrical motors, semiconductor physics, electromagnetic devices, mechanical actuators, and control systems. It is concerned with designing, controlling, and transforming power into an electric form through a power electronic system. The main component of power electronic system is a switching power converter. There are several applications of power electronics systems, such as power generation, power transmission, power distribution, and power control. Power electronics has a number of advantages such as improved efficiency in power conversion, high power density supplies, wireless power transfer, and providing power according to the specifications. This book outlines the processes and applications of power electronics in detail. The various studies that are constantly contributing towards advancing technologies and evolution of this field are examined in detail. With state-of-the-art inputs by acclaimed experts of this field, this book targets students and professionals.

The researches compiled throughout the book are authentic and of high quality, combining several disciplines and from very diverse regions from around the world. Drawing on the contributions of many researchers from diverse countries, the book's objective is to provide the readers with the latest achievements in the area of research. This book will surely be a source of knowledge to all interested and researching the field.

In the end, I would like to express my deep sense of gratitude to all the authors for meeting the set deadlines in completing and submitting their research chapters. I would also like to thank the publisher for the support offered to us throughout the course of the book. Finally, I extend my sincere thanks to my family for being a constant source of inspiration and encouragement.

Editor

A Novel Zero Dead-Time PWM Method to Improve the Current Distortion of a Three-Level NPC Inverter

Jin-Wook Kang [1], Seung-Wook Hyun [2], Yong Kan [2], Hoon Lee [3] and Jung-Hyo Lee [4],*

[1] Hanwha Defense, Seongnam 13488, Korea; kjw2171@naver.com
[2] LG Electronics, Seoul 07336, Korea; hahama19@hanmail.net (S.-W.H.); ykan88@naver.com (Y.K.)
[3] Department of Electrical and Computer Engineering, Sungkyunkwan University, Suwon 16419, Korea; dlgns520@skku.edu
[4] Department of Electrical Engineering, Kunsan National University, Gunsan, Jeollabuk-do 54150, Korea
*   Correspondence: jhlee82@kunsan.ac.kr

Abstract: This paper proposes a novel pulse width modulation (PWM) for a three-level neutral point clamped (NPC) voltage source inverter (VSI). When the conventional PWM method is used in three-level NPC VSI, dead time is required to prevent a short circuit caused by the operation of complementary devices on the upper and lower arms. However, current distortion is increased because of the dead time and it can also cause a voltage unbalance in the dc-link. To solve this problem, we propose a zero dead-time width modulation (ZDPWM) which does not require dead time used in complementary operation. The proposed technique applies the offset voltage to the space vector pulse width modulation (SVPWM) reference voltage for the same modulation index (MI) as the conventional SVPWM, but any complementary switching operation needs dead time. In addition, the proposed method is divided into four operation sections using the reference voltage and phase current to operate switching devices which flow the current depending on the section. This ZDPWM method is simply implemented by carrier and reference voltage that reduce the current distortion, because complementary operation that needs dead time is not implemented. However, the operation section is delayed due to the sampling delay that occurs during the experiment. Therefore, in this paper, we conduct a modeling of sampling delay to improve the delay of operation section. To verify the principle and feasibility of the proposed ZDPWM method, a simulation and experiment are implemented.

Keywords: three-level NPC VSI; zero dead-time PWM; sampling delay; current distortion

1. Introduction

In high power applications such as medium voltage motor drives, solar cells, vehicles, and most recently, wind generation system, etc., multilevel topologies introduced in [1] have been widely applied to reduce harmonic in the grid current, to downsize the physical filter size, and to mitigate the switching losses of the used devices as compared with the conventional two-level pulse width modulation (PWM) inverter. These multilevel topologies allow the output voltage to be closer to the sinusoidal wave by increasing the number of voltage levels, and reducing the harmonic distortion, as reported in the literature [2–4]. The major multilevel topologies that have been studied are neutral point clamped (NPC) [5,6], active net point clamped (ANPC) [7–9], flying capacitor (FC) [10–12], and cascaded multilevel inverter (CMLI) [13,14]. The ANPC application has a high cost since it has a relatively large number of switching devices. The capacitor used in FC requires precharging, and the high number of flying capacitors required with increasing output levels reduces system reliability. The disadvantages of CMLI are the complexity of synchronization and the unbalanced power losses between power modules. Among these topologies, the three-level neutral point clamped (NPC) voltage source inverter

(VSI) is the most popular topology as compared with other three-level topologies because it has some advantages such as simple operating sequence, low voltage stress, and low switching losses [15,16]. Figure 1 illustrates a three-level NPC VSI topology.

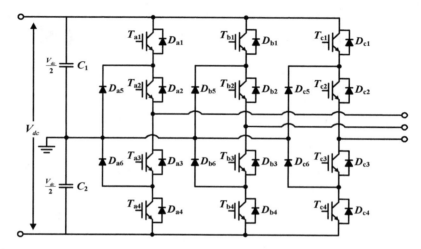

Figure 1. Configuration of the three-level neutral point clamped (NPC) voltage source inverter (VSI).

Each leg consists of four switching devices which are connected in series, two diodes, and a neutral point of dc-link as shown in Figure 1.

The performance including the efficiency and total harmonic distortion (THD) of the three-level NPC VSI is affected by the PWM method such as sinusoidal pulse width modulation (SPWM), space vector pulse width modulation (SVPWM), and discontinuous pulse width modulation (DPWM), etc.; the available output voltage area is determined depending on the PWM method [17–20]. The previously explained PWM techniques all require dead time when transitioning from positive to zero, or reverse, switching transition conditions. In this transition time, the switching device, T_{x1}, is turned off and the switching device, T_{x3}, is turned on, in parallel. During the transition time, the dead time is applied to prevent a short circuit caused by the on/off delay of the switching devices.

However, the dead time in the complementary switching method causes distortion in the output voltage, and therefore increases the THD in the phase current. For this reason, dead-time compensation methods are widely studied to reduce the THD in phase current [21–24]. In [21], the authors proposed a method to improve the harmonic components caused by dead time by applying the offset voltage which consisted of the reference voltage and distorted output voltage in an inverter. Selective harmonic elimination (SHE) modulation is also one of the ways to solve harmonic component, considering dead-time effect [25]. However, since this method is based on the PWM which operates complementarily, these methods have the characteristic of turning on the switching device even when there is no current flowing into the switching devices.

Another research integrated the sixth harmonic of the synchronous d-axis proportional-integral (PI) current regulator [22]. This method imposes high computational burden and is difficult to realize, because of the detection of harmonics by integrated operation with integral controller. A previous study used a parameter adjustment mechanism for adaptive dead-time compensation [23]. Because this method utilized the dc-bus voltage and three-phase voltage, it was difficult to achieve an accurate effect if an imbalance of three-phase voltage and dc-bus voltage existed. In [26], dead-time compensation was performed with a dead-time correction state machine. This method was based on the error from measured three-phase voltage and current, and sensitiveness existed from the disturbance, which was a disadvantage.

In the case of the previous studies, there are limitations to the compensation of current disturbance due to dead time, because of the modulation or complicated control configuration and mechanism, and based on various infarction signals.

In this paper, to solve these limitations, the zero dead-time width modulation (ZDPWM) method is proposed which does not need the dead-time, and therefore there improve the distortion of output current. A short circuit which is caused by the on/off delay of switching devices does not appear when the complementary switching operation is not used. Thus, in this paper, we describe the ZDPWM method that operates without dead time, and how this PWM is implemented by comparing the carrier and reference voltage. We also suggest an improved method that can solve the sampling delay of a microcontroller unit (MCU) that is seen in the experiment.

This paper is organized as follows: In Section 2, we describe the basic structure and theory of the existing three-level NPC topology; in Section 3, we describe the analysis of the proposed ZDPWM method and how to implement the proposed technique; in Section 4, we describe the impact of sampling delay on the proposed ZDPWM method and the technique to improve this; in Section 5, we describe the simulation and experimental results for validation of the proposed control; and in Section 6, we provide the conclusions.

2. Configuration of the Three-level Neutral Point Clamped (NPC) Inverter and Operation Principle

In this section, first, we describe the basic configuration and operating principles of NPC topology and the effects of dead time before describing the proposed techniques.

The three-level NPC inverter, as shown in Figure 1, is connected to the neutral point, the neutral point clamp diode, and the dc-link capacitor, each leg consists of four switches, and two neutral point clamp diodes. Since the two neutral point clamp diodes are connected to the neutral point of the dc-link, unlike the two-level inverter, an output state of "zero" is possible.

Table 1 shows the switching state of the three-level NPC inverter according to the output voltage. Switching states can be divided into three states depending on the output voltage. Figure 2 shows the flow of current with switching status in a three-level NPC inverter. The state, in Figure 2a, outputs a voltage of "positive" with current flowing under load from the top capacitor and switches T_{x1} and T_{x2} on one leg. T_{x1} and T_{x3}, T_{x2} and T_{x4} perform reciprocal actions, therefore, T_{x3} and T_{x4} are off. The state, in Figure 2b, is that the current is flowing from the neutral point to the load, the voltage output is "zero", with switches T_{x2} and T_{x3} on and switches T_{x1} and T_{x4} off. The state, in Figure 2c, is that the current is flowing from the lower capacitor to the load, the voltage output is "negative", with switches T_{x3} and T_{x4} on and switches T_{x1} and T_{x2} off.

Table 1. Switching state of three-level NPC inverter depending on output voltage.

State	T_{x1}	T_{x2}	T_{x3}	T_{x4}
Positive	1	1	0	0
Zero	0	1	1	0
Negative	0	0	1	1

These three-level NPC inverters can be designed to have a lower voltage on each switch than conventional two-level inverters because one leg consists of four switches. In addition, the level of voltage is higher than that of a two-level inverter, therefore, the THD is lower. All elements are switched to the fundamental frequency, which gives the inverter a high efficiency advantage [27].

The three-level NPC is generally adapting SVPWM, the same as the two-level inverters [28]. With SVPWM, the three-level NPC inverter consists of a total of 27 switching states and 19 voltage vectors, as shown in Figure 3.

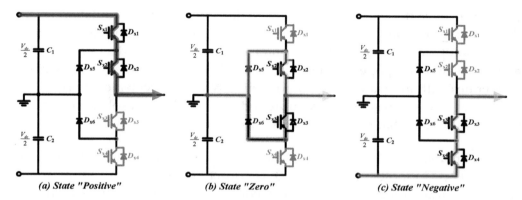

Figure 2. Current flow of three-level NPC inverter depending on the switching state.

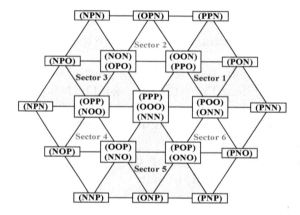

Figure 3. Vector diagram of three-level NPC inverter.

At this time, vectors requiring dead time are divided into three forms, as shown in Figure 4, depending on the state of the load. First, a unit vector, representing 120 degrees phase displacement, is as shown in Equation (1), in order to express the phase voltage of the load as:

$$a = e^{j2\pi/3} = -\frac{1}{2} + j\sqrt{3/2} \tag{1}$$

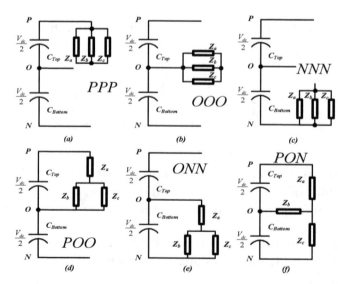

Figure 4. Voltage vector depending on the switching state: **(a)** PPP; **(b)** OOO; **(c)** NNN; **(d)** POO; **(e)** ONN; **(f)** PON.

The space vector of the phase voltage of the load can be expressed as shown in Equation (2):

$$V = \frac{2}{3}(v_{a0} + a v_{b0} + a^2 v_{c0}) \tag{2}$$

With equation (2), the voltage vector of the zero vector in Figure 4a–c can be expressed as (3):

$$\begin{aligned} V_{(PPP)} &= \frac{2}{3}\left(\frac{V_{dc}}{2} + a\frac{V_{dc}}{2} + a^2\frac{V_{dc}}{2}\right) = 0 \\ V_{(OOO)} &= \frac{2}{3}\left(0 + a + a^2\right) = 0 \\ V_{(NNN)} &= \frac{2}{3}\left(-\frac{V_{dc}}{2} - a\frac{V_{dc}}{2} - a^2\frac{V_{dc}}{2}\right) = 0 \end{aligned} \tag{3}$$

Figure 4d,e shows a small vector. When switching states are (POO) and (ONN), voltage vectors can be expressed in the following ways:

$$\begin{aligned} V_{(POO)} &= \frac{2}{3}\left(\frac{V_{dc}}{2} + a0 + a^20\right) = \frac{V_{dc}}{3} \\ V_{(oNN)} &= \frac{2}{3}\left(0 - a\frac{V_{dc}}{2} - a^2\frac{V_{dc}}{2}\right) = \frac{V_{dc}}{3} \end{aligned} \tag{4}$$

In addition, Figure 4f shows the medium vector, and if the switching state is (PON), the dc-link neutral point is connected to the load stage, which affects the voltage variation of the neutral point. In the above state, the voltage vector may be shown as follows:

$$V_{(PON)} = \frac{2}{3}\left(\frac{V_{dc}}{2} + a0 - a^2\frac{V_{dc}}{2}\right) = \frac{V_{dc}}{3}(1 - a^2) = \frac{V_{dc}}{\sqrt{3}}e^{j\pi/6} \tag{5}$$

In order to make the above load conditions, the three-level NPC inverter must perform a complementary operation of four insulated gate bipolar transistor (IGBT) and two reverse parallel diodes, as described earlier.

Figure 5 shows the characteristics of power switching devices that use dead time, when performing complementary operation. The upper side represents switching signals, and the bottom side represents collector-emitter voltage and output current. As shown in Figure 5, the transition between the on/off states may cause a short circuit, due to the time difference between rising time and falling time, and therefore dead time is applied to the on/off signals to prevent such short circuit accidents. However, due to the effect of dead time, an error occurs between the commanded voltage and the actual voltage. These voltage errors vary with the inverter current direction.

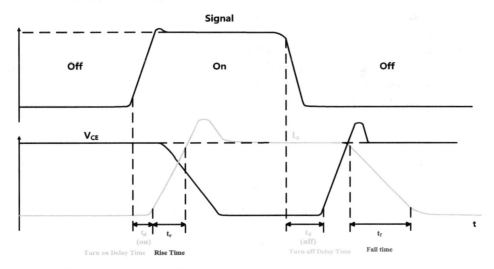

Figure 5. Turn-on/off characteristics of a switching power device.

Figure 6 shows the switching state and output voltage during dead time. Switch T_{x2} is always on, and T_{x4} is always off. When the direction of the current is $i_o > 0$, applying the dead time by t_{dead} to switch T_{x1} causes an error between the command switching and the actual switching, resulting in an error voltage equal to V_{error} in the output voltage. Similarly, when the direction of the current is $i_o < 0$, applying dead time to T_{x3} produces a voltage error as much as V_{error} at the output voltage. The error voltage generated by such dead time causes distortion of the output current, higher distortion of the electric wave, and lower stability of the system, due to the distorted output current [29].

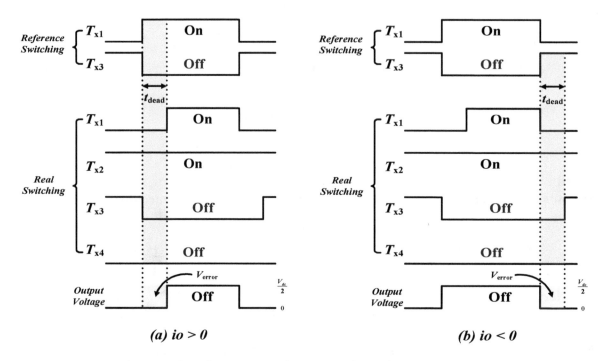

Figure 6. Switching state and output voltage during dead time.

3. Proposed Zero Dead-Time PWM Method

In this section, we describe the proposed zero dead-time PWM (ZDPWM) which totally removes dead-time utilization for PWM generation, in order to improve the previously explained current distortion caused by dead time in a three-level NPC inverter.

Figure 7 shows the switching mode according to the reference voltage and phase current. The proposed method focuses on turning on the switching devices which are flowing the current, and is divided into four operation areas depending on the direction of the reference voltage and phase current as shown in Figure 7b–e. The reference voltage and phase current conditions in each section are as follows:

$$In\ Secion\ 1,\ v_{x\_ref}\&i_{x\_out} > 0$$
$$In\ Secion\ 2,\ v_{x\_ref} < 0,\ i_{x\_out} > 0$$
$$In\ Secion\ 3,\ v_{x\_ref}\&i_{x\_out} < 0 \tag{6}$$
$$In\ Secion\ 4,\ v_{x\_ref} > 0,\ i_{x\_out} < 0$$

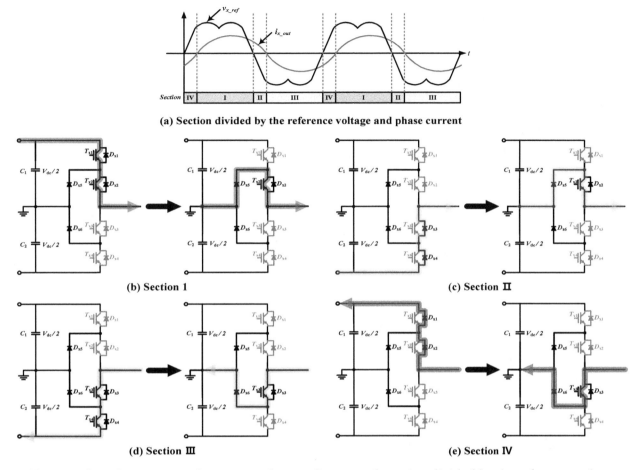

(a) Section divided by the reference voltage and phase current

(b) Section 1

(c) Section II

(d) Section III

(e) Section IV

Figure 7. Switching state and current path according to each section divided by the reference voltage and phase current.

First, Figure 7b illustrates when the direction of the current is in the "+" direction. In section I, the switch T_{x3}, which is affected by dead time, does not turn on when switching from "positive" to "zero" state. Therefore, because it does not operate the more than three switching devices, there is no dead-time interval without implementing the complementary operation. Figure 7c illustrates when the direction of the current is in the "+" direction, as in case of Figure 7b. In Section II, switch T_{x3}, which is affected by dead time, does not turn on when switching from the "negative" to "zero" state. Therefore, the complementary operation is also not performed in Section II. Figure 7d shows when the direction of the current is in the "−" direction. In Section III, the switch T_{x2}, which is affected by dead time does not turn on when switching from the "negative" to "zero" state. Thus, because it does not operate the more than three switching devices, there is also no implementation of the complementary operation. Figure 7e illustrates when the direction of the current is in the "+" direction, as in case of the Figure 7d. In Section IV, the switch T_{x2} does not turn on when switching from the "positive" to "zero" state. Therefore, the complementary operation is also not implemented in Section IV. As a result, the switching states can remove dead time from any of the operation modes as there is no short circuit fault resulted from turn on/off delay between the switching transient [30].

Figure 8 shows the switching signal and reference voltage waveforms of phase A of the proposed ZDPWM method used in the three-level NPC VSI. Depending on the direction of the output current, two offset voltages $V_{a\_offsetx}$ are generated, and the magnitude of the offset voltage is determined by the maximum modulation index (MI), as shown in Figure 8 and Equation (7) as:

$$if \ i_{x\_out} > 0, \ then \ V_{a\_offset} = V_{a\_offset1} = +1.1547 \ (modulation \ index)$$
$$if \ i_{x\_out} < 0, \ then \ V_{a\_offset} = V_{a\_offset2} = -1.1547 \ (modulation \ index)$$

(7)

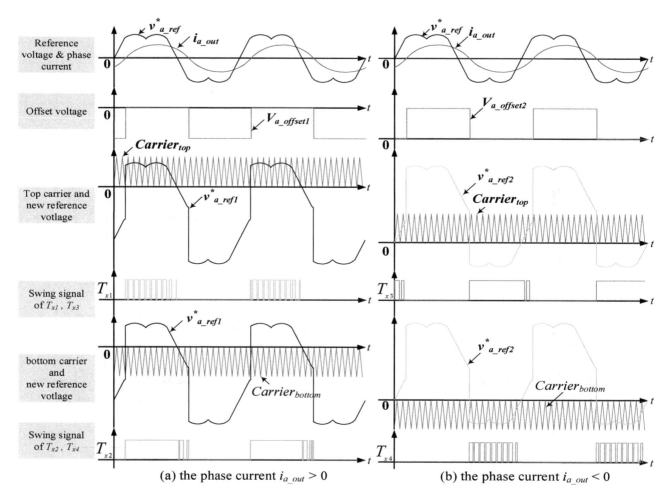

Figure 8. The switching signal and reference voltage waveforms of the proposed zero dead-time width modulation (ZDPWM) method.

The proposed technique is based on the SVPWM method; the maximum MI value of SVPWM is 1.1547. When the SVPWM generates a voltage reference from the output of the PI current controller, the offset value is designated as the maximum MI value, as described in Equation (7), in order to use the same maximum capable output voltage as the maximum voltage of SVPWM.

Two new reference voltages are generated by adding two offset voltage to the reference voltage, as shown in Figure 8. On the one hand, when $V^*_{a\_ref1}$, which is generated by merging with the "−" offset voltage, is greater than the upper carrier $Carrier_{top}$, the switching signal is generated by switch T_{x1}, and when $V^*_{a\_ref1}$ is greater than the lower carrier $Carrier_{bottom}$, the switching signal is generated by switch T_{x2}.

On the other hand, when $V^*_{a\_ref2}$, which is generated by merging with the "+" offset voltage, is less than the upper carrier $Carrier_{top}$, the switching signal is generated by switch T_{x3}, and when $V^*_{a\_ref2}$ is less than the lower carrier $Carrier_{bottom}$, the switching signal is generated by switch T_{x4}. The switching condition are expressed as follows:

$$
\begin{aligned}
&1)\ if\ v^*_{ref1} > Carrier_{Top},\ then\ T_{x1} = 1 \qquad 2)\ if\ v^*_{ref1} > Carrier_{Bottom},\ then\ T_{x2} = 1 \\
&\qquad\qquad\qquad\qquad\qquad else\ T_{x1} = 0 \qquad\qquad\qquad\qquad\qquad\qquad\qquad else\ T_{x2} = 0 \\[6pt]
&3)\ if\ v^*_{ref2} < Carrier_{Top},\ then\ T_{x3} = 1 \qquad 4)\ if\ v^*_{ref2} < Carrier_{Bottom},\ then\ T_{x4} = 1 \\
&\qquad\qquad\qquad\qquad\qquad else\ T_{x3} = 0 \qquad\qquad\qquad\qquad\qquad\qquad\qquad else\ T_{x4} = 0
\end{aligned}
\tag{8}
$$

Therefore, in the proposed ZDPWM method, since each switching signal is compared only in each area, switches T_{x1} and T_{x3}, and switches T_{x2} and T_{x4} do not perform the complementary operation to each other, unlike conventional SPWM, SVPWM method.

Figure 9 also shows that, when using the proposed ZDPWM method, there is no need for dead time at the transition of each section. As shown in Figure 9, unlike conventional techniques, a switch can be seen to perform switching operations in each area, in contrast to conventional complementary operation. The switching state in the proposed ZDPWM technique can be defined as in Table 2.

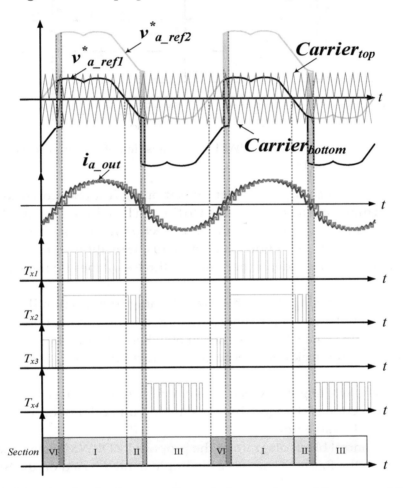

Figure 9. The switching signal and reference voltage whole waveforms of the proposed ZDPWM method.

Table 2. Switching state of the proposed ZDPWM depending on the section.

		Section I		Section II		Section III		Section IV	
	Reference Voltage	Positive		Negative		Negative		Positive	
	Phase current	Positive		Positive		Negative		Negative	
	Phase Voltage	Positive	Zero	Negative	Zero	Negative	Zero	Positive	Zero
State	Switch T_{x1}	On	Off	Off	Off	Off	Off	Off	Off
	Switch T_{x2}	On	On	Off	On	Off	Off	Off	Off
	Switch T_{x3}	Off	Off	On	Off	On	On	Off	On
	Switch T_{x4}	Off	Off	On	Off	On	Off	Off	Off

Figure 10 illustrates the determination of short-circuit accidents at the transition of a three-level NPC inverter using the proposed ZDPWM technique. Figure 10a shows the switching state of the four switches when transitioning from Section I to Section II. In Section I, T_{x1} and T_{x2} are determined by the top carrier, the bottom carrier, and the first reference voltage $v^*_{a\_ref1}$, and T_{x3} and T_{x4} are determined by the second reference voltage $v^*_{a\_ref2}$. Due to this, transition from Section I to section II is not affected by short circuit accidents.

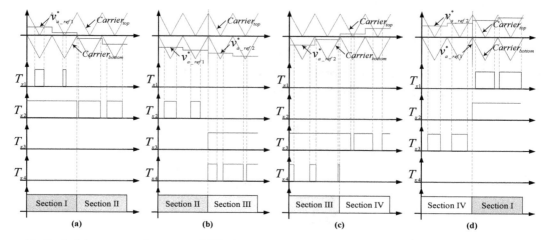

Figure 10. Switching state of proposed ZDPWM when transition of section: **(a)** from I to II; **(b)** from II to III; **(c)** from III to IV; **(d)** from IV to I.

Figure 10b shows the switching state when transitioning from Section II to Section III. Because switch T_{x1} was off when transitioning from Section I to Section II, it does not affect the state in Figure 10b. Switch T_{x2} is off before transitioning from Section II to Section III, and switches T_{x3} and T_{x4} are on at the transitioning from Section II to Section III. Due to this, transition from section II to section III is also not affected by short circuit accidents, because the three switches do not simultaneously change the state.

Figure 10c shows the switching state when transitioning from Section III to Section IV. Switches T_{x1} and T_{x2} are off state, switches T_{x3} are determined by the top carrier, bottom carrier, and second reference voltage $v^*_{a\_ref2}$, and switch T_{x4} are switched off when transitioning from Section III to Section IV. Therefore, no short circuit accident is occurred because the three switches do not change the state simultaneously.

Figure 10d shows the switching state when transitioning from Section IV to Section I. Three switches are turned on when the section is changed, but no short circuit accident occurs because switches T_{x1} and T_{x2} do not change at the same time.

Figure 11 shows a control block diagram of the proposed ZDPWM technique based on previous illustration. As shown on the figure, it can be easily implemented based on the SVPWM technique.

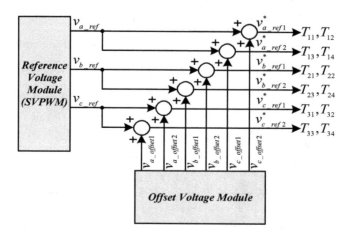

Figure 11. Control block diagram of proposed ZDPWM method.

However, since this method determines the offset of reference voltage according to the polarity of the phase current, unexpected distortion may occur if the delay occurs in the analog-to-digital converter (ADC) modules that senses the phase current. Therefore, in this paper, simple phase delay modeling was performed, and it was applied to the proposed ZDPWM method.

4. Phase Delay Modeling and Compensation Method

In this section we describe the problems that can arise when the proposed ZDPWM technique is actually implemented and how to optimize it.

Figure 12 shows the delay in sampling data caused by the ADC module in MCU. Since the sampling period of ADC modules is determined using the interrupt of the PWM modules in MCU, the initiation of the ADC for real-time data sampling is associated with the switching period, as shown in Figure 12. The ADC modules which sense the current needs more than a certain amount of time to converter analog-to-digital data. Therefore, the phase current data are not immediately converted at the beginning of the sampling, and the data sampled at the previous time is used at the start of the next sampling. The sampled phase current is calculated as the lagging current relative to the real current and, consequently, the current controller or voltage controller of the inverter is operated through these sampled current data.

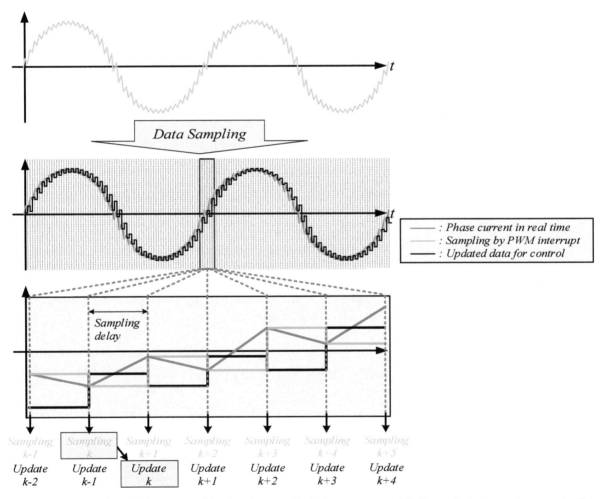

Figure 12. Sampling delay caused by analog-to-digital converter (ADC) module in a microcontroller unit (MCU).

When the conventional PWM is used, it does not significantly affect switching operation even if it is sensed as a lagging current, because the delayed phase is compensated by using the d-axis current control. However, in the case of the proposed ZDPWM, because the phase of voltage and current are shifted by the sampling delay, it causes a delayed transition of the operation section. This delay of section transition delays the recognition of the zero-crossing of the phase current and causes distortion of the phase current. This delay of section transition also cannot be improved by only the d-axis control used in the conventional SPWM, SVPWM method. Therefore, the sampling delay should be improved to have the same phase with real current for the ideal operation of the proposed ZDPWM method.

The sampling delay t_{delay} can be calculated by Equation (9) as:

$$t_{delay} = \frac{1}{f_{sw}} \tag{9}$$

where f_{sw} is the switching frequency of the inverter. The relation formula both the delay angle of phase current, which is output AC phase, and the sampling delay are calculated through proportional trigonometric equation as follows:

$$2\pi : \frac{1}{f_{ref}} = \theta_{delay} : t_{delay} \tag{10}$$

In Equation (10), f_{ref} is the frequency of the reference voltage. In this paper, we propose the sampling delay compensation method through phase angle of d-q transformation using the delay angle which is calculated from the Equation (11) as:

$$\theta_{delay} = \frac{2\pi f_{ref}}{f_{sw}} \tag{11}$$

The delayed phase current through sampling can be defined as follows:

$$\begin{bmatrix} i_a \\ i_b \\ i_c \end{bmatrix} = \begin{bmatrix} I_{peak}\cos(\theta - \theta_{delay}) \\ I_{peak}\cos(\theta - 2\pi/3 - \theta_{delay}) \\ I_{peak}\cos(\theta + 2\pi/3 - \theta_{delay}) \end{bmatrix} \tag{12}$$

Equation (12) can be expressed as a d-q axis current using the d-q transformation as follows:

$$\begin{bmatrix} i_a \\ i_b \\ i_c \end{bmatrix} T(0)R(\theta - \theta_{delay}) = \begin{bmatrix} I_q^e \\ I_d^e \end{bmatrix} = \begin{bmatrix} I_{peak}\cos(0) \\ I_{peak}\sin(0) \end{bmatrix} \left(T(0) = \frac{2}{3} \begin{bmatrix} 1 & -1/2 & -1/2 \\ 0 & \sqrt{3}/2 & -\sqrt{3}/2 \\ 1/\sqrt{2} & 1/\sqrt{2} & 1/\sqrt{2} \end{bmatrix}, R(\theta) = \begin{bmatrix} \cos\theta & \sin\theta \\ -\sin\theta & \cos\theta \end{bmatrix} \right) \tag{13}$$

In general, the phase angle which is used in d-q transformation is obtained by phase locked loop (PLL). $T(0)$ and $R(\theta)$ are transfer matrix of the Park transformation and Clark Transformation. As shown in the Equation (13), finally, the current of d and q axis does not affect the sampling delay of the ADC modules.

Therefore, to prevent the phase delay caused by the sampling delay, as shown in Equation (14), the delayed phase angle obtained by Equation (11) is compensated when performing the reverse d-q transformation as:

$$\begin{bmatrix} I_q^e \\ I_d^e \end{bmatrix} R^{-1}(\{\theta - \theta_{delay}\} + \theta_{delay})T^{-1}(0) = \begin{bmatrix} I_{peak}\cos(\theta) \\ I_{peak}\cos(\theta - 2\pi/3) \\ I_{peak}\cos(\theta + 2\pi/3) \end{bmatrix} \tag{14}$$

Figure 13 shows a flowchart to compensate for the phase angle delay of the phase current. When sensing the current, as shown in Figure 13, the phase angle of the current is delayed by the sampling delay. This delayed current is converted to a synchronous reference frame via d-q transformation, and harmonics within this value are removed by digital filters such as low pass filter. Then, the values of these converted synchronous reference frame are converted to a stationary reference frame again using the compensated phase angle shown in the Equation (14) via reverse d-q transformation. Therefore, the proposed ZDPWM method can use the phase current, which is not delayed, to calculate the offset voltage. The phase angle used in this time is compensated as much as it is delayed.

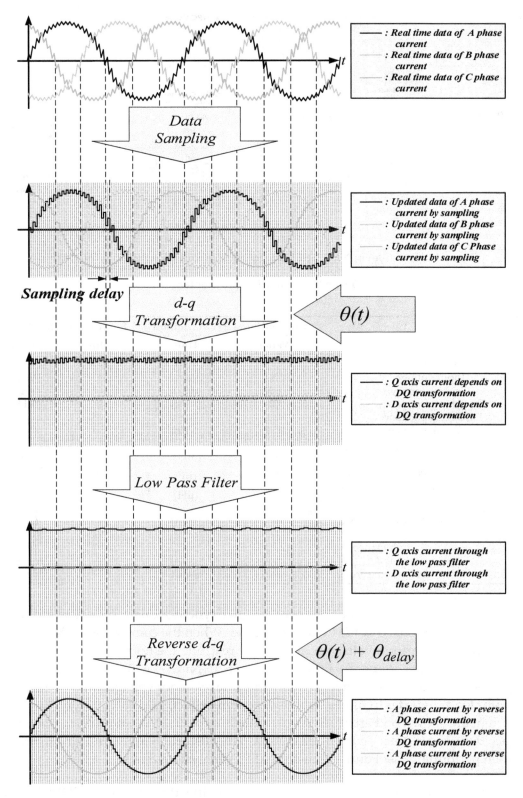

Figure 13. Flowchart of the compensating of the delayed phase angle of the phase current.

5. Simulation and Experimental Results

In order to verify the principle and feasibility of the proposed ZDPWM method, a simulation was developed using the PSIM software program. The simulation schematic in which the three-level NPC VSI back-to-back system is illustrated in Figure 14. The systems parameters of the simulation are shown in Table 3.

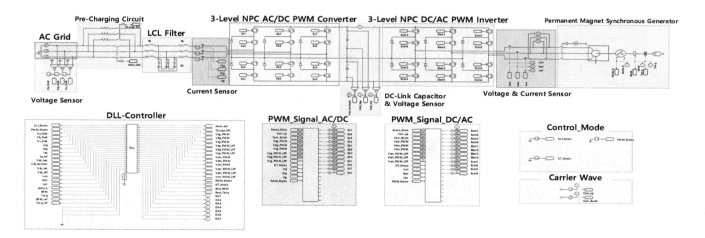

Figure 14. Three-level NPC VSI back-to-back schematic used in simulation.

Table 3. Simulation parameters.

Parameter		Conventional PWM Method		Proposed PWM Method	
		Value	Unit	Value	Unit
	Dead time	5	(µs)	0	(µs)
	Input voltage	380	(Vrms)	380	(Vrms)
	Grid frequency	60	(Hz)	60	(Hz)
	Switching frequency	10	(kHz)	10	(kHz)
	DC-link capacitance	6000	(µF)	6000	(µF)
	DC-link voltage	650	(Vdc)	650	(Vdc)
LCL filter	Grid side filter inductance	700	(µH)	700	(µH)
	Filter capacitance	10	(µF)	10	(µF)
	Converter side filter inductance	1000	(µH)	1000	(µH)
PMSG	Maximum speed	190	(rpm)	190	(rpm)
	Resistance	0.466	(ohm)	0.466	(ohm)
	Inductance	12.975	(mH)	12.975	(mH)
	Pole	24	(pole)	24	(pole)
	Maximum torque	670	(Nm)	670	(Nm)

Figure 15 shows the results of a permanent magnet synchronous motor (PMSG) torque control simulation using a conventional SVPWM with dead time. When the control starts, the grid side inverter operates as a converter to perform voltage control. After reaching the normal state, the torque command is applied to the load side inverter at 0 → 670 (Nm) → −670 (Nm) → 670 (Nm). From the top, it shows dc-link voltage, torque, grid side measured current, grid side reference voltages, and current and reference voltages to motor load. In order to figure out the dead-time effect, grid side current, LCL filter current, and motor current are shown on Figure 16. As previously explained, dead time is applied to prevent a short circuit accident, but it causes current distortion. This distortion appears in the form of low-order harmonics when performing fast Fourier transform (FFT) analysis, as shown in Figure 17.

Figure 18 shows the simulation results when applying the proposed ZDPWM in the same conditions without dead time as the previous SVPWM. The applied load is the same as the previous condition 0 → 670 (Nm) → −670 (Nm) → 670 (Nm). As shown in Figure 19, the proposed ZDPWM does not use dead time, and current distortion cannot be seen caused by dead-time effect of the current in the grid side, the LCL filter, and the load motor. Thus, the FFT analysis, as shown in Figure 20, indicate that the low-order harmonics produced by dead time are significantly reduced as compared with SVPWM.

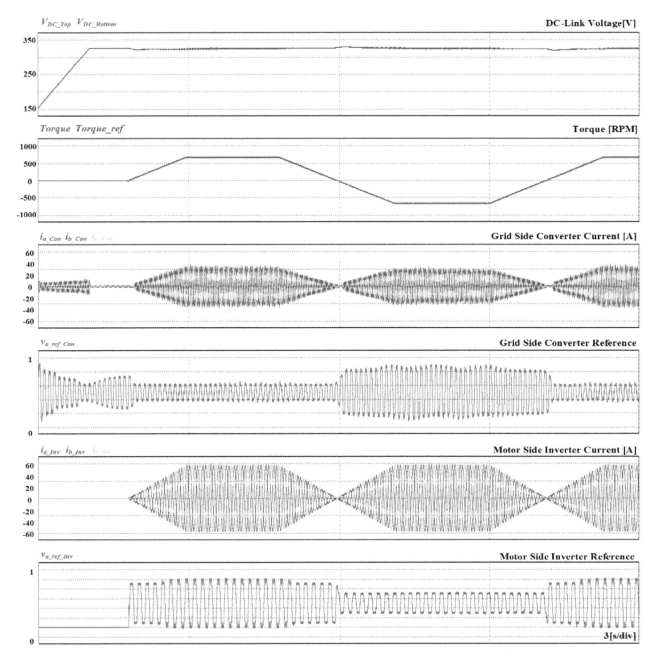

Figure 15. Simulation results of PMSG torque control using the conventional space vector pulse width modulation (SVPWM) with dead time.

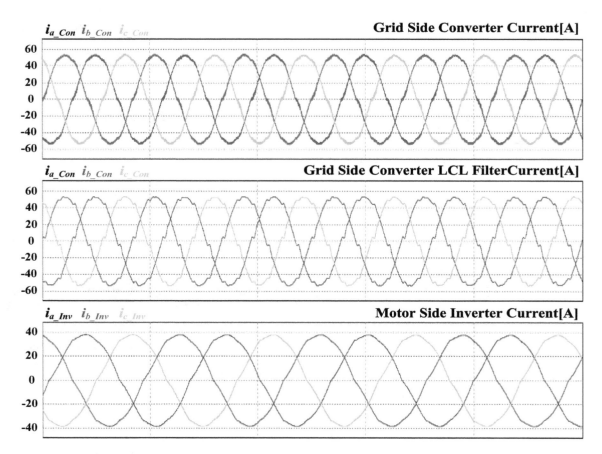

Figure 16. Simulation results of grid side current, LCL filter current, and three-phase motor current using the conventional SVPWM with dead time.

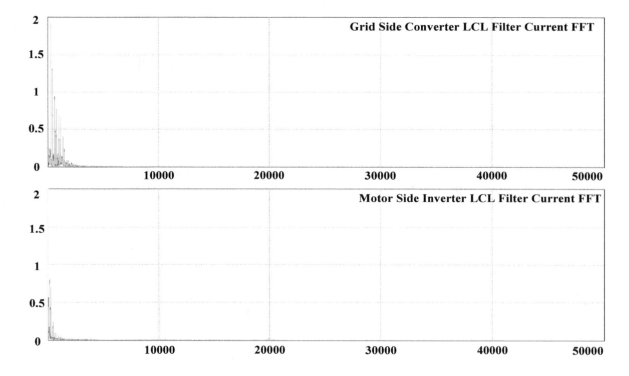

Figure 17. Fast Fourier transform (FFT) analysis results of grid side LCL filter current and motor side current using the conventional SVPWM with the dead time.

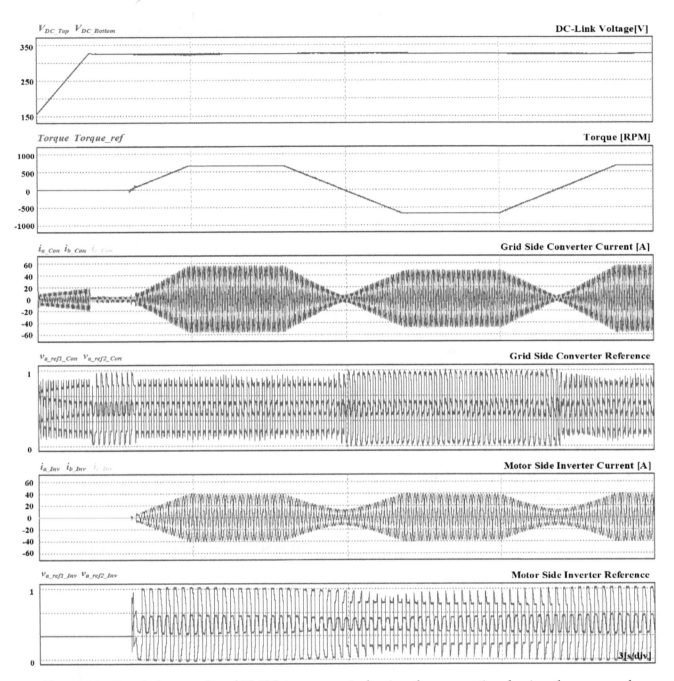

Figure 18. Simulation results of PMSG torque control using the conventional using the proposed ZDPWM method.

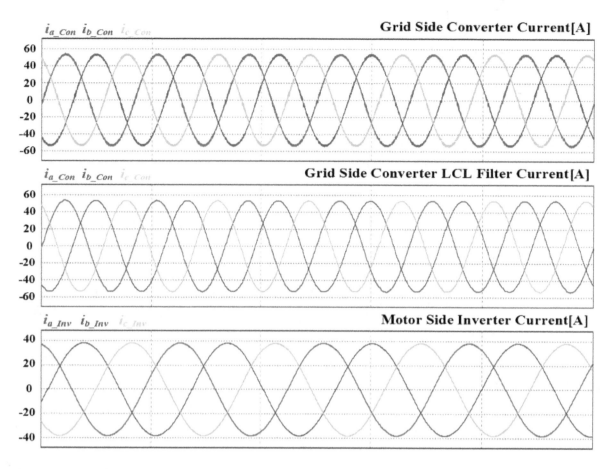

Figure 19. Simulation results of grid side current, LCL filter current and three-phase motor current using the proposed ZDPWM method.

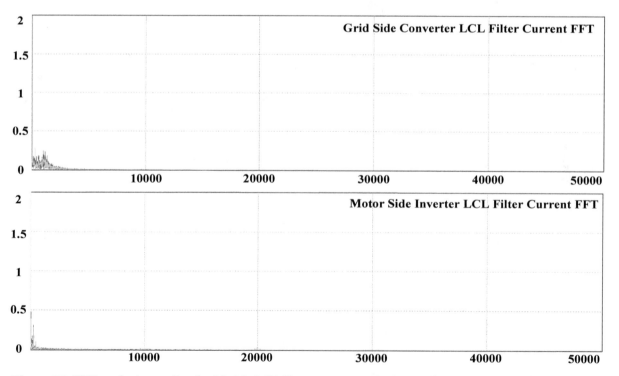

Figure 20. FFT analysis results of grid side LCL filter current and motor side current using the proposed ZDPWM method.

Figure 21 shows the comparison simulation result of the conventional SVPWM and proposed ZDPWM method. As previously mentioned, when using the conventional SVPWM method, dead time increases the distortion in the phase current, which causes an increase in the THD. However, the proposed ZDPWM method does not require dead time because it does not perform a complementary operation, as shown in Figure 21, and therefore it reduces the distortion in the phase current. Consequently, when using the conventional SVPWM method with dead time, the THD of the current is about 2.4%, but the THD of the current is about 1.2% when using the proposed method.

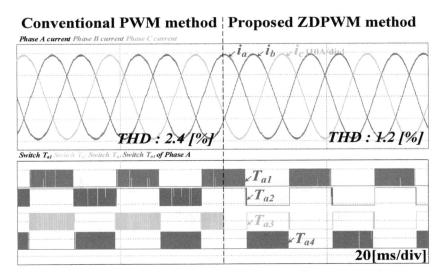

Figure 21. Simulation results of the conventional SVPWM method and proposed ZDPWM method with delay compensation applied in three-level NPC VSI (20 ms/div).

Figure 22 also shows the simulation waveforms before and after applying proposed ZDPWM with sampling delay compensation. We can see that the proposed ZDPWM method is more distorted by the sampling delay of the phase current at the section where it is transmitted, as shown in Figure 22a. At this time, the THD of current is about 2.8% and is higher than the conventional SVPWM method. Figure 22b shows that by compensating the delay, the distortion caused by the delay is reduced.

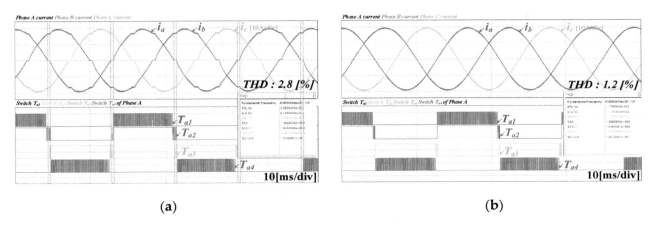

(a)   (b)

Figure 22. Simulation results of the proposed ZDPWM method applied in three-level NPC VSI: (**a**) Without delay compensation (10 ms/div); (**b**) With delay compensation (10 ms/div).

In this case, the compensated phase angle, which is calculated by using the Equation (11), is shown in Figure 23. Therefore, although the phase of the sensed current is delayed more than the real current, the phase current used in the proposed ZDPWM method has the same phase as the real current because delay is compensated during the reverse d-q transformation.

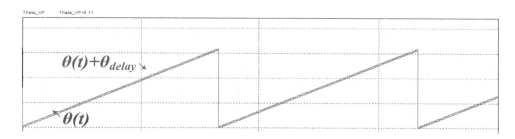

Figure 23. Simulation results of the real phase angle and compensated phase angle.

Simulation results of three methods are summarized as Table 4. The THD characteristics of the conventional SVPWM, proposed ZDPWM without delay compensation and proposed ZDPWM with delay compensation are 2.4%, 2.8%, 1.2% respectively. As a result, proposed ZDPWM with delay compensation is the best method to improve current distortion of a three-level NPC inverter.

Table 4. Comparison of the THD characteristics of the conventional SVPWM and proposed methods.

Control Method	THD (%)
Conventional SVPWM (5μs dead time)	2.4
Proposed ZDPWM without sampling delay compensation (no dead time)	2.8
Proposed ZDPWM with sampling delay compensation (no dead time)	1.2

In addition, in the NPC topology, in this paper, voltage unbalancing occurs in the dc-link due to load condition, difference in charging/discharging status according to current direction, etc. Therefore, as illustrated in [31,32], voltage balancing in dc-link is generally achieved by applying offset voltage to the PWM output from SVPWM, PSPWM, or LSPWM. In this paper, we also applied offset voltage to compensate dc-link unbalancing at the output end of ZDPWM, and conducted simulation by applying offset voltage the same as the conventional SVPWM.

Figures 24 and 25 show the simulation results applying SVPWM and the proposed ZDPWM in a three-level inverter, respectively. As shown in the figure, all applications in this paper perform voltage balance control, and therefore the dc-link voltage imbalance in the upper/lower capacitors does not occur.

An experiment was performed to verify the feasibility of the proposed ZDPWM method applied in a three-level NPC VSI. The configurations of the experimental system and power stacks are as follows in Figures 26 and 27.

The controller is implemented on TMS320F28346 and that of a floating point microcontroller unit at 300 MHz rate frequency. The switching and sampling frequency is 10 kHz. The power is supplied to the three-level NPC inverter through the three-level NPC PWM converter.

Figure 28a shows the output phase current waveforms of each phase when the conventional SVPWM method is used in a three-level NPC VSI. In this experiment, the dead time was 5 us. As shown in Figure 28, we can see that dead time causes the distortion of the output current and, in particular, large distortion of the current, at zero crossing points. Figure 28b shows the phase A output current and reference voltage waveform, and the switching signal according to the reference voltage is generated, as shown in Figure 28c. As a result, it can be seen that the THD of the output current has deteriorated to about 2.5% due to the dead time.

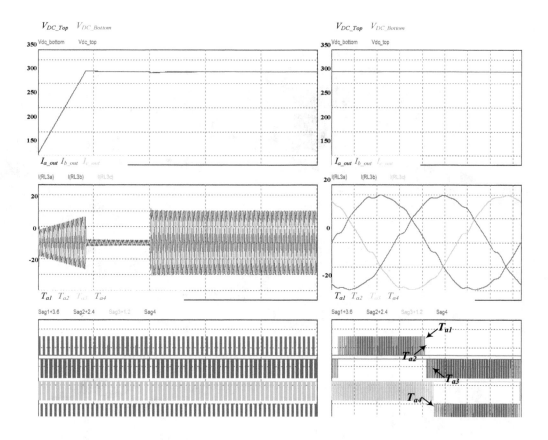

Figure 24. Simulation results using the dc-link balancing control applied in the SVPWM method.

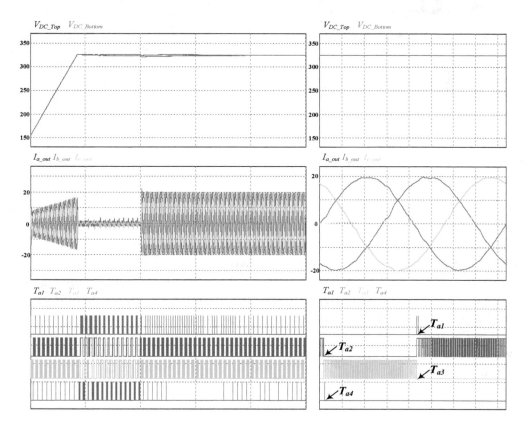

Figure 25. Simulation results using the dc-link balancing control applied in the proposed ZDPWM method.

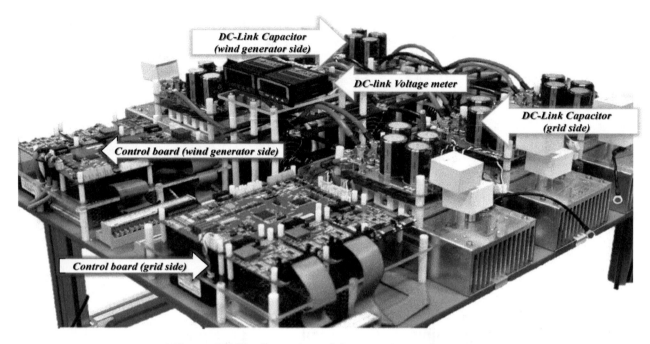

Figure 26. Configuration of the experimental system.

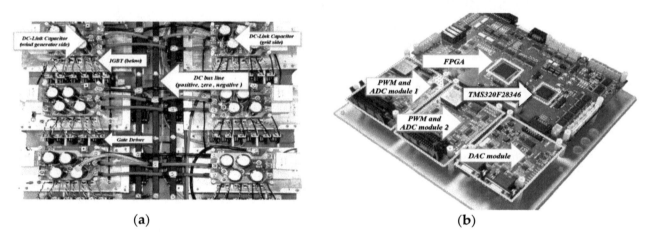

(a)   **(b)**

Figure 27. Configuration of the experimental setup: (**a**) The dc-link, switching devices, and gate driver; (**b**) The control board.

Figure 29a shows the output phase current waveforms of each phase when the proposed ZDPWM method with delay compensation is used in a three-level NPC VSI. In this experiment, the dead time was not applied. Figure 29b shows the phase A output current and reference voltage waveform. Figure 29c shows the switching signal generated by two reference voltage which are adding the offset voltage to reference voltage. In the conventional SVPWM method, each switching signal is generated by applying the dead time in order to perform the complementary operation and prevent the short circuit, but in the proposed ZDPWM method, it is possible to verify that the switches T_{a1} and T_{a3}, and the switches T_{a2} and T_{a4} do not perform the complementary operation, as shown Figure 29d. As a result, we can see that the THD of the output current is about 1.2% better than the conventional SVPWM method by about 1.3%.

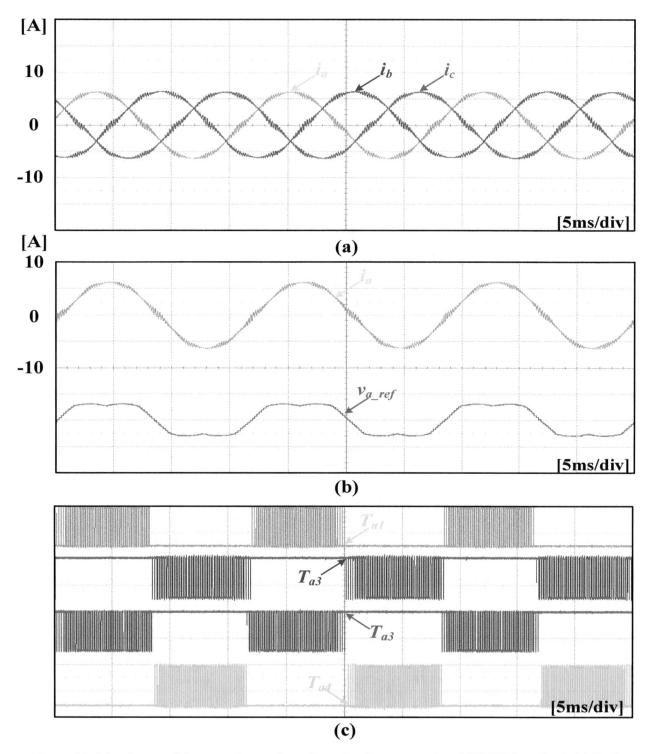

Figure 28. Waveforms of the experimental results using the conventional SVPWM method: (**a**) Each phase current; (**b**) Reference voltage; (**c**) Switching signal of the one leg switching devices.

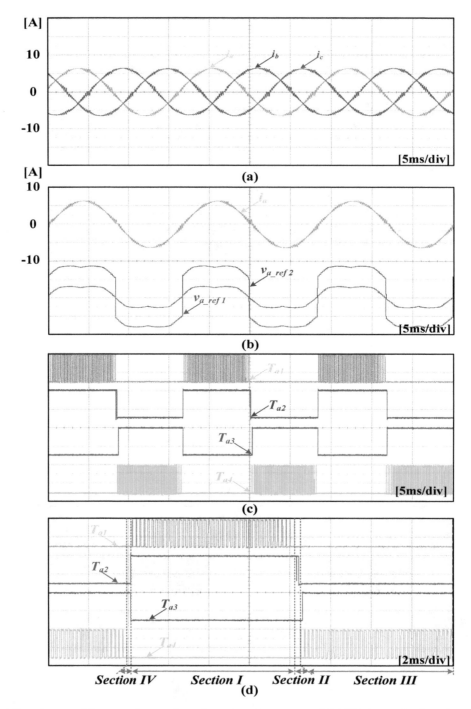

Figure 29. Waveforms of the experimental results using the proposed ZDPWM method with compensation method: (**a**) Each phase current; (**b**) Reference voltage; (**c**) Switching signal (5 ms/div); (**d**) Switching signal (2 ms/div).

6. Conclusions

In this paper, we have proposed the ZDPWM method with compensation of the sampling delay in three-level NPC VSI. It can reduce the current distortion caused by the dead time and it is easy to implement.

We discovered that complementary switching operation of the NPC VSI was the reason dead time was needed and we described how to operate the NPC VSI without dead time. In addition, the sampling delay modeling was described to compensate the delay of the phase current.

In this paper, a simulation and an experiment were performed and presented to verify the proposed ZDPWM method as well as their feasibility. From them, the proposed method resulted in better performance of lower current harmonic distortion.

From the simulation result, the current regulated by conventional SVPWM has a THD of about 2.4%, the current regulated by proposed ZDPWM without sampling delay compensation has a THD of about 2.8%, and the current regulated by proposed ZDPWM with sampling delay compensation has a THD of about 1.2%. Additionally, in experimental result shows that the current regulated by proposed ZDPWM without sampling delay compensation has a THD of about 2.5% and the current regulated by proposed ZDPWM with sampling delay compensation has a THD of about 1.2%, similar to the simulation result. These results indicate that THD can be significantly reduced by the proposed method. As a result, the proposed ZDPWM is a good PWM strategy for three-level NPC VSI systems.

Author Contributions: J.-W.K. and S.-W.H. conceived and designed the experiment; J.-W.K. and S.-W.H. performed the experiment; J.-W.K. and Y.K. analyzed the theory; J.-W.K. wrote the manuscript; H.L. reviewed the manuscript and search state-of-the art literature; J.-H.L. participated in research plan development and revised the manuscript. All authors have read and agreed to the published version of the manuscript.

Acknowledgments: This research was supported by the National Research Foundation of Korea (grant number NRF-2018R1C1B6008895); and the Basic Science Research Program through the National Research Foundation of Korea (NRF), funded by the Ministry of Education (NRF-2016R1A6A1A03013567).

References

1. Ying, J.; Gan, H. High power conversion technologies and trend. In Proceedings of the 2012 7th International Power Electronics and Motion Control Conference (IPEMC), Harbin, China, 2–5 June 2012; pp. 1766–1770.

2. Marzoughi, A.; Burgos, R.; Boroyevich, D.; Xue, Y. Investigation and comparison of cascaded H-bridge and modular multilevel converter topologies for medium-voltage drive application. In Proceedings of the 40th Annual Conference of IEEE Industrial Electronics Society-IECON, Dallas, TX, USA, 29 October–1 November 2014; pp. 1562–1568.

3. Qashqai, P.; Sheikholeslami, A.; Vahedi, H.; Al-Haddad, K. A review on multilevel converter topologies for electric transportation applications. In Proceedings of the IEEE-Vehicular Power and Propulsion Conference, Montréal, QC, Canada, 19–22 October 2015; pp. 1–6.

4. Kang, J.W.; Lee, H.; Hyun, S.W.; Kim, J.; Won, C.Y. An Enhanced Control Scheme Based on New Adaptive Filters for Cascaded NPC/H-Bridge System. *Energies* **2018**, *11*, 1034. [CrossRef]

5. Beniak, R.; Górecki, K.; Paduch, P.; Rogowski, K. Reduced Switch Count in Space Vector PWM for Three-Level NPC Inverter. *Energies* **2020**, *13*, 5945. [CrossRef]

6. Hakami, S.S.; Lee, K.-B. Four-Level Hysteresis-Based DTC for Torque Capability Improvement of IPMSM Fed by Three-Level NPC Inverter. *Electronics* **2020**, *9*, 1558. [CrossRef]

7. Feng, Z.; Zhang, X.; Wang, J.; Yu, S. A High-Efficiency Three-Level ANPC Inverter Based on Hybrid SiC and Si Devices. *Energies* **2020**, *13*, 1159. [CrossRef]

8. Teston, S.A.; Vilerá, K.V.; Mezaroba, M.; Rech, C. Control System Development for the Three-Ports ANPC Converter. *Energies* **2020**, *13*, 3967. [CrossRef]

9. Halabi, L.M.; Mohd Alsofyani, I.; Lee, K.-B. Open-Circuit Fault Tolerance Method for Three-Level Hybrid Active Neutral Point Clamped Converters. *Electronics* **2020**, *9*, 1535. [CrossRef]

10. Hammami, M.; Ricco, M.; Ruderman, A.; Grandi, G. Three-Phase Three-Level Flying Capacitor PV Generation System with an Embedded Ripple Correlation Control MPPT Algorithm. *Electronics* **2019**, *8*, 118. [CrossRef]

11. Rana, R.A.; Patel, S.A.; Muthusamy, A.; Lee, C.; Kim, H.-J. Review of Multilevel Voltage Source Inverter Topologies and Analysis of Harmonics Distortions in FC-MLI. *Electronics* **2019**, *8*, 1329. [CrossRef]

12. Estévez-Bén, A.A.; Alvarez-Diazcomas, A.; Rodríguez-Reséndiz, J. Transformerless Multilevel Voltage-Source Inverter Topology Comparative Study for PV Systems. *Energies* **2020**, *13*, 3261. [CrossRef]

13. Rodriguez, J.; Jih-Sheng, L.; Fang, P. Multilevel inverters: A survey of topologies, controls, and applications. *IEEE Trans. Ind. Electron.* **2002**, *49*, 724–738. [CrossRef]

14. Kang, J.-W.; Hyun, S.-W.; Ha, J.-O.; Won, C.-Y. Improved Neutral-Point Voltage-Shifting Strategy for Power Balancing in Cascaded NPC/H-Bridge Inverter. *Electronics* **2018**, *7*, 167. [CrossRef]

15. Marzoughi, A.; Burgos, R.; Boroyevich, D.; Xue, Y. Design and Comparison of Cascaded H-Bridge, Modular Multilevel Converter, and 5-L Active Neutral Point Clamped Topologies for Motor Drive Applications. *IEEE Trans. Ind. Appl.* **2018**, *54*, 1404–1413. [CrossRef]

16. Song, Z.; Tian, Y.; Yan, Z.; Chen, Z. Direct power control for three-phase two-level voltage-source rectifiers based on extended-state observation. *IEEE Trans. Ind. Electron.* **2016**, *63*, 4593–4603. [CrossRef]

17. Younis, M.A.; Rahim, N.A.; Mekhilef, S. High Efficiency THIPWM Three-Phase Inverter for Grid Connected System. In Proceedings of the 2010 IEEE Symposium on Industrial Electronics and Applications (ISIEA), Penang, Malaysia, 3–6 October 2010; pp. 88–93.

18. Ben-Brahim, L. A Discontinuous PWM Method for Balancing the Neutral Point Voltage in Three-Level Inverter-Fed Variable Frequency Drives. *IEEE Trans. Energy Convers.* **2008**, *23*, 1057–1063. [CrossRef]

19. Holmes, D.G.; Lipo, T.A. *Pulse Width Modulation for Power Converters*; Wiley: New York, NY, USA, 2003.

20. Hashempour, M.M.; Yang, M.Y.; Lee, T.L. An Adaptive Control of DPWM for Clamped-Three-Level Photovoltaic Inverters with Unbalanced Neutral-Point Voltage. *IEEE Trans. Ind. Appl.* **2018**, *54*, 6133–6148. [CrossRef]

21. Mese., H.; Ersak., A. Compensation of dead-time effects in three-level neutral point clamped inverters based on space vector PWM. In Proceedings of the International Aegean Conference on Electrical Machines and Power Electronics and Electromotion, Joint Conference 2011, Istanbul, Turkey, 8–10 September 2011; pp. 101–108.

22. Hwang, S.H.; Kim, J.M. Dead Time Compensation Method for Voltage-Fed PWM Inverter. *IEEE Trans. Energy Convers.* **2010**, *25*, 1–10. [CrossRef]

23. Herran, M.A.; Fischer, J.R.; Gonzalez, S.A.; Judewicz, M.G.; Carrica, D.O. Adaptive Dead-Time Compensation for Grid-Connected PWM Inverters of Single-Stage PV Systems. *IEEE Trans. Power Electron.* **2013**, *28*, 2816–2825. [CrossRef]

24. Ji, Y.; Yang, Y.; Zhou, J.; Ding, H.; Guo, X.; Padmanaban, S. Control Strategies of Mitigating Dead-time Effect on Power Converters: An Overview. *Electronics* **2019**, *8*, 196. [CrossRef]

25. Cheng, J.; Chen, D.; Chen, G. Modeling and Compensation for Dead-Time Effect in High Power IGBT/IGCT Converters with SHE-PWM Modulation. *Energies* **2020**, *13*, 4348. [CrossRef]

26. Nabae, A.; Takahashi, I.; Akagi, H. A New Neutral-Point-Point-Clamped PWM Inverter. *IEEE Trans. Ind. Appl.* **1981**, *IA-17*, 518–523. [CrossRef]

27. Iqbal, S.; Xin, A.; Jan, M.U.; Abdelbaky, M.A.; Rehman, H.U.; Salman, S.; Aurangzeb, M.; Rizvi, S.A.A.; Shah, N.A. Improvement of Power Converters Performance by an Efficient Use of Dead Time Compensation Technique. *Appl. Sci.* **2020**, *10*, 3121. [CrossRef]

28. Lai, Y.-S.; Chou, Y.-K.; Pai, S.-Y. Simple PWM Technique of Capacitor Voltage Balance for Three-Level Inverter with DC-link Voltage Sensor Only. In Proceedings of the IECON 2007—33rd Annual Conference of the IEEE Industrial Electronics Society, Taipei, Taiwan, 5–8 November 2007; pp. 5–8.

29. Jeong, S.-G.; Park, M.-H. The analysis and compensation of dead-time effects in PWM inverters. *IEEE Trans. Ind. Electron.* **1991**, *38*, 108–114. [CrossRef]

30. Hyun, S.W.; Hong, S.W.; Won, C.Y. A compensation method to reduce sampling delay of zero dead-time PWM using 3-level NPC PWM inverter. In Proceedings of the 2016 IEEE Transportation Electrification Conference and Expo, Asia-Pacific (ITEC Asia-Pacific), Busan, Korea, 1–4 June 2016; pp. 465–469.

31. Chaturvedi, P.; Jain, S.; Agarwal, P. Carrier-Based Neutral Point Potential Regulator with Reduced Switching Losses for Three-Level Diode-Clamped Inverter. *IEEE Trans. Ind. Electron.* **2014**, *61*, 613–624. [CrossRef]

32. Malakondareddy, B.; Kumar, S.S.; Gounden, N.A.; Anand, I. An adaptive pi control scheme to balance the neutral-point voltage in a solar pv fed grid connected neutral point clamped inverter. *Int. J. Electr. Power Energy Syst.* **2019**, *110*, 318–331. [CrossRef]

2

A Fast and Accurate Maximum Power Point Tracking Approach Based on Neural Network Assisted Fractional Open-Circuit Voltage

Ahmad Alzahrani 🆔

Electrical Engineering Department, Faculty of Engineering, Najran University, Najran 66446, Saudi Arabia; asalzahrani@nu.edu.sa

Abstract: This paper presents an enhanced maximum power point tracking approach to extract power from photovoltaic panels. The proposed method uses an artificial neural network technique to improve the fractional open-circuit voltage method by learning the correlation between the open-circuit voltage, temperature, and irradiance. The proposed method considers temperature variation and can eliminate the steady-state oscillation that comes with conventional algorithms, which improves the overall efficiency of the photovoltaic system. A comparison with the traditional and most widely used algorithms is discussed and shows the difference in performance. The presented algorithm is implemented with a Ćuk converter and tested under various weather and irradiance conditions. The results validate the competitiveness of the algorithm against other algorithms.

Keywords: DC-DC; MPPT; P&O; FOCV; PV; renewable; power electronics; solar

1. Introduction

The advancement of photovoltaic (PV) technologies and the drop in manufacturing costs have led to a sudden increase in the number of mega-watt-sized solar projects. Developed countries have planned to wean themselves off their reliance on fossil-fuels, as solar investment becomes an appealing option [1–3]. PV energy is superior to other sources because it does not require periodic maintenance, and PV panels have a lifetime of two and a half decades [4–6]. PV systems can be either grid-tied systems or off-grid systems [7]. In the grid-tied systems, the load is powered by the PV system, the main electric grid, or both [8]. A grid-tied system is less complicated than an off-grid system, which is more commonly seen since it has fewer components and is generally more cost effective; for example, there are usually no batteries in grid-tied systems [9]. The off-grid system can operate without a battery if the load is set to be powered during peak hours of the day, like a water pump system [10]. However, if the load is a home appliance, it is better to have enough energy storage devices with enough capacity for a couple of days [11].

The output power of the PV systems is intermittent and highly depends on weather and sun radiation levels. The sun irradiance affects the PV current directly, and the temperature affects mostly the output voltage [12–14]. Therefore, it is essential to have power processing units regulate the output voltage and extract the most out of the PV panels. The load seen by the PV determines the operating point on the power-voltage curve (P-V curve) and often does not match the resistance at the maximum power [15]. The resistance mismatch results in a loss of power and a reduction of the system's overall efficiency. The resistance seen by the PV needs to be varied to match the maximum power resistance at various temperatures and solar irradiance levels [16–19]. Power processing units enable the operation at the highest power if a maximum power point tracker (MPPT) controls the power processing unit.

MPPT is an algorithm or method that tracks the maximum operating point, which is necessary to extract the most out of the PV panel, enhance the operation of the overall system, and extend the lifetime of the PV panels [20].

Many MPPT methods have been developed and utilized [21–25] during the last two decades. Several studies evaluated the performance of such methods [26–30] based on both uniform solar irradiance and partially shaded conditions. The essential aspects of the comparisons are the implementation complexity, reaching the real maximum power point (MPP), the required number of sensors, convergence speed, cost, and efficiency [15,31]. Several other aspects are considered, such as the dependency on the PV panel or array and whether the control system is analog or digital. The most widely used method is perturbation and observation (P&O). In P&O, the power is calculated from the current and voltage sensors and then compared to the previous power reading. If the current reading is higher than the previous, the reference voltage should be increased until it reaches the maximum power. However, if the current reading is less than the previous reading, the reference voltage is higher, the maximum power voltage (Vmpp). Therefore, the reference voltage needs to be decreased. The P&O conversion speed depends on the perturbation size (Δ). That is, if Δ is small, the convergence speed will be slow. On the other hand, when Δ is large, the convergence speed will be fast. Very large Δ might cause the control system's instability, and it might never reach the real maximum power.

Several modifications to P&O have been suggested and experimented with, such as in [32–35], where the dynamic perturbation size was introduced. Similar methods to P&O, such as hill-climbing (HC) and incremental conductance (IC), were presented in [16,36], respectively, where their speed depends on the size of the step size. Several other methods were introduced, such as fuzzy logic, genetic algorithms, fractional open-circuit voltage (FOCV), fractional short-circuit current, neural networks, ripple correlation control, and many others. The most common methods will be discussed in detail and will be illustrated based on their performance in the next sections.

FOCV is the simplest and easiest to implement tracking approach. However, under the variation of temperature, the tracker cannot reach the true maximum power point. This paper presents an enhancement modification to the FOCV. The feed-forward neural network (FFNN) is used to learn the correlation between the temperature, irradiance, voltage at maximum power, and the open-circuit voltage and enhance the performance by providing the accurate voltage reference for the pulse width modulation controller. That is, the FFNN is used to find the fraction of the voltage at the maximum power and the open-circuit voltage at specific irradiance and temperature instead of guessing or estimating the value of the fraction. Therefore, the advantages of the proposed method are:

- It provides accurate results in all weather and irradiance conditions.
- It has a stable operation with no oscillation around the MPP.
- It is a cost-effective method, which does not require expensive sensing components.
- It has a faster transient response and can track the MPP during a sudden decrease in the solar irradiance.

The rest of this paper is organized as follows: Section 2 discusses several PV models and the models' performance under weather and temperature variations. Section 3 illustrates several MPPT methods and compares them to their performance, cost, and implementation. Section 4 presents the proposed modification to the fractional open-circuit and the implementation details. The proposed method results under several conditions and the performance comparison with other commonly used methods are presented in Section 5. Finally, the conclusions are presented in Section 6.

2. PV Characteristics and Performance

2.1. PV Mathematical Models

The simplest model of a PV cell can be represented by a light generated current source connected in parallel to an ideal diode. The current source depends on the level of solar irradiance, and the current generated is called the photocurrent I_{ph}. The equivalent circuit of this model is shown in Figure 1a. The PV current I_{PV} of the ideal model is given by:

$$I_{PV} = I_{ph} - I_0(e^{\frac{V_{PV}}{V_T}} - 1) \tag{1}$$

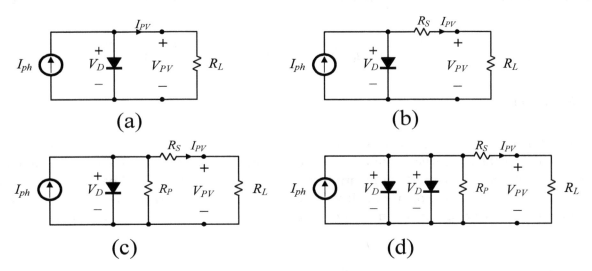

Figure 1. Several PV models: (a) the ideal model, which is lossless; (b) single diode with series resistance; (c) single diode with series and parallel resistances; (d) the two-diode model.

The V_T is the thermal voltage of the diode, which can be calculated by:

$$V_T = \frac{n k_{Boltzmann} T_{cell}}{q} \tag{2}$$

where n is called the ideality factor. The ideality factor indicates how close the diode is from the ideal diode. Therefore, $n = 1$ if the diode is ideal. T_{Cell} is the temperature of the cell in Kelvin, and $k_{Boltzmann}$ is the Boltzmann constant, which has a value of $k_{Boltzmann} = 1.38 \times 10^{-23}$ J/K. The amount of charge in one electron is $q \approx 1.602 \times 10^{-19}$ C . The ideal model takes into account the direct relationship between the photocurrent and the solar insolation. However, it is not enough to model the nonideality, such as the losses. The simple model is similar to the ideal model with the addition of a series resistance R_s, as shown in Figure 1b. The series resistance represents the losses in the low-doped semiconductor material, as well as the conduction losses in wires and contacts. The PV current is calculated by:

$$I_{pv} = I_{ph} - I_0(e^{\frac{V+I \times R_S}{V_T}} - 1) \tag{3}$$

The series resistance is not enough to accurately model all the losses. The PV cell suffers from the impurities of the p-n junction and the leakage current across the cell borders, which lead to power loss. Therefore, parallel resistances model the loss and improve the temperature sensitivity. Figure 1c shows the equivalent circuit of the PV cell with the parallel resistance. The current equation of a single diode model with series and parallel resistance is given by:

$$I_{pv} = I_{ph} - I_0(e^{\frac{V+I \times R_S}{V_T}} - 1) - \frac{V + I \times R_S}{R_P} \tag{4}$$

A higher level of modeling accuracy can be achieved using the two-diode model, as shown in Figure 1d. The second diode represents the junction recombination, which corrects the ideality factor at low voltages. The current of the two-diode model is given by:

$$I_{pv} = I_{ph} - I_{0_1}(e^{\frac{V+I\times R_S}{V_{T_1}}} - 1) - I_{0_2}(e^{\frac{V+I\times R_S}{V_{T_2}}} - 1) - \frac{V+I\times R_S}{R_P} \tag{5}$$

Several studies also introduced a three-diode model [37]. The current of the third diode represents recombinations in the blemished regions and grain boundaries. The current equation is given by:

$$I_{pv} = I_{ph} - I_{0_1}(e^{\frac{V+I\times R_S}{V_{T_1}}} - 1) - I_{0_2}(e^{\frac{V+I\times R_S}{V_{T_2}}} - 1) - I_{0_3}(e^{\frac{V+I\times R_S}{V_{T_3}}} - 1) - \frac{V+I\times R_S}{R_P} \tag{6}$$

2.2. PV Performance

The performance of a PV cell is related to its characteristic and power curves. The characteristic curve is also called the I-V curve, which shows the relationship between the PV voltage and current under specific solar irradiance and temperature, as shown in Figure 2a. Several parameters are known from the I-V curve, such as the open-circuit voltage, the short-circuit current, the voltage and current at the maximum power, and the maximum power point. Furthermore, the fill factor can be calculated, and it is given by:

$$FF = \frac{P_{mpp}}{I_{SC}\times V_{OC}} = \frac{V_{mpp}\times V_{mpp}}{I_{SC}\times V_{OC}} \tag{7}$$

The performance also can be indicated using a power curve (P-V) that shows the power versus the voltage. Figure 2b shows the power curve under standard conditions. Under normal operation, the curve has a single peak, which represents the maximum power point. The irradiance level and temperature influence both the I-V and P-V curves. The effect of temperature on the P-V curve is illustrated in Figure 3a. As the temperature increases, the maximum output power decreases. On the other hand, the effect of irradiance on the PV curve is shown in Figure 3b. As the solar irradiance increases, the maximum power increases. Similarly, with the I-V curve, the increase in temperature reduces the open-circuit voltage, as shown in Figure 4. The short-circuit current increases as the solar irradiance increases. The effects of temperature on the open-circuit voltage and short-circuit current are given by:

$$V_{oc}(T) = V_{oc,STC} + \Delta_v(T - T_{STC}) \tag{8}$$

$$I_{sc}(T) = I_{sc,STC}(1 + \Delta_i(T - T_{STC})) \tag{9}$$

where Δ_v and Δ_i the temperature coefficients for open-circuit voltage and the short-circuit current, respectively. The temperature of the standard test conditions (STCs) is given by T_{STC}. The open-circuit voltage $V_{oc,STC}$ and short-circuit current $I_{sc,STC}$ are measured in the STCs. The effect of solar irradiation on the short-circuit current can be characterized by:

$$I_{sc}(G, T) = I_{sc}(T)\frac{G}{G_{STC}} \tag{10}$$

where G_{STC} is the solar irradiation of the STCs, which is 1000 W/m^2:

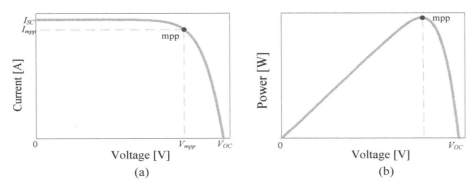

Figure 2. The performance curves of PV cells: (**a**) I-V characteristic curve. From this curve, one can find the short-circuit current, open-circuit voltage, voltage and current at maximum power, fill factor, and resistance at maximum power. (**b**) P-V curve, which shows the maximum output power.

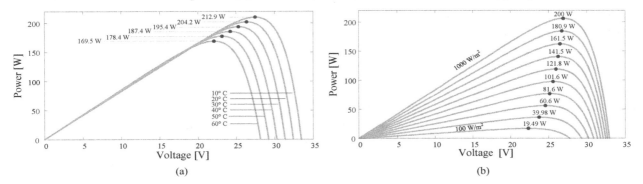

Figure 3. The P-V under variable conditions: (**a**) Under different temperature degrees. (**b**) Under different solar irradiance levels.

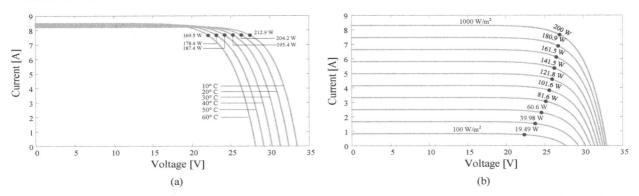

Figure 4. The I-V under variable conditions: (**a**) Under different temperature degrees. (**b**) Under different solar irradiance levels.

3. Maximum Power Point Tracking Approaches

A single PV cell has a unique mpp, corresponding to the solar irradiance and temperature point. The resistance at that point is represented by R_{mpp}. The R_{mpp} at the STCs can be calculated by:

$$R_{mpp,STC} = \frac{V_{mpp,STC}}{I_{mpp,STC}} \tag{11}$$

The variation in solar irradiance and temperature values makes R_{mpp} variable. Therefore, a tracker is needed to allow a PV panel to operate at R_{mpp} and extract the maximum power. Many approaches have been introduced to track the MPP. The commonly used technique in the solar industry is the P&O because of the reasonable trade-off between simplicity and performance. The P&O approach flowchart is seen in Figure 5. This approach perturbs the duty cycle and observes the power if it increases or decreases. Then, a step-change to the duty cycle takes place. The step-change determines the conversion speed and accuracy. P&O with a large step-change lacks accuracy and might experience instability issues. Making the step-change small reduces the conversion speeds, lacks speed, and might perform poorly at fast transitioning solar irradiance. The adaptive step can be applied, but without the guarantee of stable operation [36,38–40]. Similarly, hill-climbing and incremental conductance have similar performance to P&O, with no significant advantages. The ripple correlation control approach takes advantage of ripples imposed by switching devices on the PV panel [41,42]. In RCC, the derivative of the voltage is multiplied by the derivative of the power. If the product is zero, then the PV operates at its maximum power point. If the product is higher or lower than the MPP, then the PV operates on the left or the right of the MPP, respectively. The drawback of using RCC is the required multiplied devices, which are power-consuming devices.

Intelligent tracking approaches, such as fuzzy logic and artificial neural networks, were used to track the MPP. In fuzzy logic, the PV voltage and current are used to calculate the slope of the P-I characteristic and the change in the slope, which will be the input of the fuzzification stage. Fuzzification converts the numeric to linguistic variables to be used in the inference stage to make a decision. After a decision is made by a set of conditionals statements, the defuzzification stage converts the decision into numerical values [43–45]. Although the performance of the fuzzy controller is higher than the traditional P&O, the rules of the membership function selection are based on trial and error, which increases the implementation time. The neural network method was used to track the MPP. The technique uses acquired data, usually irradiance and temperature, to train the neurons and provide the optimal weight that makes the correct control decision about the duty [46,47]. The NN can track the true MPP and has solid performance, but it uses an expensive pyranometer to read the irradiance values.

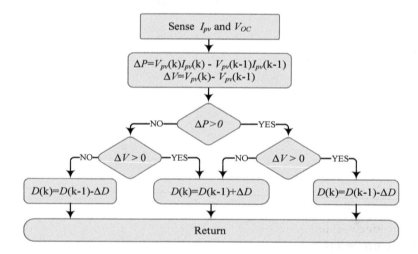

Figure 5. The flowchart of the P&O tracking method.

The fractional open-circuit voltage (FOCV) method is the simplest MPPT method. In this method, the MPP voltage can be approximated by multiplying the open-circuit voltage with a fraction number k. The k is a constant value predetermined by either experimentation or assumption, and it has a typical value ranging between 0.6 and 0.93. The open-circuit voltage must be sensed periodically. The challenge is that the open-circuit voltage cannot be measured while the system operates without disconnecting the system. Disconnecting the system momentarily to measure the open-circuit voltage leads to power loss and overall efficiency degradation. In order to find the open-circuit voltage without PV panel disconnection, a pilot cell with the same characteristics as the main PV panel is needed. Figure 6 shows the implementation of the FOCV with a pilot cell.

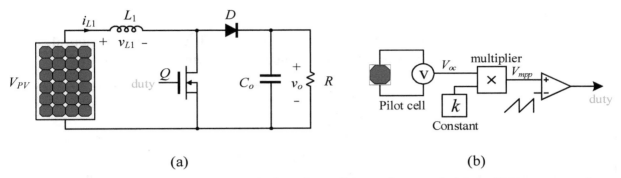

(a)   (b)

Figure 6. An example of how FOCV with the pilot cell is implemented: (**a**) the PV is connected to a conventional boost converter, which is controlled by the duty cycle (**b**) the pilot cell provides the reading of the V_{OC} to the control circuit.

4. The Proposed Method

4.1. Theory of Operation

The conventional FOCV is unable to reach the actual MPP, and neural network methods require expensive sensors to give accurate results. The proposed method uses a neural network to correlate the open-circuit voltage with temperature, therefore giving the right and precise reference voltage or duty cycle. The proposed FFNN is fed input from an external pilot circuit and a temperature sensor. The external pilot cell provides implicit information about the solar irradiance level and independent open-circuit voltage without system disconnection. Figure 7a shows the maximum power of the PV plotted versus the voltage and current of the PV panel. We use a neural network to learn and approximate maximum power functions perfectly under specific solar irradiance and temperature. Therefore, the reading of the irradiance and temperature is needed to find the real MPP. The ratio between the maximum power voltage V_{mpp} and the open-circuit voltage varies depending on the irradiance level and the temperature, as shown in Figure 7b. The PV panel's open-circuit voltage directly relates to the irradiance and an inverse relationship with the temperature, as illustrated in Figure 7c. Similarly, the effects of solar irradiance and temperature on the maximum power and the relationship between the maximum power and maximum voltage and current are shown in Figure 7d and Figure 7e, respectively. The figures show that the open-circuit voltage can provide information about the solar irradiance, and by using only the temperature sensor, the true MPPs can be found.

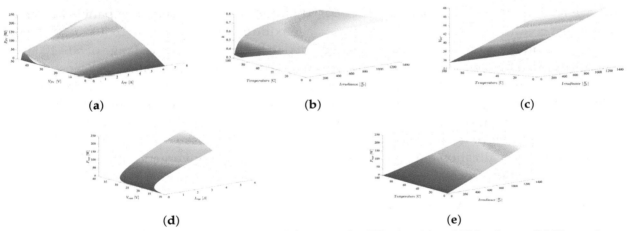

(**a**)   (**b**)   (**c**)

(**d**)   (**e**)

Figure 7. (**a**) The output power of the PV panel versus the PV current and PV voltage. (**b**) The ratio between the maximum voltage versus the irradiance and temperature. (**c**) The effects of solar irradiance and temperature on the open-circuit voltage of the PV panel. (**d**) The effects of solar irradiance and temperature on the maximum power of the PV panel. (**e**) The maximum power versus the maximum current and the maximum voltage of the PV.

4.2. Neural Network Based FOCV Setup

The setup has three stages; preprocessing, setting and training the neural network, and implementation and testing. Figure 8 shows the stages of implementing the controller.

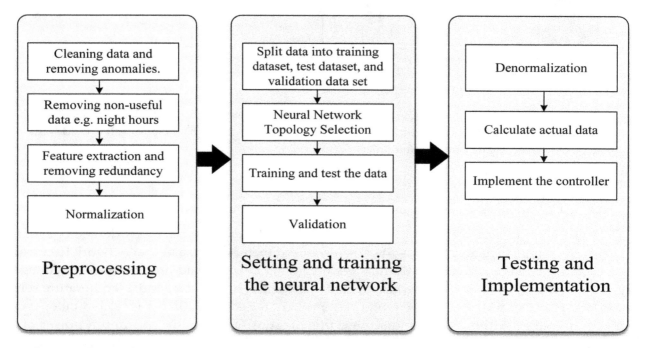

Figure 8. Stages for implementing the neural network based FOC controller. The implementation process has three stages; preprocessing, setting and training the neural network, and testing and implementation.

4.2.1. Preprocessing

The data can be obtained using voltage sweep at different solar irradiance, either experimentally or by simulation. Figure 9 shows the circuit used to obtain the dataset, which can be either simulation or experimental. A large dataset is needed to train the network and achieve higher test accuracy. The process can be automated in simulation using an iterative loop with a small step size, which allows for higher resolution data. The data obtained must be correctly labeled. Each maximum power point and open-circuit voltage correspond to specific solar irradiance and temperature values. In this study, the Kyocera KC200GT solar panel was used to perform the sweep. The panel was tested at standard test conditions (STCs), which had 200 W maximum power, $V_{mpp} = 26.3$ V, $I_{mpp} = 26.3$ A, $V_{OC} = 32.9\tilde{V}$, and $I_{SC} = 26.3$ A. The data were recorded in the solar irradiance range of $50-1000\tilde{W}/m^2$ and temperature range of $15-65$ °C. The recorded data require cleaning the incorrect sensor readings or other values that are out of PV range, for example negative values of power. Missing data points can be filled with the average of the row or the mode. Because the data are obtained by the voltage sweep, some data are redundant. The redundancy needs to be removed to improve the training process. Table 1 shows a sample of the recorded data after cleaning. The final step in the preprocessing stage is normalization. Normalization enhances the learning speed and convergence time [48]. The values after the normalization process lie between 0 and 1.

4.2.2. Setting the Network And Training

In this stage, the data are split into a training dataset and a testing dataset, either by using a holdout set or a cross-validation set. The validation dataset is optional, and the controller can be validated eventually after simulation and implementation. The holdout set used is 70% for training and 30% for the test. The selection of the neural network topology can be difficult due to the wide variety of neural network topologies and hyperparameters. A feedforward neural network is used in this paper to map labeled input data to labeled output data and learn the relationship. Figure 10 shows the structure of the training network. The training consists of two neural networks. The first neural network NN_1 is used to approximate the relationship between the open-circuit voltage and the solar irradiance under variable temperature values. NN_1 eliminates the need for an expensive pyranometer, as the solar irradiance can be estimated with the value of open-circuit voltage. The structure of NN_1 contains an input layer with two neurons, a hidden layer with 8 neurons, and a single output layer. The activation function of the hidden layer is *sigmoid* [49], which is given by:

$$f(z) = \frac{1}{1 + e^{-z}} \tag{12}$$

where z is nothing but the summation of the bias and the product of the weight and input examples $z = b + WX$. The output layer has a linear function where $f(z) = z$. The inputs to NN_1 are $V_{OC}(t)$ and the temperature $T(t)$, and the target is the solar irradiance $G(t)$. Similarly, NN_2 has an input layer that takes two inputs, the predicted irradiance and the temperature $T(t)$. The hidden layer consists of 10 neurons with the *sigmoid* activation function. The output layer has a linear activation function, and the output is k.

After selecting the topology, the weight is randomized, and the hyperparameters are set. Both networks are trained using the Levenberg–Marquardt backpropagation algorithm [50] to update the weight. The training took about 50 epochs to reach the goal set, a low mean squared error (MSE).

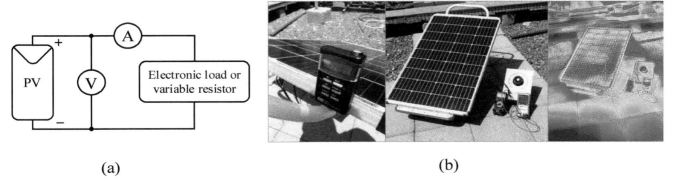

(a)   (b)

Figure 9. The approach used to produce the dataset for neural network training: (**a**) schematic diagram of the circuit (**b**) experimental setup. The electronic load used to sweep the voltage and get the I-V characteristic curve at specific irradiance and temperature values.

Table 1. Sample of the recorded data.

G	T	V_{mpp}	I_{mpp}	MPP	V_{OC}	K
400	20	26.3106	3.0091	79.1704	31.8532	0.826
800	45	23.7686	6.0453	143.6872	29.9841	0.7927
1000	50	23.2755	7.5521	175.7788	29.8026	0.781
600	25	26.0688	4.5389	118.323	31.9547	0.8158
100	30	22.896	0.714	16.3469	27.9174	0.8201
200	25	24.7221	1.4769	36.5124	29.9076	0.8266
300	35	23.9895	2.2404	53.747	29.3799	0.8165
900	40	24.4561	6.8125	166.607	30.8412	0.793
700	30	25.5538	5.3	135.436	31.602	0.8086
500	40	23.9789	3.7658	90.2998	29.7159	0.8069

4.2.3. Testing and Implementation of The Controller

After the training process is complete, the weight and bias are adjusted to map the relationship between the inputs and the output correctly. Then, the data are denormalized; the reverse of the normalization process. To evaluate the performance of the neural network, the mean squared error is used, which is given by:

$$MSE = \frac{1}{n} \sum_{t=1}^{n} (k - \hat{k})^2 \tag{13}$$

where $\hat{k}$ is the predicted output value and n is the number of samples. The learning curve of the training process and testing is shown in Figure 11. The MSE error is about 2.2×10^{-10} at the end of the 50th epoch. The neural network block is implemented using the following equation.

$$k = W_2 \times sigmoid(W_1 X + b_1) + b_2 \tag{14}$$

where W_1, W_2, b_1, b_2 are the weight matrix for the hidden layer, the weight matrix for the output layer, the bias for the hidden layer, and the output layer's bias, respectively. A multiplier is then used to find the voltage reference by multiplying the open-circuit voltage and the fraction k. A further safe operation can be considered by limiting the value of k to be between 0 and 1.

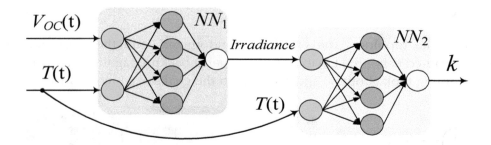

Figure 10. The two neural networks used in training: NN_1 used to approximate the relationship between the open-circuit voltage and the solar irradiance under variable temperature values, and NN_2 used to map the relationship between the irradiance and fractional open-circuit factor.

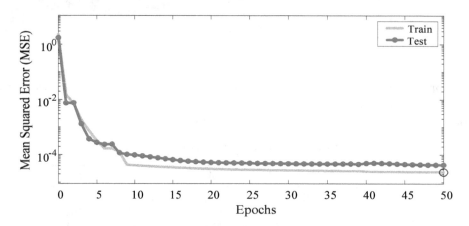

Figure 11. The learning curve for training NN_1. The performance of the test is similar to the training.

5. Simulations

The simulation setup of the proposed method was implemented using MATLAB and PLECS Blockset, as shown in Figure 12. The model KC200GT was used as an input source to supply a resistive load. The PV system was interfaced using a Ćuk converter, as shown in Figure 12a. The Ćuk converter was selected because of the non-pulsating input and output currents, which enhance the PV cell's lifetime and obtain better reading values of the PV current, as well as providing the ability to step-up and step-down the voltage. Then, the resistance of the maximum power point (Rmpp) can be tracked from zero to infinity ∞. The Ćuk converter is designed to operate in continuous conduction mode, where the inductor current at the input level must be above zero in the steady-state. In the continuous conduction mode, the converter goes into two subintervals. In Subinterval 1, the MOSFET is ON, and the diode is in reverse bias mode. The inductor L_1 is charging from the input source, and the capacitor C_1 is discharging to the output. The equivalent circuit of this subinterval is illustrated in Figure 13a. In Subinterval 2, the MOSFET is OFF, and the diode is in forwarding conduction mode. The capacitor C_1 is charging from the input source and the inductor L_1. The equivalent circuit of Subinterval 2 is shown in Figure 13b. The output voltage of the Ćuk converter is inverted and given by:

$$V_o = \frac{d}{1-d} V_{PV} \tag{15}$$

The parameters used in the simulation are $L_1 = 100$ µH, $L_2 = 10$ µH, $C = 10$ µF, $V_f = 0.8$ V, $R_{on} = 15$ mΩ. The MPPT tracker shown in Figure 12b consists of the pilot PV cell and the neural network. The pilot PV cell is connected directly to a voltage sensor to obtain the open-circuit voltage. The controller uses a pilot cell for open-circuit voltage measurements and a temperature sensor to provide temperature values. Then, the trained neural network takes the V_{OC} and $T(t)$ values at time t and provides the corresponding k. The value of k will be multiplied by V_{OC} to provide the reference voltage to the pulse width modulator (PWM). The PWM operates at a 100 kHz switching frequency, and the control bandwidth does not exceed one-tenth of the switching frequency.

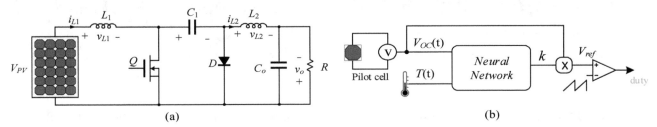

(a)   (b)

Figure 12. The experimental setup: (**a**) Ćuk converter (**b**) the proposed controller.

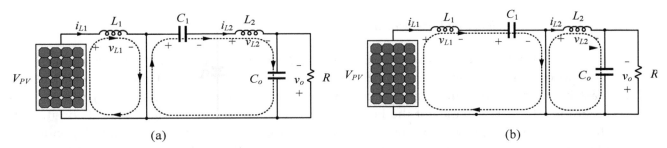

(a)   (b)

Figure 13. The subintervals in continuous conduction mode. (**a**) is when the active switch is ON; (**b**) is the when the active switch is OFF.

5.1. Case1: Variation of Solar Irradiance with a Constant Temperature Value

In this case, the output load and temperature are assumed constant for two reasons. First, the relationship between the output power and solar irradiance is stronger than the relationship between the output load and temperature, which gives an acceptable estimation of the tracking algorithm's performance if the temperature value is constant. Second, all algorithms need an ideal test, where the load is constant to find their best performance. Figure 14 shows the performance of the algorithms under the variation of solar irradiance. The proposed method outperforms the P&O algorithm in terms of accuracy and can reach the MPP faster. P&O oscillates around the MPP without reaching the exact MPP. The FOCV method has a similar dynamic performance to the proposed method. However, the accuracy of the FOCV is less than the proposed method.

Figure 15 shows a new way to compare the performance of the two algorithms. Therefore in Figure 15a, the power difference between the FOCV and the proposed method is shown to indicated positive and negative regions. The positive region is when the difference in power is higher than zero, which indicates that the power produced using the proposed method is higher than the power produced using the FOCV. The negative region is when the difference in power is less than zero and indicates that the FOCV performs better. The comparison with P&O is shown in Figure 15b.

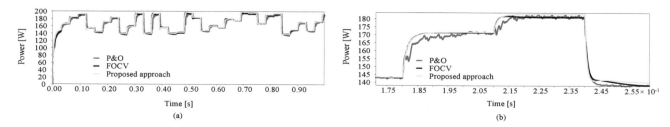

Figure 14. The output waveform of the PV panel for Case 1 using various tracking algorithm: (**a**) The performance over a long time period. (**b**) Zoomed area.

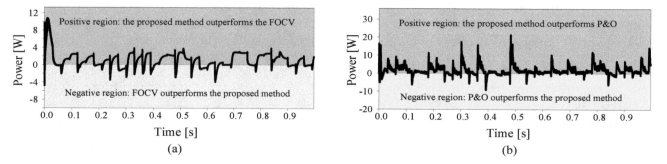

Figure 15. Case 1: comparison between the proposed method with (**a**) the FOCV method and (**b**) the P&O method.

5.2. Variation of Both Irradiance and Temperature

This case simulates a more realistic situation, where the temperature and irradiance vary. The output power, in this case, is shown in Figure 16. It can be seen that the proposed method exhibits faster dynamic speed and higher accuracy than the other methods. The varying temperature greatly degrades the performance of the FOCV since the open-circuit voltage depends on it. P&O is also affected by the temperature variation. However, since the P&O algorithm is independent of the panel information, the accuracy is higher than the FOCV method. The proposed method is faster during transitions from one irradiance level to another.

5.3. Energy Analysis

It can be seen from Figures 14–17 that the proposed converter outperforms the other tracking algorithms. The overall comparison can be determined using the average of the difference between the two tracking methods. In the case of fixed temperature, the extracted energy by the proposed method is about 4.27 kW/y higher than the FOCV and 5.1 kWh/y higher than P&O. In the case of variable temperature, the energy extracted by the proposed method is 12.1 kWh/y higher than the extracted energy by the FOCV and 8.9 kWh/y higher than the extracted energy by P&O. Note that the previous analysis is per panel; the difference in energy extraction would be much higher for a large PV system. The performance comparison is summarized in Table 2. The energy and efficiency of the proposed are higher than the other methods, with better transitioning time and stability.

Table 2. Performance comparison of the proposed method with commonly used methods.

Method	Case 1		Case 2		Performance			
	Energy (kW/year)	$\eta\%$	Energy (kW/year)	$\eta\%$	Oscillation	Transition	True MPP	Implementation
FOCV	358.6	98.2	349.6	95.8	No	Fast	NO	Digital and analog
P&O	375.1	98	352.9	96.6	Yes	Slow	Yes	Digital and analog
Proposed	362.8	99.4	361.7	99.1	No	Fast	Yes	Digital

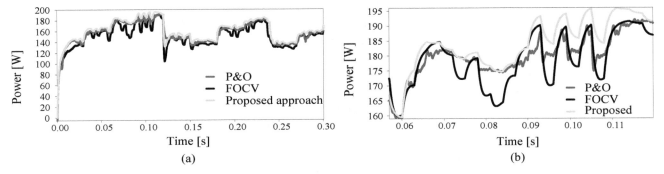

Figure 16. The output waveform of the PV panel for Case 2 using various tracking algorithms: (**a**) The performance over a long time period. (**b**) Zoomed area.

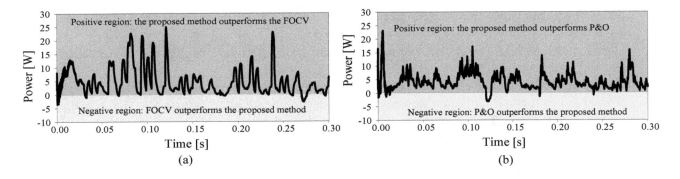

Figure 17. Case 2: comparison between the proposed method with (**a**) the FOCV method and (**b**) the P&O method.

6. Conclusions

A novel approach for finding the maximum power point of a photovoltaic system is developed. The approach is based on the fractional open-circuit voltage to find the true maximum power point using an artificial neural network approach. The neural network can correlate the reference voltage of the maximum power to the open-circuit voltage and temperature. Unlike other neural network approaches, the presented approach does not require expensive irradiance sensor circuitry. The proposed method is used to control the Ćuk converter to interface with a 200 W solar panel. Both the simulation and experimental results confirm the validation of the new approach. The results show that the proposed method has a fast transitioning time, better efficiency, and higher extracted energy than other methods.

Acknowledgments: The author is thankful to the Deanship of Scientific Research at Najran University for funding this work through Grant Research Code NU/ESCI/17/075.

Abbreviations

The following abbreviations are used in this manuscript:

PV   Photovoltaics
MPPT   Maximum power point tracking
FFNN   Feedforward neural network
FOCV   Fractional circuit voltage

References

1.   Fayaz, H.; Rahim, N.; Saidur, R.; Solangi, K.; Niaz, H.; Hossain, M. Solar energy policy: Malaysia vs developed countries. In Proceedings of the 2011 IEEE Conference on Clean Energy and Technology (CET), Kuala Lumpur, Malaysia, 27–29 June 2011; pp. 374–378.

2.   Mendoza, J.M.F.; Gallego-Schmid, A.; Rivera, X.C.S.; Rieradevall, J.; Azapagic, A. Sustainability assessment of home-made solar cookers for use in developed countries. *Sci. Total Environ.* **2019**, *648*, 184–196. [CrossRef] [PubMed]

3.   Liu, Z. What is the future of solar energy? Economic and policy barriers. *Energy Sources Part Econ. Planning Policy* **2018**, *13*, 169–172. [CrossRef]

4.   Irfan, M.; Zhao, Z.Y.; Ikram, M.; Gilal, N.G.; Li, H.; Rehman, A. Assessment of India's energy dynamics: Prospects of solar energy. *J. Renew. Sustain. Energy* **2020**, *12*, 053701. [CrossRef]

5.   Messenger, R.A.; Abtahi, A. *Photovoltaic Systems Engineering*; CRC Press: Boca Raton, FL, USA, 2017.

6.   Tyagi, V.; Rahim, N.A.; Rahim, N.; Jeyraj, A.; Selvaraj, L. Progress in solar PV technology: Research and achievement. *Renew. Sustain. Energy Rev.* **2013**, *20*, 443–461. [CrossRef]

7.   Villalva, M.G.; Gazoli, J.R.; Ruppert Filho, E. Comprehensive approach to modeling and simulation of photovoltaic arrays. *IEEE Trans. Power Electron.* **2009**, *24*, 1198–1208. [CrossRef]

8.   Shi, Y.; Li, R.; Xue, Y.; Li, H. High-frequency-link-based grid-tied PV system with small DC-link capacitor and low-frequency ripple-free maximum power point tracking. *IEEE Trans. Power Electron.* **2015**, *31*, 328–339. [CrossRef]

9.   Ghenai, C.; Bettayeb, M. Design and optimization of grid-tied and off-grid solar PV systems for super-efficient electrical appliances. *Energy Effic.* **2020**, *13*, 291–305. [CrossRef]

10.   Rafique, M.M.; Bahaidarah, H.M.; Anwar, M.K. Enabling private sector investment in off-grid electrification for cleaner production: Optimum designing and achievable rate of unit electricity. *J. Clean. Prod.* **2019**, *206*, 508–523. [CrossRef]

11.   Kavya Santhoshi, B.; Mohana Sundaram, K.; Padmanaban, S.; Holm-Nielsen, J.B.; KK, P. Critical review of PV grid-tied inverters. *Energies* **2019**, *12*, 1921. [CrossRef]

12.   Qing, X.; Niu, Y. Hourly day-ahead solar irradiance prediction using weather forecasts by LSTM. *Energy* **2018**, *148*, 461–468. [CrossRef]

13.   Khosravi, A.; Koury, R.; Machado, L.; Pabon, J. Prediction of hourly solar radiation in Abu Musa Island using machine learning algorithms. *J. Clean. Prod.* **2018**, *176*, 63–75. [CrossRef]

14.   Dong, N.; Chang, J.F.; Wu, A.G.; Gao, Z.K. A novel convolutional neural network framework based solar irradiance prediction method. *Int. J. Electr. Power Energy Syst.* **2020**, *114*, 105411. [CrossRef]

15.   Esram, T.; Chapman, P. Comparision of Photovoltaic Array MPPT Techniques. *IEEE Trans. Energy Convers.* **2007**, *22*, 439–449. [CrossRef]

16.   Liu, F.; Duan, S.; Liu, F.; Liu, B.; Kang, Y. A variable step size INC MPPT method for PV systems. *IEEE Trans. Ind. Electron.* **2008**, *55*, 2622–2628.

17.   Subudhi, B.; Pradhan, R. A comparative study on maximum power point tracking techniques for photovoltaic power systems. *IEEE Trans. Sustain. Energy* **2012**, *4*, 89–98. [CrossRef]

18.   Abdelsalam, A.K.; Massoud, A.M.; Ahmed, S.; Enjeti, P.N. High-performance adaptive perturb and observe MPPT technique for photovoltaic-based microgrids. *IEEE Trans. Power Electron.* **2011**, *26*, 1010–1021. [CrossRef]

19.   Mei, Q.; Shan, M.; Liu, L.; Guerrero, J.M. A novel improved variable step-size incremental-resistance MPPT method for PV systems. *IEEE Trans. Ind. Electron.* **2010**, *58*, 2427–2434. [CrossRef]

20.   Lin, L.; Zhang, J.; Shao, S. Differential Power Processing Architecture With Virtual Port Connected in Series and MPPT in Submodule Level. *IEEE Access* **2020**, *8*, 137897–137909. [CrossRef]

21. Motahhir, S.; El Hammoumi, A.; El Ghzizal, A. The most used MPPT algorithms: Review and the suitable low-cost embedded board for each algorithm. *J. Clean. Prod.* **2020**, *246*, 118983. [CrossRef]

22. Bollipo, R.B.; Mikkili, S.; Bonthagorla, P.K. Critical Review on PV MPPT Techniques: Classical, Intelligent and Optimisation. *IET Renew. Power Gener.* **2020**, *14*, 1433–1452.

23. Pathak, P.K.; Yadav, A.K.; Alvi, P. Advanced Solar MPPT Techniques Under Uniform and Non-Uniform Irradiance: A Comprehensive Review. *J. Sol. Energy Eng.* **2020**, *142*. [CrossRef]

24. Mansoor, M.; Mirza, A.F.; Ling, Q. Harris hawk optimization-based MPPT control for PV Systems under Partial Shading Conditions. *J. Clean. Prod.* **2020**, *274*, 122857. [CrossRef]

25. Abdel-Rahim, O.; Alamir, N.; Orabi, M.; Ismeil, M. Fixed-frequency phase-shift modulated PV-MPPT for LLC resonant converters. *J. Power Electron.* **2020**, *20*, 279–291. [CrossRef]

26. Pradhan, R.; Panda, A. Performance evaluation of a MPPT controller with model predictive control for a photovoltaic system. *Int. J. Electron.* **2020**, *107*, 1543–1558. [CrossRef]

27. De Brito, M.A.G.; Galotto, L.; Sampaio, L.P.; e Melo, G.d.A.; Canesin, C.A. Evaluation of the main MPPT techniques for photovoltaic applications. *IEEE Trans. Ind. Electron.* **2012**, *60*, 1156–1167. [CrossRef]

28. Chen, W.; Shen, H.; Shu, B.; Qin, H.; Deng, T. Evaluation of performance of MPPT devices in PV systems with storage batteries. *Renew. Energy* **2007**, *32*, 1611–1622. [CrossRef]

29. Uoya, M.; Koizumi, H. A calculation method of photovoltaic array's operating point for MPPT evaluation based on one-dimensional Newton–Raphson method. *IEEE Trans. Ind. Appl.* **2014**, *51*, 567–575. [CrossRef]

30. Rezk, H.; Mazen, A.O.; Gomaa, M.R.; Tolba, M.A.; Fathy, A.; Abdelkareem, M.A.; Olabi, A.; Abou Hashema, M. A novel statistical performance evaluation of most modern optimization-based global MPPT techniques for partially shaded PV system. *Renew. Sustain. Energy Rev.* **2019**, *115*, 109372. [CrossRef]

31. Garcia, M.; Maruri, J.M.; Marroyo, L.; Lorenzo, E.; Pérez, M. Partial shadowing, MPPT performance and inverter configurations: observations at tracking PV plants. *Prog. Photovoltaics Res. Appl.* **2008**, *16*, 529–536. [CrossRef]

32. Femia, N.; Petrone, G.; Spagnuolo, G.; Vitelli, M. Optimization of perturb and observe maximum power point tracking method. *IEEE Trans. Power Electron.* **2005**, *20*, 963–973. [CrossRef]

33. Hussein, K.; Muta, I.; Hoshino, T.; Osakada, M. Maximum photovoltaic power tracking: An algorithm for rapidly changing atmospheric conditions. *IEE Proc. Gener. Transm. Distrib.* **1995**, *142*, 59–64. [CrossRef]

34. Salas, V.; Olias, E.; Barrado, A.; Lazaro, A. Review of the maximum power point tracking algorithms for stand-alone photovoltaic systems. *Sol. Energy Mater. Sol. Cells* **2006**, *90*, 1555–1578. [CrossRef]

35. Elgendy, M.A.; Zahawi, B.; Atkinson, D.J. Assessment of perturb and observe MPPT algorithm implementation techniques for PV pumping applications. *IEEE Trans. Sustain. Energy* **2011**, *3*, 21–33. [CrossRef]

36. Xiao, W.; Dunford, W.G. A modified adaptive hill climbing MPPT method for photovoltaic power systems. In Proceedings of the 2004 IEEE 35th Annual Power Electronics Specialists Conference (IEEE Cat. No. 04CH37551), Aachen, Germany, 20–25 June 2004; Volume 3, pp. 1957–1963.

37. Nishioka, K.; Sakitani, N.; Uraoka, Y.; Fuyuki, T. Analysis of multicrystalline silicon solar cells by modified 3-diode equivalent circuit model taking leakage current through periphery into consideration. *Sol. Energy Mater. Sol. Cells* **2007**, *91*, 1222–1227. [CrossRef]

38. Brunton, S.L.; Rowley, C.W.; Kulkarni, S.R.; Clarkson, C. Maximum power point tracking for photovoltaic optimization using ripple-based extremum seeking control. *IEEE Trans. Power Electron.* **2010**, *25*, 2531–2540. [CrossRef]

39. Mohammad, A.N.M.; Radzi, M.A.M.; Azis, N.; Shafie, S.; Zainuri, M.A.A.M. An Enhanced Adaptive Perturb and Observe Technique for Efficient Maximum Power Point Tracking Under Partial Shading Conditions. *Appl. Sci.* **2020**, *10*, 3912

40. Liu, F.; Kang, Y.; Zhang, Y.; Duan, S. Comparison of P&O and hill climbing MPPT methods for grid-connected PV converter. In Proceedings of the 2008 3rd IEEE Conference on Industrial Electronics and Applications, Singapore, 3–5 June 2008; pp. 804–807.

41. Esram, T.; Kimball, J.W.; Krein, P.T.; Chapman, P.L.; Midya, P. Dynamic maximum power point tracking of photovoltaic arrays using ripple correlation control. *IEEE Trans. Power Electron.* **2006**, *21*, 1282–1291. [CrossRef]

42. Hammami, M.; Grandi, G. A single-phase multilevel PV generation system with an improved ripple correlation control MPPT algorithm. *Energies* **2017**, *10*, 2037. [CrossRef]

43. Kottas, T.L.; Boutalis, Y.S.; Karlis, A.D. New maximum power point tracker for PV arrays using fuzzy controller in close cooperation with fuzzy cognitive networks. *IEEE Trans. Energy Convers.* **2006**, *21*, 793–803. [CrossRef]

44. Alajmi, B.N.; Ahmed, K.H.; Finney, S.J.; Williams, B.W. Fuzzy-logic-control approach of a modified hill-climbing method for maximum power point in microgrid standalone photovoltaic system. *IEEE Trans. Power Electron.* **2010**, *26*, 1022–1030. [CrossRef]

45. Bendib, B.; Krim, F.; Belmili, H.; Almi, M.; Boulouma, S. Advanced Fuzzy MPPT Controller for a stand-alone PV system. *Energy Procedia* **2014**, *50*, 383–392. [CrossRef]

46. Lin, W.M.; Hong, C.M.; Chen, C.H. Neural-network-based MPPT control of a stand-alone hybrid power generation system. *IEEE Trans. Power Electron.* **2011**, *26*, 3571–3581. [CrossRef]

47. Zečević, Ž.; Rolevski, M. Neural Network Approach to MPPT Control and Irradiance Estimation. *Appl. Sci.* **2020**, *10*, 5051. [CrossRef]

48. Mueller, A.V.; Hemond, H.F. Extended artificial neural networks: Incorporation of a priori chemical knowledge enables use of ion selective electrodes for in-situ measurement of ions at environmentally relevant levels. *Talanta* **2013**, *117*, 112–118. [CrossRef] [PubMed]

49. Simpson, P.K. *Artificial Neural Systems: Foundations, Paradigms, Applications, and Implementations*; Elsevier Science Inc.: Amsterdam, The Netherlands, 1989.

50. Hagan, M.T.; Menhaj, M.B. Training feedforward networks with the Marquardt algorithm. *IEEE Trans. Neural Netw.* **1994**, *5*, 989–993. [CrossRef] [PubMed]

3

Design, Simulation and Hardware Implementation of Shunt Hybrid Compensator using Synchronous Rotating Reference Frame (SRRF)-Based Control Technique

R. Balasubramanian, K. Parkavikathirvelu, R. Sankaran and Rengarajan Amirtharajan *

School of Electrical & Electronics Engineering (SEEE)S, SASTRA Deemed University, Thirumalaisamudram, Thanjavur 613401, India; rbalu@eee.sastra.edu (R.B.); to_parkavi@eee.sastra.edu (K.P.); rs@eee.sastra.edu (R.S.)
* Correspondence: amir@ece.sastra.edu

Abstract: This paper deals with the design, simulation, and implementation of shunt hybrid compensator to maintain the power quality in three-phase distribution networks feeding different types balanced and unbalanced nonlinear loads. The configuration of the compensator consists of a selective harmonic elimination passive filter, a series-connected conventional six-pulse IGBT inverter, acting as the active filter terminated with a DC link capacitor. The theory and modelling of the compensator based on current harmonic components at the load end and their decomposition in d-q axis frame of reference are utilized in the reference current generation algorithm. Accordingly, the source current waveform is made to follow the reference current waveform using a high-frequency, carrier-based controller. Further, this inner current control loop is supported by a slower outer voltage control loop for sustaining desirable DC link voltage. Performance of the compensator is evaluated through MATLAB simulation covering different types of loads and reduction of harmonic currents and THD at the supply side along with excellent regulation of DC link voltage are confirmed. The performance of a hybrid compensator designed and fabricated using the above principles is evaluated and corroborated with the simulation results.

Keywords: harmonics; hybrid power filter; active power filter; power quality; total harmonic distortion

1. Introduction

One major area of research that has gained attention in recent times is maintaining power quality of distribution systems. The power quality issues arise due to the widespread usage of processed power in industrial applications and commercial/domestic applications [1–3]. For example, variable speed drives are implemented through power modulators which consist of high-power controlled/uncontrolled rectifiers feeding variable voltage and variable frequency multiphase inverters. Similarly, commercial power consumption is characterised by appliances like computers, photocopiers, and fax machines, along with fluorescent and CFL lamps. All the above represent nonlinear loads, resulting in lower supply-side power factor and waveform distortion, indicated by harmonic components in voltage and current. The adverse effect of harmonics includes heating and extra losses, saturation and malfunctioning of distribution transformers, interference with communication signals, damages to consumer utilities, and in extreme cases, the failure of supply-side equipment [4].

The initial steps towards mitigation of the above problems were focused on the low power factor at the supply side only, whereby a passive power filter (PPF) connected in shunt compensates for the lagging reactive current, which was extended with selective harmonic elimination. The disadvantage of this approach is insensitivity with load current changes and fluctuations in supply-side voltage.

The filter performance also depends on the load power factor, which may be variable. For example, in [5], the load is a motor that can work at different conditions. As a means of overcoming these problems, various active compensator topologies comprising of series and shunt elements have been proposed which have gained wide acceptance. However, some of the disadvantages, like high initial and operating cost due to the use of high-rating semiconductors and also the need for maintaining a high DC link voltage, have limited the application of pure shunt active power filter in medium- and high-power installations [6,7]. As a result, the stage was set for development of shunt hybrid power filters (HPFs), which represent a judicious combination of both passive and active power filters (APFs). The various hybrid compensator configurations are: (i) series active power filter and shunt passive power filter, (ii) shunt active power filter and shunt passive power filter and (iii) a series combination of passive power filter and active power compensator connected in shunt with the system. The series element in configuration (i) has to be rated for maximum load current and is not flexible for many applications. The configuration (ii) with two independent compensators in parallel requires both blocks to be rated at the supply voltage leading to high DC link capacitor voltage rating [8–10].

In comparison, the third topology where the passive power filter elements appear in series with the standard active power filter circuitry poses important advantages in terms of reflecting nearly zero impedance of the passive power filter for load current harmonics and at the same time high impedance for system side voltage harmonics. Further, it leads to absorbing the fundamental voltage component across the passive power filter, thereby reducing the voltage rating of the DC link capacitor to only the harmonic components which are to be suppressed. Accordingly, the rating and cost of the capacitor and the power semiconductor switches in the compensator are considerably reduced [8]. Further, this topology is effective in preventing system resonance, reducing switching noise and avoidance of any circulating current in the compensator.

The control requirements for all configurations involving active power filter boil down to the generation of reference current waveform, switching and triggering timings for semiconductor switches to match the source current and also for maintaining the desired voltage across the DC link capacitor. Accordingly, a variety of control schemes supported by related algorithms have been reported in the literature [11–20]. Time domain methods provide fast response, compared with frequency domain methods. Accordingly, many authors have proposed control techniques such as instantaneous reactive power theory, synchronous rotating reference frame (SRRF) theory, sliding mode controller, neural network techniques and feedforward control to improve the performance of both active as well as hybrid filters.

In this paper, the third topology has been utilised, where the passive power filter is designed with the aim of 5th and 7th selective harmonic elimination along with an active power compensator connected in series. The active component of the compensator is modelled in the stationary abc frame of mains and further transformed to the rotating dq frame to avoid time dependence of parameters to reduce the control complexity. A control technique using PI controller, based on decoupled currents, is used to inject currents from the compensator to ensure tracking of the reference waveform [13,14,21]. An independent outer control loop using another PI controller regulates the DC link voltage for sustained operation of the compensator. The parallel combination of 5th and 7th tuned passive harmonic filters connected in series with active filter configuration, the fundamental voltage of the system mainly drops on the PPF capacitor, not in the APF. Hence, the APF DC link voltage has been reduced with an objective of APF voltage rating reduction. The passive filter parameters present in this topology not only function as harmonic filter but also act as a filter for the switching ripples present in the system. Finally, the entire power and control circuits are fabricated, where the FPGA development kit SPARTAN-6 has been employed as the controller. Further, the performance improvement of the supply system along with the fabricated compensator has been evaluated by carrying out by a series of experiments on a prototype system and the results are presented covering different loading conditions.

2. System Configuration of Hybrid Compensator

The system shown in Figure 1 consists of a standard three-phase 400 V, 50 Hz mains connected to a pair of three-phase uncontrolled bridge rectifiers feeding individual R-L, R-C loads, which introduce harmonic currents in the supply system due to the nonlinearity. The shunt compensator contains 5th and 7th tuned passive filters, which are in parallel and the combination is connected in series with low voltage rated active power filter. The parallel connected passive filters having selective harmonic elimination are meant for effective reactive power compensation. This system eliminates the inherent disadvantages of both active and passive filters. The hybrid compensator is operated such that the distortion currents generated due to nonlinear loads are confined within the PPF and do not flow in the AC supply mains. The fundamental voltage drop across passive power filter components permits shunt APF to operate at low DC link voltage. The three-phase source voltage, the voltage at the point of common coupling, load and compensator currents are denoted as V_{abc}, V_{s123}, i_{L123} and i_{c123}, respectively, and shown individually in Figure 1.

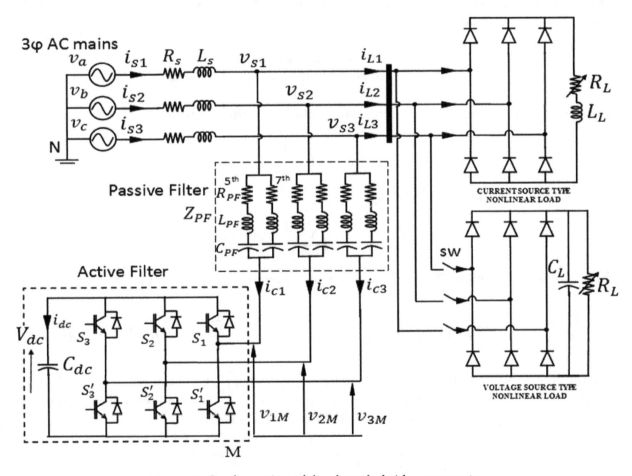

Figure 1. Configuration of the shunt hybrid compensator.

2.1. Modelling of Shunt Hybrid Power Compensator

By applying Kirchhoff's laws in Figure 1, the following equations in differential form in stationary three-phase reference frame are obtained, where the R_{PFe}, L_{PFeq} and C_{PFeq} are equivalent parameter values of the 5th and 7th selective harmonic filters.

$$v_{sk} = L_{PFeq}\frac{di_{ck}}{dt} + R_{PFeq}i_{ck} + (1/C_{PFeq})\int i_{ck}dt + v_{kM} + v_{MN} \qquad (1)$$

where $k = 1, 2, 3$ represent the three phases.

Differentiation of Equation (1) to eliminate the integral term yields

$$\frac{dv_{sk}}{dt} = L_{PFeq}\frac{d^2i_{ck}}{dt^2} + R_{PFeq}\frac{di_{ck}}{dt} + \frac{1}{C_{PFeq}}i_{ck} + \frac{dv_{kM}}{dt} + \frac{dv_{MN}}{dt} \tag{2}$$

Assuming balanced three-phase supply voltage yields

$$v_{s1} + v_{s2} + v_{s3} = 0$$

Summing the three equations included in (1) for k = 1, 2, 3 and assuming nonexistence of the zero-sequence current into three-wire system [16] results in

$$v_{MN} = -\frac{1}{3}\sum_{k=1}^{3} v_{kM} \tag{3}$$

The switching function C_k [12] of the kth leg of the inverter is the state of the power semiconductor devices S_k and S_k' and is defined as

$$C_k = \begin{cases} 1, \ if \ S_k \ is \ On \ and \ S_k' \ is \ Off \\ 0, \ if \ S_k \ is \ Off \ and \ S_k' \ is \ On \end{cases} \tag{4}$$

Thus, with $v_{kM} = C_k V_{dc}$ and differentiation of the same, this leads to

$$\frac{dv_{kM}}{dt} = C_k \frac{dV_{dc}}{dt} \tag{5}$$

By differentiating Equation (3), we get

$$\frac{dv_{MN}}{dt} = -\frac{1}{3}\sum_{k=1}^{3}\frac{d}{dt}(C_k V_{dc}) \tag{6}$$

Substitution of Equations (5) and (6) into (2) and rearranging the same yields

$$\frac{d^2i_{ck}}{dt^2} = -\frac{R_{PFeq}}{L_{PFeq}}\frac{di_{ck}}{dt} - \frac{1}{C_{PFeq}L_{PFeq}}i_{ck} - \frac{1}{L_{PFeq}}\left(C_k - \frac{1}{3}\sum_{m=1}^{3}C_m\right)\frac{dV_{dc}}{dt} + \frac{1}{L_{PFeq}}\frac{dv_{sk}}{dt} \tag{7}$$

defining the switching state function as $q_{nk} = \left(C_k - \frac{1}{3}\sum_{m=1}^{3}C_m\right)_n$, where, n = 0 or 1.

In other words, the vector q_n epends on the parameters C_1, C_2, C_3 through a matrix transformation given below, indicating the interaction among the three phases [15].

$$\begin{bmatrix} q_{n1} \\ q_{n2} \\ q_{n3} \end{bmatrix} = \frac{1}{3}\begin{bmatrix} 2 & -1 & -1 \\ -1 & 2 & -1 \\ -1 & -1 & 2 \end{bmatrix}\begin{bmatrix} C_1 \\ C_2 \\ C_3 \end{bmatrix} \tag{8}$$

$$\frac{d^2i_{ck}}{dt^2} = -\frac{R_{PFeq}}{L_{PFeq}}\frac{di_{ck}}{dt} - \frac{1}{C_{PFeq}L_{PFeq}}i_{ck} - \frac{1}{L_{PFeq}}q_{nk}\frac{dV_{dc}}{dt} + \frac{1}{L_{PFeq}}\frac{dv_{sk}}{dt} \tag{9}$$

The capacitor current $i_{dc} = i_{c1} + i_{c2} + i_{c3}$ and is related to V_{dc} by

$$\frac{dV_{dc}}{dt} = \frac{1}{C_{dc}}i_{dc} \tag{10}$$

Expressing the DClink capacitor current i_{dc} in terms of the switching and compensator currents, the following equation is obtained

$$\frac{dV_{dc}}{dt} = \frac{1}{C_{dc}} \sum_{k=1}^{3} q_{nk} i_{ck} = \frac{1}{C_{dc}} [q_{n123}]^T [i_{c123}] \tag{11}$$

In the nonexistence of zero-sequence currents, the variables i_{c3} and q_{n3} can be eliminated by the substitution of $i_{c3} = -(i_{c1} + i_{c2})$ and $q_{n3} = -(q_{n1} + q_{n2})$ so that Equation (11) for the modelling of the capacitor is modified as follows.

$$\frac{dV_{dc}}{dt} = \frac{1}{C_{dc}} [2q_{n1} + q_{n2}] i_{c1} + \frac{1}{C_{dc}} [q_{n1} + 2q_{n2}] i_{c2} \tag{12}$$

The complete model of the shunt hybrid power compensator in abc reference frame is indicated by Equations (9) and (12).

2.2. Equations in dq Frame

The model given by Equations (9) and (12) is transformed into synchronous orthogonal frame using the transformation matrix [13]

$$T_{dq}^{123} = \sqrt{\frac{2}{3}} \begin{bmatrix} \cos\theta & \cos\left(\theta - \frac{2\pi}{3}\right) & \cos\left(\theta - \frac{4\pi}{3}\right) \\ -\sin\theta & -\sin\left(\theta - \frac{2\pi}{3}\right) & -\sin\left(\theta - \frac{4\pi}{3}\right) \end{bmatrix} \tag{13}$$

where $\theta = \omega t$ and ω represents the mains frequency.

Since T_{dq}^{123} is orthogonal, $\left(T_{dq}^{123}\right)^{-1} = \left(T_{dq}^{123}\right)^T$ and Equation (12) can be written as

$$\frac{dV_{dc}}{dt} = \frac{1}{C_{dc}} \left(T_{dq}^{123} [q_{ndq}]\right)^T \left(T_{dq}^{123} [i_{dq}]\right) = \frac{1}{C_{dc}} [q_{ndq}][i_{dq}] \tag{14}$$

On the other hand, Equation (9) can be written as

$$\frac{d^2}{dt^2} [i_{c12}] = -\frac{R_{PFeq}}{L_{PFeq}} \frac{d}{dt} [i_{c12}] - \frac{1}{C_{PFeq} L_{PFeq}} [i_{c12}] - \frac{1}{L_{PFeq}} [q_{n12}] \frac{dV_{dc}}{dt} + \frac{1}{L_{PFeq}} \frac{d}{dt} [v_{s12}] \tag{15}$$

The three-phase current with the absence of zero-sequence components can be converted into d-q frame using reduced transformation matrix. Applying the transformations in Equation (15), the complete d-q frame dynamic model of the system is obtained as follows;

$$L_{PFeq} \frac{d^2 i_d}{dt^2} = -R_{PFeq} \frac{di_d}{dt} + 2\omega L_{PFeq} \frac{di_q}{dt} - \left(-\omega^2 L_{PFeq} + \frac{1}{C_{PFeq}}\right) i_d + \omega R_{PFeq} i_q - q_{nd} \frac{dV_{dc}}{dt} + \frac{dv_d}{dt} - \omega v_q \tag{16}$$

$$L_{PFeq} \frac{d^2 i_q}{dt^2} = -R_{PFeq} \frac{di_q}{dt} - 2\omega L_{PFeq} \frac{di_d}{dt} - \left(-\omega^2 L_{PFeq} + \frac{1}{C_{PFeq}}\right) i_q - \omega R_{PFeq} i_d - q_{nq} \frac{dV_{dc}}{dt} + \frac{dv_q}{dt} + \omega v_d \tag{17}$$

$$C_{dc} \frac{dV_{dc}}{dt} = q_{nd} i_d + q_{nq} i_q \tag{18}$$

The role of i_d in Equation (16) is interpreted as the component for meeting the switching losses in the compensator, whereas the component i_q is utilised to supply reactive power and maintain the DClink voltage across the capacitor for sustaining the compensator action.

It is specifically noted that this set of Equations (16)–(18) contain nonlinear terms involving the control variables q_{nd} and q_{nq}. Accordingly, the implementation of this control strategy is termed as nonlinear control technique by many authors in the literature [12,15,16].

2.3. Control of Harmonic Currents

Based on the load and compensator models presented in Section 2.2, the control problem is formulated with the objective of minimizing supply-side current harmonics and improving the power factor. Also, for maintaining the performance during load fluctuations, it is necessary to maintain a desired DC link capacitor voltage. The control law is derived using the following approach.

Rewriting Equations (16) and (17) in a more convenient form, we get

$$L_{PFeq}\frac{d^2i_d}{dt^2} + R_{PFeq}\quad\frac{di_d}{dt} + \left(-\omega^2 L_{PFeq} + \frac{1}{C_{PFeq}}\right)i_d$$
$$= 2\omega L_{PFeq}\frac{di_q}{dt} + \omega R_{PFeq}i_q - q_{nd}\frac{dV_{dc}}{dt} + \frac{dv_d}{dt} - \omega v_q \tag{19}$$

$$L_{PFeq}\frac{d^2i_q}{dt^2} + R_{PFeq}\quad\frac{di_q}{dt} + \left(-\omega^2 L_{PFeq} + \frac{1}{C_{PFeq}}\right)i_q$$
$$= -2\omega L_{PFeq}\frac{di_q}{dt} - \omega R_{PFeq}i_d - q_{nq}\frac{dV_{dc}}{dt} + \frac{dv_q}{dt} + \omega v_d \tag{20}$$

The control variables u_d and u_q are defined as

$$u_d = 2\omega L_{PFeq}\frac{di_q}{dt} + \omega R_{PFeq}i_q - q_{nd}\frac{dV_{dc}}{dt} + \frac{dv_d}{dt} - \omega v_q \tag{21}$$

$$u_q = -2\omega L_{PFeq}\frac{di_d}{dt} - \omega R_{PFeq}i_d - q_{nq}\frac{dV_{dc}}{dt} + \frac{dV_q}{dt} + \omega v_d \tag{22}$$

Using the idea of decoupling the current harmonic components for the purpose of tracking the reference current, the error signals $\bar{i}_d = i_d^* - i_d$ and $\bar{i}_q = i_q^* - i_q$ are generated and processed through a pair of PI controllers [12,14] to obtain u_d and u_q signals which are given below.

i_d^* and i_q^* are the reference currents deduced from the load current, i_d and i_q are the actual compensator currents. Load current in d-q coordinate is processed using a pair of fourth-order Butterworth low-pass filters with cut-off frequency set at 60 Hz to extract the harmonic current references alone.

$$u_d = K_P\bar{i}_d + K_I\int\bar{i}_d dt$$

$$u_q = K_P\bar{i}_q + K_I\int\bar{i}_q dt$$

From Equations (19) and (20), we obtain the transfer function as follows

$$\frac{I_d(s)}{U_d(s)} = \frac{I_q(s)}{U_q(s)} = \frac{1}{L_{PFeq}\left(s^2 + \frac{R_{PFeq}}{L_{PFeq}}s + \frac{1}{C_{PFeq}L_{PFeq}} - \omega^2\right)} \tag{23}$$

The current control of the closed-loop system is shown in Figure 2 and represents the signal flow of the variables $i_q(s)$; the other component $i_d(s)$ is obtained concurrently in a similar manner. The transfer function of the full closed-loop control module is derived as

$$\frac{I_q(s)}{I_q^*(s)} = \frac{K_P}{L_{PFeq}}\left[\frac{s + \frac{K_I}{K_P}}{s^3 + \frac{R_{PFeq}}{L_{PFeq}}s^2 + \left(\frac{1}{C_{PFeq}L_{PFeq}} - \omega^2 + \frac{K_P}{L_{PFeq}}\right)s + \frac{K_I}{L_{PFeq}}}\right] \tag{24}$$

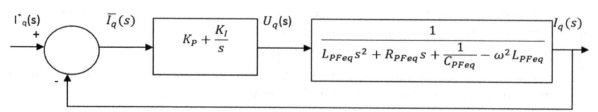

Figure 2. Structure of current control loop for i_q and i_d.

From Equations (21) and (22), the control variables of the proposed system are defined by the following equations.

$$q_{nd} = \frac{2\omega L_{PFeq}\frac{di_q}{dt} + \omega R_{PFeq}i_q + \frac{dV_d}{dt} - \omega v_q - u_d}{\frac{dV_{dc}}{dt}} \tag{25}$$

$$q_{nq} = \frac{-2\omega L_{PFeq}\frac{di_d}{dt} + \omega R_{PFeq}i_d + \frac{dV_q}{dt} - \omega v_d - u_q}{\frac{dV_{dc}}{dt}} \tag{26}$$

The above equations represent both the decoupled linear compensation part and cancellation of nonlinearity.

2.4. DC Link Voltage Control

The capacitor in the APF does not need any external DC source but gets charged through the rectifier action of the built-in reverse diodes across the six Insulated-gate bipolar transistors (IGBTs). The power loss in the capacitor and switching losses in the inverter have to be met by the active component id of the compensator current from the mains, while the component i_q supplies the reactive power stored in the capacitor. The power losses in this circuit can reduce the DC link capacitor voltage, thereby weakening the function of the active filter. Hence, it is necessary to maintain the voltage across the DC link capacitor at a designed reference value by an additional voltage regulator, which modifies the PWM signals appropriately. This regulator is implemented by using a PI controller [12], which processes the error between the reference voltage V_{dc}^* and the actual capacitor voltage V_{dc}. The parameters of the PI regulator are chosen in such a way that the DC voltage is maintained around its desired value. The design values for the PI controller parameters have been obtained following the approach suggested by Salem Rahmani et al. [13,14]. The overall transfer function of this controller is incorporated as a subsystem in the simulation schematic.

The control scheme of the proposed hybrid power compensator is shown in Figure 3, where the various signals such as V_{dc}, V_{dc}^*, i_d, i_q, i_d^* and i_q^* are processed using Equations (21) to (26) to obtain the gate trigger signals.

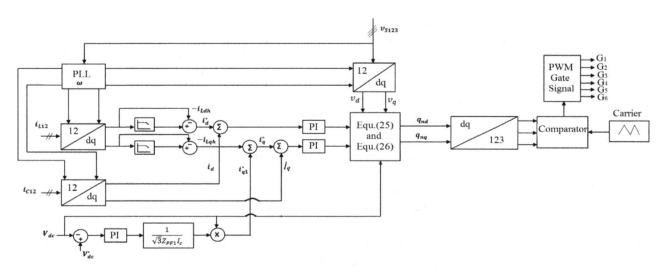

Figure 3. Control structure of the shunt hybrid compensator.

2.5. Simulink Model of the Shunt Hybrid Power Compensator

The Simulink schematic of the power distribution system along with the proposed shunt hybrid power compensator, which translates the entire system equations presented in Section 2 into functional blocks in the Simulink model, has been developed using MATLAB software. The nonlinear load has been modelled using a three-phase full bridge diode rectifier feeding RL and RC loads separately as subsystems and connected to the three phase mains. At the source side, a resistor–inductor combination is used to represent the line impedance of each phase before PCC. The unbalance case in the nonlinear load has been formed by connecting a single-phase diode rectifier feeding RL load between phase1 and ground. The synchronization of the subsystem with main supply frequency is accomplished by using a three-phase discrete PLL block [22–25]. The PLL block detects the supply frequency; the detected supply frequency is used to synchronize the compensator d-q axis current and the distorted load d-q axis current to extract harmonic component needs to be compensated. During load disturbance and supply voltage distortions, the variation in the supply frequency is detected by the PLL block and synchronizes the subsystems accordingly.

The hybrid filter consists of a parallel connection of selective 5th and 7th harmonic elimination passive filters as depicted in Figure 1 along with APF. The entire system simulation is carried out in a discrete mode, variable step size with ode45 (Dormand–Prince) solver.

3. Simulation Results

To evaluate the performance of the shunt HPF controlled by the proposed control algorithm, a Simulink-based schematic was created so as to operate from a 400 V, 50 Hz supply. The load consists of a set of balanced and unbalanced nonlinear loads, which are selectively connected for successive simulation runs. In this work, the performances of the compensator corresponding to the following loads are analysed: (i) three-phase rectifier feeding RL load, (ii) three-phase rectifier feeding RC load, (iii) dynamic load variation and (iv) unbalanced load. Table 1 indicates the specification of the system parameters used in the simulation.

Table 1. Parameters of the system.

Phase Voltage and Frequency	Vsrms = 230 V and fs = 50 Hz
Impedance of the line	Rs = 0.1 Ω, Ls = 4 mH
Nonlinear load of current source type	RL = 50 Ω, LL = 10 mH
Nonlinear load of voltage source type	RL = 32 Ω, CL = 1000 μF
5th tuned PPF parameters	R = 0.1 Ω, L = 10 mH, C = 40 μF
7th tuned PPF parameters	R = 0.1 Ω, L = 7 mH, C = 30 μF
DClink voltage and capacitance	V_{dc} = 25 V, C_{dc} = 6600 μF
Parameters of the outer loop PI controller	k1 = 0.22 and k2 = 15.85
Inner loop PI controller parameters	K_P = 0.6 and K_I = 1.2

3.1. Performance of Shunt HPF to the Nonlinear Load of Current Source Type

The three-phase mains supply a diode bridge rectifier, whose output is wired to a balanced R-L load, which imposes a typical nonlinearity, resulting in harmonic currents at the supply end. The proposed hybrid shunt compensator is wired at PCC as shown in Figure 1. The results are shown in Figure 4a–e covering the waveforms of the supply voltage, currents of the load, source current after compensation, currents of the compensator and DC link voltage, respectively. The waveform depicts that the supply current waveform is almost sinusoidal after compensation and the DC link capacitor voltage is maintained constant. Figure 5 depicts the simulation results covering the harmonic spectrum of the supply current before and after compensation. The results indicate that the THD of the supply current is reduced from 25.74% without compensator to 4.68% with compensator. It is seen that the proposed control strategy with shunt HPF can mitigate the current harmonics present in the supply system within the limit specified by IEEE 519-1992 standards.

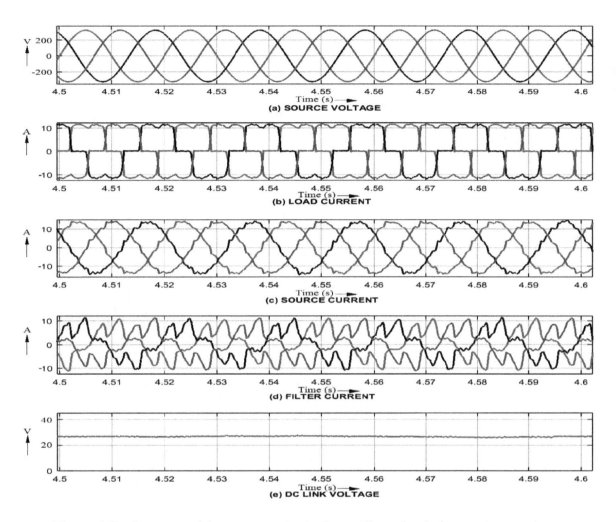

Figure 4. Performance of the compensator to the nonlinear load of current source type.

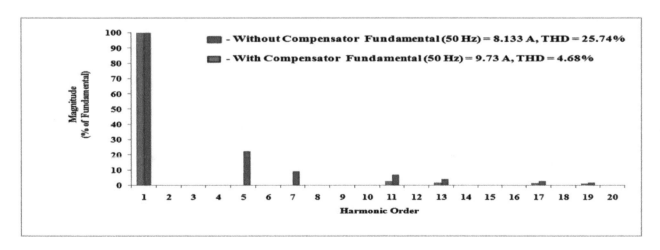

Figure 5. FFT analysis of the supply current with and without compensator.

3.2. Varying Three-Phase Rectifier-Fed RL Load

Since the load in a distribution system can vary concerning time, it is essential to verify the suitability of the designed filter under varying load conditions. The simulation results for sudden change of load current from 9.74 A (rms) to 17.17 A at $t = 4$ s and restoration of the same to the previous value at t = 4.1 s are presented in Figure 6a–e covering source voltage, load current, source current, compensator current and DC link voltage waveforms, respectively. The THD of the supply current before step change in the load current has been improved from 25.74% without compensator to 4.55% with compensator. Similarly, after a step change in the load current, the supply current THD has been reduced from 23.06% to 4.14% with compensator. In both the cases, the compensator is capable of maintaining the THD of the source-side current within the limit specified by the IEEE 519-1992 standards. It is observed from Figure 6e,c that the compensator is able to maintain the DC link voltage as constant and supply current waveforms to be sinusoidal.

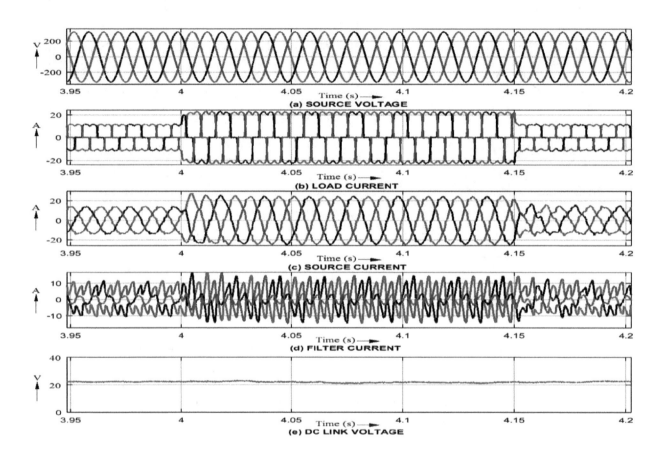

Figure 6. Response of the compensator to sudden change of load current.

3.3. Three-Phase Rectifier-Fed RC Load

The load in the simulation schematic is replaced by an R-C circuit with parameter values as shown in Table 1 and the simulation is executed to obtain the performance of the compensator. Figure 7a–e shows the source voltage, load current, supply current after compensation, compensator current and DC link voltage waveforms, respectively. It is observed from the figures that the DC link voltage of the compensator is maintained constant during compensation, and the supply current variation approaches a sinusoidal waveform. Figure 8 shows the harmonic spectrum of line currents from mains before and after compensation. It shows that the THD of the source current is reduced considerably from 27.49% to 4.08%. The system is highly nonlinear and subjected to disturbances due to loading. Hence, in practice, the DC link voltage varies slightly due to ripples and disturbances inherent in the operation of the compensator.

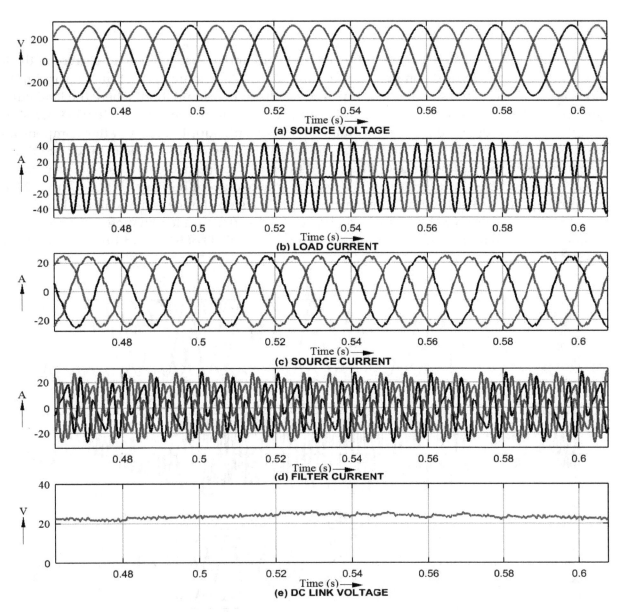

Figure 7. Performance of compensator for voltage source type nonlinear load.

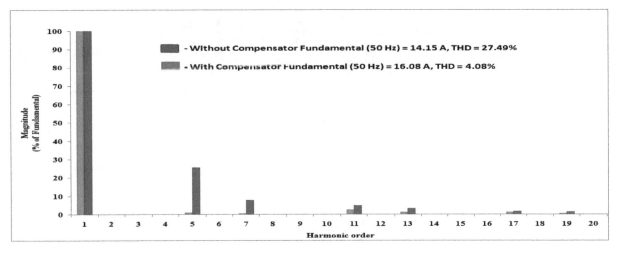

Figure 8. Harmonic spectrum of supply current before and after compensation.

3.4. Varying Three-Phase Rectifier-Fed RC Load

By introducing a 100% step increase followed by a similar decrease in the load current of the R-C circuit, the ability of the hybrid compensator to mitigate the harmonic current under varying load conditions is examined in the simulation setup. The results of this simulation, where the load current changes from 16.05 A (rms) to 30.32 A at $t = 2$ s and restoration of the same to 16.05 A are shown in Figure 9a–e. These depict the supply voltage, load current, supply current after compensation, compensator current and the DC link voltage, respectively. The obtained results indicate that the desirable features in the performance of the hybrid filter are maintained even after the step changes in the load within a very short time. The THD of the source current has been reduced from 27.49% without compensator to 3.46% with compensator before the step change is made in the load current. Similarly, the THD has been reduced from 20.91% to 4.77% after the step change. It is seen that the compensator operation reduces the harmonic content within a THD of 5%, in addition to maintaining a steady DC link voltage.

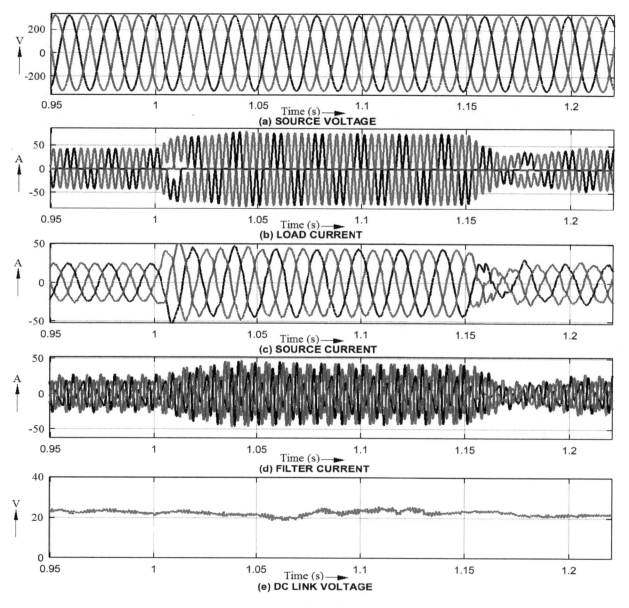

Figure 9. Performance of the compensator for variation in three-phase rectifier-fed RC load.

3.5. Unbalanced Loading Condition

To study the performance of a shunt hybrid power compensator employing synchronous rotating reference-based nonlinear control method, an unbalance in load is created by connecting a single-phase rectifier feeding R-L load across the phase 'a' and phase 'b' of the supply system. The compensation performance at steady state of the proposed compensator, such as the supply voltage, load current, supply current, compensator current and DC link voltage waveforms, is depicted in Figure 10a–e. The THD of supply phase 'a' current is reduced from 9.46% to 2.54% after compensation. Similarly, supply phase 'b' current has been reduced from 7.04% to 1.91%, while phase 'c' current has been reduced from 25.47% to 5.37%.

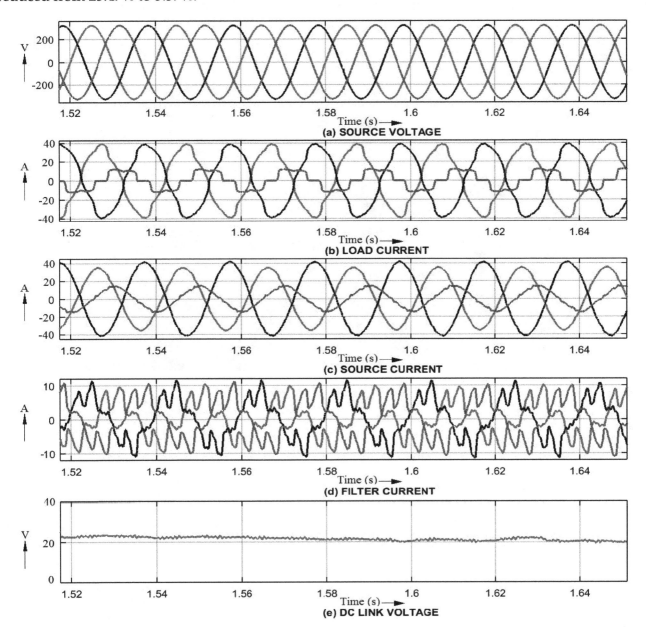

Figure 10. Performance of the compensator to the unbalanced nonlinear load.

It is seen from the obtained results that the proposed control method with shunt hybrid compensator effectively compensates the harmonic distortion present in the supply current under unbalanced loading conditions.

Since the load in a distribution system can vary with respect to time, it is essential to verify the suitability of the designed filter under varying load conditions. During dynamic variations in the load current and unbalance, as shown in Figures 6, 9 and 10, the compensator is able to maintain constant DC link voltage as well as reduce the amount of harmonics present in the system.

4. Hardware Fabrication

A prototype of the distribution system along with the passive and active parts of the hybrid compensator has been fabricated and experimental work on the same was carried out. Figure 11 shows the overall setup of the experimental work. The active filter part of the prototype has been developed using intelligent power module PEC16DSMO1. The IPM consists of a three-phase six-pulse inverter with six IGBT semiconductor power switches and is controlled in real time based on the software control program from the SPARTAN-6 FPGA development board. These IGBT switches are capable of operating at high frequencies of the carrier waveform up to 20 kHz.

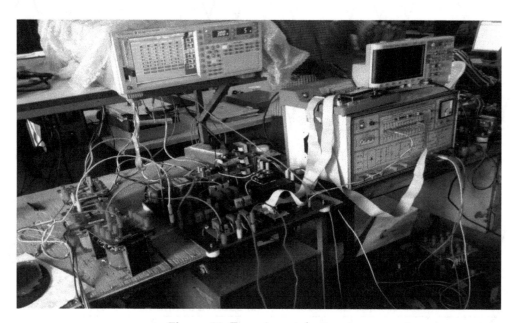

Figure 11. Experimental setup.

4.1. Hardware Resultsand Discussion

A series of experiments involving different types of loads and their dynamic variations have been conducted, and the performance of the compensator has been evaluated. A six-channel YOKOGAWA Power Quality Analyser has been used to capture the signals such as supply voltage, source current, load current, compensator current and DC link voltage. The hardware system parameters are shown in Table 2

Table 2. Experimental circuit parameters.

Active Power Filter	Intelligent Power Module (IPM) PEC16DSMO1 with 6 IGBT Switches
IGBT rating	25 A, 1200 V
Switching frequency of APF switches	2 kHz
Current sensors	LTS 25-NP
Voltage sensors	LV 25-P
Filter inductors	7 mH and 10 mH
Filter capacitors	30 μF and 40 μF
DC link capacitor	6400 μF

4.1.1. Performance of the Compensator for Current Source Type Nonlinear Load

Figure 12 shows the experimental response of the compensator for the current source type nonlinear load. The compensation performance of the hybrid filter has been verified by turn on the compensator at the point 'S' as shown in Figure 12. It is seen from the figure that the source current becomes sinusoidal after the operation of compensator at point 'S'. The harmonic spectrum of the source current without and with compensator is shown in Figure 13a,b and it is observed that the THD of the source current has been reduced from 26.225% to 2.212%.

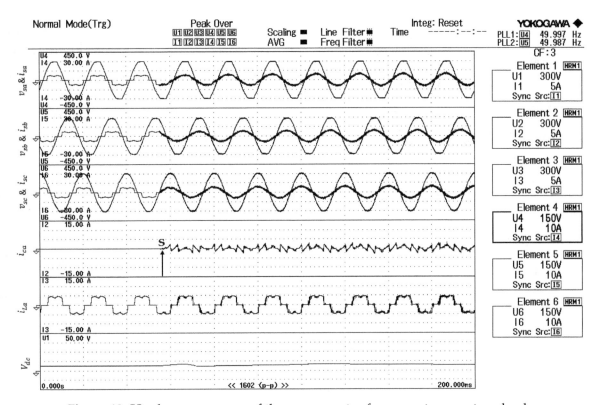

Figure 12. Hardware response of the compensator for current source type load.

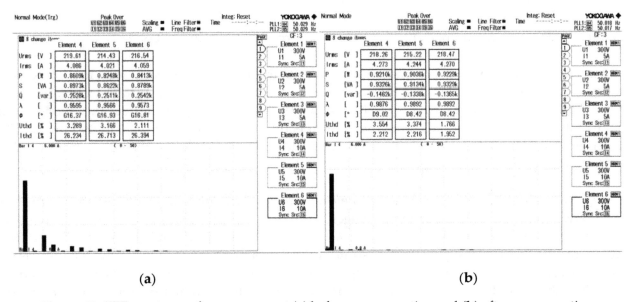

(a)   (b)

Figure 13. FFT spectrum of source current (**a**) before compensation and (**b**) after compensation.

4.1.2. Performance of the Compensator for Varying Current Source Type Nonlinear Load

The performance of the hybrid compensator has been verified experimentally by creating a step change in the load current at 'A' as shown in Figure 14. It is observed that the source current in all the three phases are sinusoidal and the compensator is able compensate the harmonic distortion even after a step change in the load current. The THD of the source current after compensation has been observed as 3.26%, which is less than 5% as specified by IEEE 519-1992 standards.

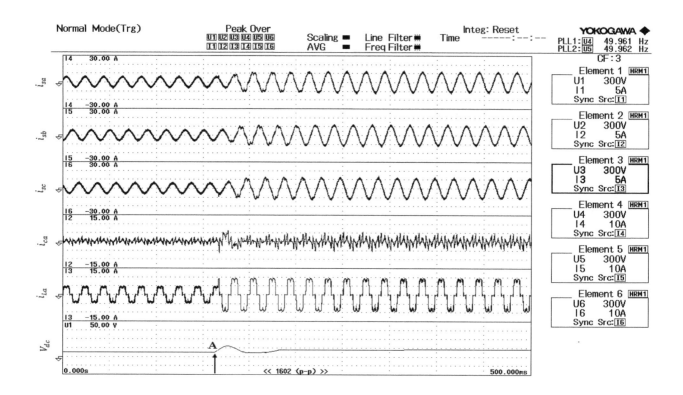

Figure 14. Performance of the compensator for time-varying current source type nonlinear load.

4.1.3. Performance of the Compensator for Voltage Source Type Nonlinear Load

The performance of the compensator for voltage source type nonlinear load has been experimentally verified by connecting the nonlinear load in the three-phase supply system. Figure 15 shows the three-phase source currents of all phases individually, compensator current of one phase, load current of one phase and DC link voltage, respectively. It is seen that the supply current after compensation is sinusoidal without distortion and the DC link voltage of the compensator is also maintained constant.

The FFT spectrum of the source current before and after compensation is shown in Figure 16a,b, respectively. It is observed from the FFT spectrum that the supply current harmonics have been reduced from 67.795% without compensator to 4.136% with compensator. It has been confirmed from the above results that the designed control strategy with proposed shunt hybrid power compensator is able to effectively compensate both the types of nonlinear load distortions present in the supply current. It is observed that the experimental work results are closely matching with simulation work results in all cases and the obtained performance waveforms are also matching with each other.

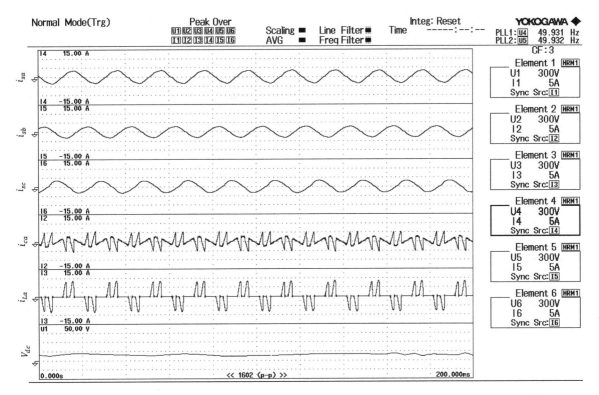

Figure 15. Performance of the compensator for voltage source type nonlinear load.

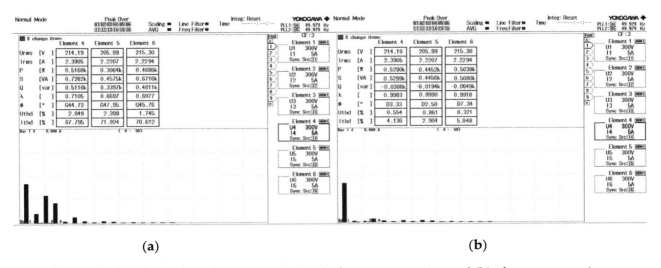

(a)   (b)

Figure 16. FFT Spectrum of source current (**a**) before compensation and (**b**) after compensation.

4.1.4. Performance Comparison and Discussion

Performance of the proposed hybrid filter with the adopted control strategy has been compared with some other control methods proposed in the literature. Table 3 shows the comparison of various control techniques proposed in the literature for harmonic reduction. It is clearly understood from Table 3 that the synchronous rotating reference theory based nonlinear control with the adopted hybrid filter topology proposed in this work gives better THD in the source current during supply feeds of different types of nonlinear loads. Most of the techniques proposed in the literature cover only current source type nonlinear load. In this work, different types of nonlinear loads and various loading conditio..ns are considered during simulation and experimental work. The experimental work has been carried out for low current rating nonlinear loads compared to the simulation work. Hence, it has been observed that the experimental THD values are better than the simulation results.

Furthermore, the hardware setup has been developed for low rating loads only due to the cost. The results of simulation indicate effectiveness of the compensator topology and versatility of the control scheme for meeting the requirements of a set of nonlinear loads indicated above. In addition, the *kVA* ratings of the designed active filter for both types of load are only a small percentage of the respective load ratings, which is a major advantage of this scheme. Although the overall performance of the hybrid compensator has been highly satisfactory, while covering alternate cases of load, their fluctuations and unbalance, it appears that the PI controller-based algorithm shows a minor deficiency, when the passive filter parameter values undergo even minor changes. This can be traced to the presence of fixed values of the PI controller gain and time constant parameters. A possible and feasible solution to this issue of filter parameter variations is to adaptively adjust the PI parameters' value. Although a three-phase distribution system for power quality enhancement has been considered in this paper, it is limited to a three-wire system only with isolated neutral. Hence, issues arising out of zero-sequence currents due to load unbalance have not been considered. This research work can be extended for a three-phase, four-wire system. Further, incorporation of controllers utilizing various other soft computing techniques like neural, fuzzy-neural and genetic algorithms can be carried out as further work.

Table 3. Comparison of the performance of various control techniques.

Control Methods	SRF Theory-Based Nonlinear Control for SHAPF [13]	p-q Theory-Based Control for SHAPF [26]	SRF Theory-Based Control for SHAPF [27]	Parallel Connected SHAPF [28]
THD% Three-phase rectifier-fed RL load	4.6	4.32	-	4.3
THD% Three-phase rectifier-fed RC load	4.08	4.15	-	-
Unbalanced load % THD	2.29 to 4.80	-	1.18 to 2	4.5 to 4.7
DC link voltage	25 V APF rating is less	50 V APF rating is moderate	220 V APF rating is more	26 V APF rating is less

5. Conclusions

A three-phase hybrid compensator scheme discussed in this paper is configured with 5th and 7th selective harmonic elimination passive filter in series with an active filter in the form of an IGBT-based PWM inverter, triggered using a control algorithm, based on a synchronously rotating reference frame (SRRF) in the *d-q* axis. The overall objective is to improve the steady state power factor and to ensure dynamic harmonic compensation at the supply side of three-phase mains. Accordingly, the entire system is simulated using Simulink and the results are presented, covering a set of nonlinear loads. The simulation results indicate the effectiveness of the compensator topology and versatility of the control scheme for meeting the requirements of a set of nonlinear loads. It is seen that significant reduction of source current THD is obtained within limits specified by IEEE 519-1992 standard along with excellent regulation of the DC link voltage.

A series of experiments were carried out using the above prototype along with alternate nonlinear loads and the overall performance of the hybrid filter was recorded using six channels DSO. The THD values as measured by the DSO over a series of experiments were found to be closely matching with corresponding THD values obtained during simulation for all loading conditions. It is further verified that the experimental results compared favourably with the simulation waveforms, thereby validating the simulation model. The experimental results confirm the effectiveness of the SRRF-based control strategy for the hybrid compensator scheme to achieve source side current harmonics reduction and power factor improvement to near unity.

Author Contributions: Conceptualization, R.B. and K.P.; methodology, R.B. and K.P.; software, R.B. and K.P.; validation, R.B., K.P.; and R.S.; formal analysis, R.B.; investigation, R.B. and K.P.; data curation, R.B., and K.P.; writing—original draft preparation, R.B., K.P., R.S. and R.A.; writing—review and editing, R.S. and R.A.; supervision, R.S. and R.A.

Acknowledgments: Authors wish to express their sincere thanks to acknowledge SASTRA Deemed University, Thanjavur, India 613401 for extending infrastructural support to carry out this work.

References

1. Peng, F.Z. Harmonic sources and filtering approaches. *IEEE Ind. Appl. Mag.* **2001**, *7*, 18–25. [CrossRef]
2. Dugan, R.C.; Mark, F.M.; Surya, S.; Wayne, H.B. *Electrical Power Systems Quality*, 3rd ed.; Tata McGraw-Hill Education: Noida, India, 2012; ISBN 9780071761550.
3. Singh, B.; Singh, B.N.; Chandra, A.; Al-Haddad, K.; Pandey, A.; Kothari, D.P. A review of three-phase improved power quality AC-DC converters. *IEEE Trans. Ind. Electron.* **2004**, *51*, 641–660. [CrossRef]
4. Wagner, V.E.; Balda, J.C.; Griffith, D.C.; McEachern, A.; Barnes, T.M.; Hartmann, D.P.; Phileggi, D.J.; Emannuel, A.E.; Horton, W.F.; Reid, W.E.; et al. Effects of harmonics on equipment. *IEEE Trans. Power Deliv.* **1993**, *8*, 672–680. [CrossRef]
5. Das, J.C. Passive filters-potentialities and limitations. *IEEE Trans. Ind. Appl.* **2004**, *40*, 232–241. [CrossRef]
6. Singh, B.; Al-Haddad, K.; Chandra, A. A review of active filters for power quality improvement. *IEEE Trans. Ind. Electron.* **1999**, *46*, 960–971. [CrossRef]
7. Akagi, H. New trends in active filters for power conditioning. *IEEE Trans. Ind. Appl.* **1996**, *32*, 1312–1322. [CrossRef]
8. Lam, C.-S.; Wong, M.-C. *Design and Control of Hybrid Active Power Filters*; Springer: Berlin, Germany, 2014.
9. Singh, B.; Verma, V.; Chandra, A.; Al-Haddad, K. Hybrid filters for power quality improvement. *IEE Proc.-Gener. Transm. Distrib.* **2005**, *152*, 365–378. [CrossRef]
10. Rivas, D.; Morán, L.; Dixon, J.W.; Espinoza, J.R. Improving passive filter compensation performance with active techniques. *IEEE Trans. Ind. Electron.* **2003**, *50*, 161–170. [CrossRef]
11. Samadaei, E.; Lesan, S.; Cherati, S.M. A new schematic for hybrid active power filter controller. In Proceedings of the 2011 IEEE Applied Power Electronics Colloquium (IAPEC), Johor Bahru, Malaysia, 18–19 April 2011; pp. 143–148.
12. Mendalek, N.; Al-Haddad, K.; Fnaiech, F.; Dessaint, L.A. Nonlinear control technique to enhance dynamic performance of a shunt active power filter. *IEE Proc.-Electr. Power Appl.* **2003**, *150*, 373–379. [CrossRef]
13. Rahmani, S.; Hamadi, A.; Mendalek, N.; Al-Haddad, K. A new control technique for three-phase shunt hybrid power filter. *IEEE Trans. Ind. Electron.* **2009**, *56*, 2904–2915. [CrossRef]
14. Rahmani, S.; Mendalek, N.; Al-Haddad, K. Experimental design of a nonlinear control technique for three-phase shunt active power filter. *IEEE Trans. Ind. Electron.* **2010**, *57*, 3364–3375. [CrossRef]
15. Zouidi, A.; Fnaiech, F.; Al-Haddad, K. Voltage source inverter based three-phase shunt active power filter: Topology, Modeling and control strategies. In Proceedings of the 2006 IEEE International Symposium on Industrial Electronics, Montreal, QC, Canada, 9–13 July 2006; Volume 2, pp. 785–790.
16. Mendalek, N.; Al-Haddad, K. Modeling and nonlinear control of shunt active power filter in the synchronous reference frame. In Proceedings of the Ninth International Conference on Harmonics and Quality of Power. Proceedings (Cat. No.00EX441), Orlando, FL, USA, 1–4 October 2000; Volume 1, pp. 30–35.
17. Balasubramanian, R.; Sankaran, R.; Palani, S. Simulation and performance evaluation of shunt hybrid power filter using fuzzy logic based non-linear control for power quality improvement. *Sadhana* **2017**, *42*, 1443–1452. [CrossRef]
18. Palandöken, M.; Aksoy, M.; Tümay, M. Application of fuzzy logic controller to active power filters. *Electr. Eng.* **2004**, *86*, 191–198. [CrossRef]
19. Dey, P.; Mekhilef, S. Current controllers of active power filter for power quality improvement: A technical analysis. *Automatika* **2015**, *56*, 42–54. [CrossRef]
20. Unnikrishnan, A.K.; Subhash Joshi, T.G.; Manju, A.S.; Joseph, A. Shunt hybrid active power filter for harmonic mitigation: A practical design approach. *Sadhana* **2015**, *40*, 1257–1272.
21. Chauhan, S.K.; Shah, M.C.; Tiwari, R.R.; Tekwani, P.N. Analysis, design and digital implementation of a shunt active power filter with different schemes of reference current generation. *IET Power Electron.* **2013**, *7*, 627–639. [CrossRef]

22. Cataliotti, A.; Cosentino, V. A Time-domain strategy for the measurement of IEEE standard 1459–2000 Power quantities in Non sinusoidal Three–phase and Single–phase systems. *IEEE Trans. Power Deliv.* **2008**, *23*, 2113–2123. [CrossRef]

23. Cataliotti, A.; Cosentino, V. Disturbing Load Identification in Power Systems: A Single-Point Time–Domain Method Based on IEEE 1459–2000. *IEEE Trans. Instrum. Meas.* **2009**, *58*, 1436–1445. [CrossRef]

24. Golestan, S.; Gurrero, J.M.; Vasquez, J.C. Three–Phase PLLs: A Review of Recent Advances. *IEEE Trans. Power Electron.* **2017**, *32*, 1894–1907. [CrossRef]

25. Golestan, S.; Ge, J.M.; Vasquez, J.; Abusorrah, A.M.; Al-Turki, Y.A. Modeling, Tuning, and Performance Comparison of Second–Order-Generalized-Integrator–Based FLLs. *IEEE Trans. Power Electron.* **2018**, *33*, 10229–10239. [CrossRef]

26. Balasubramanian, R.; Palani, S. Simulation and performance evaluation of shunt hybrid power filter for power quality improvement using PQ theory. *Int. J. Electr. Comput. Eng.* **2016**, *6*, 2603–2609. [CrossRef]

27. Dey, P.; Mekhilef, S. Synchronous reference frame based control technique for shunt hybrid active power filter under non-ideal voltage. In Proceedings of the 2014 IEEE Innovative Smart Grid Technologies-Asia (ISGT ASIA), Kuala Lumpur, Malaysia, 20–23 May 2014; pp. 481–486.

28. Bhattacharya, A.; Chakraborty, C.; Bhattacharya, S. Parallel-connected shunt hybrid active power filters operating at different switching frequencies for improved performance. *IEEE Trans. Ind. Electron.* **2012**, *59*, 4007–4019. [CrossRef]

4

Line Frequency Instability of One-Cycle-Controlled Boost Power Factor Correction Converter

Rui Zhang [1,*], Wei Ma [2,*], Lei Wang [3], Min Hu [2], Longhan Cao [4], Hongjun Zhou [2] and Yihui Zhang [2]

[1] College of Safety Engineering, Chongqing University of Science and Technology, Chongqing 401331, China
[2] College of Electrical Engineering, Chongqing University of Science and Technology, Chongqing 401331, China; minhuxzr@outlook.com (M.H.); zhjzsj007@163.com (H.Z.); zhangyihui@163.com (Y.Z.)
[3] The School of Electric Power Engineering, South China University of Technology, Guangzhou 510641, China; w.lei19@mail.scut.edu.cn
[4] Department of Electrical Egineering, Chongqing Institute of Communication, Chongqing 400035, China; lhcao1@hotmail.com
[*] Correspondence: zhangrui@cqust.edu.cn (R.Z.); me16888@163.com (W.M.)

Abstract: Power Factor Correction (PFC) converters are widely used in engineering. A classical PFC control circuit employs two complicated feedback control loops and a multiplier, while the One-Cycle-Controlled (OCC) PFC converter has a simple control circuit. In OCC PFC converters, the voltage loop is implemented with a PID control and the multiplier is not needed. Although linear theory is used in designing the OCC PFC converter control circuit, it cannot be used in predicting non-linear phenomena in the converter. In this paper, a non-linear model of the OCC PFC Boost converter is proposed based on the double averaging method. The line frequency instability of the converter is predicted by studying the DC component, the first harmonic component and the second harmonic component of the main circuit and the control circuit. The effect of the input voltage and the output capacitance on the stability of the converter is studied. The correctness of the proposed model is verified with numerical simulations and experimental measurements.

Keywords: power factor correction; line frequency instability; one cycle control; non-linear phenomena; bifurcation; boost converter

1. Introduction

Power Factor Correction (PFC) plays an important role in electrical engineering [1]. A PFC converter takes AC voltage as its input and outputs DC voltage. Different from traditional diode rectifiers, a PFC converter in average current mode has a high power factor. In electrical engineering, the average current mode Boost PFC converter is widely used. Although the topology of the Boost PFC is simple, the control circuit is complicated [2–4]. The control circuit consists of two loops. The first is the current control loop, with the aim of forcing the inductor current to be in the same phase as the reference. The second loop is the voltage control loop. The design of the voltage control loop is of great importance because its main objective is achieving a stable system and a near unity power factor [4]. The dynamics of the PFC converter depends on these two control loops. The traditional implementation for the PFC converter requires a multiplier, whose output is the reference current added to the current control loop. The existence of the multiplier increases the control complexity. The dynamics of the PFC converter has interested many researchers and some non-linear phenomena have been observed in the last few years [5–13]. In general, there are two kinds of non-linear dynamics in the PFC converter. The first is the so-called switching frequency instability, which is mainly the result of bifurcation and chaos caused by the current control loop [5]. The second is the line frequency instability, which is the result of bifurcation and chaos caused by the voltage control loop [7–11,14]. Among them, the line frequency instabilities are more detrimental to the normal operation of the PFC converter, as it changes the power factor to an unacceptable value. The power factor of the converter is much less than one due to the line frequency instabilities and thus, it is of great importance to select the appropriate parameters in the design process. In the traditional design, the researchers adopted the linear system theory and it has been shown that this design cannot predict the line frequency instabilities in the PFC converter [7]. Therefore, the researchers developed some powerful methods to compute the boundaries of line frequency stabilities. Among them, the method of harmonic balance needs an exact computation of the unstable periodic orbit of the control voltage [9]. Harmonic balance is applied to the model of the converter incorporating the multiplier and Floquet theory is adopted to decide the stability of the converter. According to Floquet theory, the stability of the converter is identified by calculating the eigenvalues of the transition matrix of the system. Another important method is the method of double averaging, which is based on the first harmonic component in the PFC converter line frequency model [14]. This method is more familiar to many researchers and engineers. In this paper, the later method is adopted to study the non-linear dynamics of the continuous conduction mode One-Cycle-Controlled (OCC) Boost PFC converter.

Different from the traditional average current mode PFC converter, the OCC PFC converter simplifies the control circuit [3]. The one-cycle control belongs to non-linear controls. When this control method is applied to the PFC converter, the current control loop is replaced by a resettable integrator. Therefore, only one voltage control loop is required and the multiplier is not needed. It has been shown that the control circuit of the OCC Boost PFC converter saves space and cost compared to the traditional PFC Boost converter. In most applications, the voltage control loop is designed based on the linear system theory and the prediction of dynamics of the converter is also based on the linear system theory. Therefore, non-linear dynamics of the converter are uncovered. In many applications, bifurcation and chaos are observed but are not addressed. The reason is that the non-linear systems theory is not applied to the OCC PFC converter. In this paper, the method of double averaging is adopted to predict the non-linear dynamics of the OCC Boost PFC converter. Although this method has been applied to the traditional PFC converter, there is still a problem when applying it to the OCC PFC converter, because the control circuits in the two converters are totally different and as a result, some new consequences will occur in the OCC PFC converter. It is important to note that in a previous study [15], the non-linear dynamics of the OCC PFC converter were observed by experiments and no effective computation was provided. In the present paper, the computation is based on the exact non-linear model of the OCC PFC converter and therefore, the conclusions are meaningful in the design process of the converter.

2. The OCC Boost PFC Converter

2.1. The OCC Boost PFC Converter and Its Control Circuit

The OCC Boost PFC converter consists of a diode rectifier and a boost converter, as shown in Figure 1. In some applications, the load of the PFC converter is another DC-DC converter. In this paper, the load of the PFC converter is a resistor, because the emphasis of this paper is on the non-linear dynamics of the PFC converter. The control circuit in Figure 1 is equivalent to a commercial control IC IR1150, which is used to verify the theoretical results in this paper. Apparently, this control circuit has fewer resistors and capacitors than the traditional average current mode PFC converter, where both the current loop and the voltage loop have at least one resistor and one capacitor. The output of the converter is divided by R_{f1} and R_{f2}. The divided voltage is connected to the input of an Operational Amplifier (OA), whose other input is the reference V_{ref}. In IR1150, the OA is a trans-conductance type amplifier. The output of the OA is v_m, which is the input to a resettable integrator. The output of the integrator is compared with another voltage composed of v_m and the voltage across the current sense resistor R_{s1}. The integrator is reset by the output $\overline{Q}$ of the flip-flop. The sensed voltage is amplified by a DC gain $G_{DC} = 2.5$. The operation of the control circuit is described here. In the beginning of every switching period T_s, the clock sets the flip-flop and the output Q of the flip-flop turns on the switch S. At the same time, the integrator outputs the value of the integral of its input signal. When the output of the integrator exceeds the sum of v_m and the sensed voltage, the comparator resets the flip-flop and the output Q turns off the switch S. Therefore, the diode D turns on. Furthermore, the output $\overline{Q}$ resets the integrator until the next clock signal.

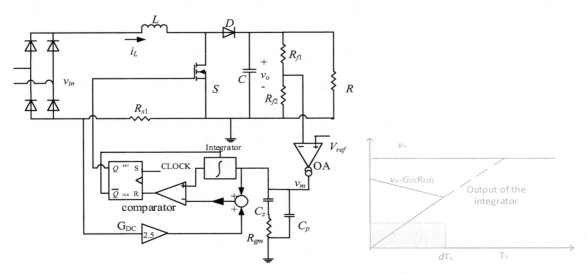

Figure 1. Circuit diagram of the OCC Boost PFC converter and its operating principle.

2.2. The OCC Boost PFC Converter Model

In this paper, the line frequency dynamics are studied. Therefore, the method of double averaging is adopted. The method is composed of two averaging processes. The first averaging is applied to the switching period [16–22]. The converter has two topology structures during one switching period, and is described by:

$$\begin{cases} \frac{di_L}{dt} = \frac{1}{L}v_{in} \\ \frac{dv_o}{dt} = \frac{-1}{RC}v_o \end{cases} \text{ (the switch S is on) or } \begin{cases} \frac{di_L}{dt} = \frac{1}{L}(v_{in} - v_o) \\ \frac{dv_o}{dt} = \frac{1}{C}\left(i_L - \frac{v_o}{R}\right) \end{cases} \text{ (the switch S is off).} \tag{1}$$

By averaging over one switching period, one obtains the following:

$$\begin{cases} (1-d)v_o = v_{in} - L\frac{di_L}{dt} \\ (1-d)i_L = C\frac{dv_o}{dt} + \frac{v_o}{R} \end{cases}. \tag{2}$$

From Equation (2), we obtain:

$$\frac{C}{2}\frac{dv_o^2}{dt} = -\frac{v_0^2}{R} + i_L v_{in} - \frac{L}{2}\frac{di_L^2}{dt}. \tag{3}$$

It is important to note that the dynamics of the inductor during one switching period can be omitted when the converter operates stably. Therefore, one has

$$\frac{C}{2}\frac{dv_o^2}{dt} = -\frac{v_0^2}{R} + i_L v_{in}. \tag{4}$$

Based on the operating principle of the converter [3–8], one has:

$$R_s i_L(t) = v_m / T(d), \tag{5}$$

where:

$$T(d) = v_o/v_{in}, \; R_s = R_{s1} \times 2.5 \tag{6}$$

From Equations (5) and (6), we can obtain:

$$i_L(t) = v_{in}v_m/(R_s v_o). \tag{7}$$

It is important to note that $v_{in} = V_m|\sin \omega_m t|$. Substituting Equation (7) into (4), we obtain:

$$\frac{C}{2}\frac{dv_o^2}{dt} = -\frac{v_0^2}{R} + \frac{v_m}{R_s v_o}V_m^2(1 - \cos 2\omega_m t). \tag{8}$$

On the other hand, the control loop includes the OA. Figure 1 provides the following transfer function of the OA:

$$H(s) = \frac{g_m(1 + sR_{gm}C_z)}{s(C_z + C_p + sR_{gm}C_z C_p)}. \tag{9}$$

As $C_z \gg C_p$, Equation (9) can be written as:

$$H(s) = \frac{g_m(1 + sR_{gm}C_z)}{sC_z}. \tag{10}$$

Therefore, the voltage control loop in Figure 1 is described by:

$$C_z\frac{dv_m}{dt} = g_m(V_{ref} - \frac{R_{f2}}{R_{f1}+R_{f2}}v_o) - g_m R_{gm} C_z \frac{R_{f2}}{R_{f1}+R_{f2}}\frac{dv_o}{dt}, \tag{11}$$

where g_m is the trans-conductance of the amplifier.

From Figure 1, one has:

$$i_L(t) = \frac{v_{in}v_m R_{f2}}{R_s\left(R_{f1}+R_{f2}\right)V_{ref}}. \tag{12}$$

Therefore, the OCC Boost PFC converter is described by:

$$\begin{cases} \frac{C}{2}\frac{dx^2}{dt} = -\frac{x^2}{R} + \frac{y}{R_s(1+\beta)V_{ref}}V_m^2(1 - \cos 2\omega_m t) \\ C_z\frac{dy}{dt} = g_m(V_{ref} - \frac{1}{1+\beta}x) - g_m R_{gm}C_z\frac{1}{1+\beta}\frac{dx}{dt} \end{cases}, \tag{13}$$

where $v_m = y$, $v_o = x$, $\beta = R_{f1}/R_{f2}$.

The next step is applying the second averaging for Equation (13). The second averaging involves taking the moving average over the main period. For any variable $u(t)$ in Equation (13), we have the following expression based on Fourier analysis [23]:

$$u(t) \approx u_0 + u_1 e^{j\omega_m t} + u_{-1}e^{-j\omega_m t} + u_2 e^{j2\omega_m t} + u_{-2}e^{-j2\omega_m t}. \tag{14}$$

where

$$u_k \approx \frac{\omega_m}{2\pi}\int_{t-\frac{2\pi}{\omega_m}}^{t} u(\tau)\exp(-jk\omega_m\tau)d\tau(k = 0, \pm1, \pm2). \tag{15}$$

It is important to note that $u_{-1} = u_1^*$, $u_{-2} = u_2^*$, where $*$ stands for complex conjugate. Taking the second averaging on Equation (13) based on Equation (14) and Equations (A1)–(A5) in Appendix A, one has:

$$\frac{C}{2}\frac{d}{dt}\left(x_0^2 + 2x_{1r}^2 + 2x_{1i}^2 + 2x_{2r}^2 + 2x_{2i}^2\right) + \frac{1}{R}\left(x_0^2 + 2x_{1r}^2 + 2x_{1i}^2 + 2x_{2r}^2 + 2x_{2i}^2\right)$$
$$= \frac{V_{in}^2}{R_s(1+\beta)V_{ref}}(y_0 - y_{2r}). \tag{16}$$

$$\frac{C}{2}\frac{d}{dt}\left(x_0 x_1 + x_{1r}x_{2r} + x_{1i}x_{2i} + j(x_{1r}x_{2i} - x_{1i}x_{2r})\right)$$
$$+\left(j\frac{\omega_m C}{2} + \frac{1}{R}\right)\left(x_0 x_1 + x_{1r}x_{2r} + x_{1i}x_{2i} + j(x_{1r}x_{2i} - x_{1i}x_{2r})\right) \tag{17}$$
$$= \frac{V_{in}^2}{R_s(1+\beta)V_{ref}}\left(\frac{y_1}{2} - \frac{y_1^*}{4}\right)$$

$$\frac{C}{2}\frac{d}{dt}\left((x_1^*)^2 + 2x_0 x_2\right) + \left(j\omega_m C + \frac{1}{R}\right)\left((x_1^*)^2 + 2x_0 x_2\right)$$
$$= \frac{V_{in}^2}{R_s(1+\beta)V_{ref}}\left(y_2 - \frac{1}{2}y_0\right) \tag{18}$$

$$C_z\frac{d}{dt}y_0 = g_m\left(V_{ref} - \frac{1}{1+\beta}x_0\right) - g_m R_{gm}C_z\frac{1}{1+\beta}\frac{d}{dt}x_0. \tag{19}$$

$$C_z\left(\frac{d}{dt}y_1 + j\omega_m y_1\right) = -g_m\frac{1}{1+\beta}x_1 - g_m R_{gm}C_z\frac{1}{1+\beta}\left(\frac{d}{dt}x_1 + j\omega_m x_1\right). \tag{20}$$

$$C_z\left(\frac{d}{dt}y_2 + j2\omega_m y_2\right) = -g_m\frac{1}{1+\beta}x_2 - g_m R_{gm}C_z\frac{1}{1+\beta}\left(\frac{d}{dt}x_2 + j2\omega_m x_2\right). \tag{21}$$

Equations (16)–(21) describe the DC component, the first harmonic component and the second harmonic component of the main circuit and the control circuit, respectively.

3. Stability of the OCC Boost PFC Converter

The stability of the OCC Boost PFC converter was studied based on Equations (16)–(21). To do this, the DC component, the first harmonic component and the second harmonic component are studied.

3.1. The First Harmonic Component

We obtain the steady-state solution by making all time-derivatives in Equations (16)–(21) equal to zero. Therefore, Equation (17) becomes:

$$\left(j\frac{\omega_m C}{2} + \frac{1}{R}\right)\left(x_0 x_{1r} + x_{1r}x_{2r} + x_{1i}x_{2i} + j(x_0 x_{1i} + x_{1r}x_{2i} - x_{1i}x_{2r})\right)$$
$$= \frac{V_m^2}{R_s(1+\beta)V_{ref}}\left(\frac{y_{1r}}{4} + j\frac{3y_{1i}}{4}\right). \tag{22}$$

Considering the real and imaginary part of Equation (22), one has:

$$\begin{cases} \frac{1}{R}(x_0 x_{1r} + x_{1r}x_{2r} + x_{1i}x_{2i}) - \frac{\omega_m C}{2}(x_0 x_{1i} + x_{1r}x_{2i} - x_{1i}x_{2r}) = \frac{V_m^2}{R_s(1+\beta)V_{ref}}\frac{y_{1r}}{4} \\ \frac{\omega_m C}{2}(x_0 x_{1r} + x_{1r}x_{2r} + x_{1i}x_{2i}) + \frac{1}{R}(x_0 x_{1i} + x_{1r}x_{2i} - x_{1i}x_{2r}) = \frac{V_m^2}{R_s(1+\beta)V_{ref}}\frac{3y_{1i}}{4} \end{cases}. \tag{23}$$

Equation (23) has another form, which is the following:

$$\begin{pmatrix} x_{1r} \\ x_{1i} \end{pmatrix} = \frac{\frac{V_m^2}{4R_s(1+\beta)V_{ref}}}{\left(\frac{1}{R^2} + \frac{\omega_m^2 C^2}{4}\right)\left(x_0^2 - (x_{2r}^2 + x_{2i}^2)\right)}$$
$$\times \begin{pmatrix} \frac{x_0 - x_{2r}}{R} + \frac{\omega_m C x_{2i}}{2} & -3\left(\frac{x_{2i}}{R} - \frac{\omega_m C(x_0 - x_{2r})}{2}\right) \\ -\frac{x_{2i}}{R} - \frac{\omega_m C(x_0 + x_{2r})}{2} & 3\left(\frac{x_0 + x_{2r}}{R} - \frac{\omega_m C x_{2i}}{2}\right) \end{pmatrix}\begin{pmatrix} y_{1r} \\ y_{1i} \end{pmatrix}. \tag{24}$$

By making all time derivatives in Equation (20) equal to zero, one has:

$$C_z j\omega_m(y_{1r} + jy_{1i}) = -g_m\frac{1}{1+\beta}(x_{1r} + jx_{1i}) - g_m R_{gm}C_z\frac{1}{1+\beta}j\omega_m(x_{1r} + jx_{1i}). \tag{25}$$

Considering the real and imaginary part of Equation (25), one obtains:

$$\begin{cases} -C_z\omega_m y_{1i} = -g_m\frac{1}{1+\beta}x_{1r} + g_m R_{gm}C_z\frac{1}{1+\beta}\omega_m x_{1i} \\ C_z\omega_m y_{1r} = -g_m\frac{1}{1+\beta}x_{1i} - g_m R_{gm}C_z\frac{1}{1+\beta}\omega_m x_{1r} \end{cases} . \tag{26}$$

Equation (26) can be written as:

$$\begin{pmatrix} y_{1r} \\ y_{1i} \end{pmatrix} = \frac{1}{C_z\omega_m}\begin{pmatrix} -g_m R_{gm}C_z\frac{1}{1+\beta}\omega_m & -g_m\frac{1}{1+\beta} \\ g_m\frac{1}{1+\beta} & -g_m R_{gm}C_z\frac{1}{1+\beta}\omega_m \end{pmatrix}\begin{pmatrix} x_{1r} \\ x_{1i} \end{pmatrix}. \tag{27}$$

Equations (24) and (27) describe the transfer function of the first harmonic component in the main circuit and the control circuit of the converter, respectively. By integrating them, one has the total transfer function as follows:

$$\begin{aligned} M = &\frac{1}{C_z\omega_m}\begin{pmatrix} -g_m R_{gm}C_z\frac{1}{1+\beta}\omega_m & -g_m\frac{1}{1+\beta} \\ g_m\frac{1}{1+\beta} & -g_m R_{gm}C_z\frac{1}{1+\beta}\omega_m \end{pmatrix} \\ &\times \frac{\frac{V_m^2}{4R_s(1+\beta)V_{ref}}}{\left(\frac{1}{R^2}+\frac{\omega_m^2 C^2}{4}\right)\left(x_0^2-\left(x_{2r}^2+x_{2i}^2\right)\right)} \\ &\times \begin{pmatrix} \frac{x_0-x_{2r}}{R}+\frac{\omega_m C x_{2i}}{2} & -3\left(\frac{x_{2i}}{R}-\frac{\omega_m C(x_0-x_{2r})}{2}\right) \\ -\frac{x_{2i}}{R}-\frac{\omega_m C(x_0+x_{2r})}{2} & 3\left(\frac{x_0+x_{2r}}{R}-\frac{\omega_m C x_{2i}}{2}\right) \end{pmatrix} \end{aligned} . \tag{28}$$

One needs the DC component and the second harmonic component before studying Equation (28).

3.2. The DC Component and the Second Harmonic Component

The DC component and the second harmonic component are computed from Equations (16) and (18). It is important to note that the first harmonic component is smaller than the DC component and the second harmonic component in Equations (16) and (18). Therefore, one has:

$$\frac{C}{2}\frac{d}{dt}\left(x_0^2+2x_{2r}^2+2x_{2i}^2\right)+\frac{1}{R}\left(x_0^2+2x_{2r}^2+2x_{2i}^2\right)=\frac{V_m^2}{R_s(1+\beta)V_{ref}}(y_0-y_{2r}). \tag{29}$$

$$\frac{C}{2}\frac{d}{dt}(2x_0x_2)+\left(j\omega_m C+\frac{1}{R}\right)(2x_0x_2)=\frac{V_m^2}{R_s(1+\beta)V_{ref}}\left(y_2-\frac{1}{2}y_0\right). \tag{30}$$

Equations (19), (21), (29) and (30) form the model describing the DC component and the second harmonic component. By making all time-derivatives in those four equations equal to zero, one obtains:

$$\frac{1}{R}\left(x_0^2+2x_{2r}^2+2x_{2i}^2\right)=\frac{V_m^2}{R_s(1+\beta)V_{ref}}(y_0-y_{2r}). \tag{31}$$

$$\left(j\omega_m C+\frac{1}{R}\right)(2x_0x_2)=\frac{V_m^2}{R_s(1+\beta)V_{ref}}\left(y_2-\frac{1}{2}y_0\right). \tag{32}$$

$$g_m\left(V_{ref}-\frac{1}{1+\beta}x_0\right)=0. \tag{33}$$

$$j2C_z\omega_m y_2 = -g_m\frac{1}{1+\beta}x_2 - j2\omega_m g_m R_{gm}C_z\frac{1}{1+\beta}x_2. \tag{34}$$

From Equations (31)–(34), one obtains the DC component and the second harmonic component. The steady-state value of the DC component is:

$$x_0=(1+\beta)V_{ref}. \tag{35}$$

When Equation (35) is satisfied, the second harmonic component is zero. Now we can study Equation (28).

3.3. Stability of the OCC Boost PFC Converter

Based on the DC component and the second harmonic component computed in Section 3.2, one can simplify Equation (28) into:

$$M = \frac{g_m}{C_z \omega_m} \frac{\frac{V_m^2}{4R_s(1+\beta)^2 V_{ref}}}{\left(\frac{1}{R^2} + \frac{\omega_m^2 C^2}{4}\right)x_0} \begin{pmatrix} -\frac{R_{gm}C_z\omega_m}{R} + \frac{\omega_m C}{2} & -\frac{3R_{gm}C_z\omega_m^2 C}{2} - \frac{3}{R} \\ \frac{1}{R} + \frac{R_{gm}C_z\omega_m^2 C}{2} & \frac{3\omega_m C}{2} - \frac{3R_{gm}C_z\omega_m}{R} \end{pmatrix}. \tag{36}$$

It is important to note that M in (36) is the round-trip signal transfer function of the first harmonic component. When all eigenvalues of M are less than 1, the first harmonic component converges to zero. At the same time, the DC component and the second harmonic component are almost constant, as shown in Section 3.2 and Figure 2. Therefore the converter operates in a stable manner. When the absolute values of one eigenvalue of M is more than 1, the first harmonic component does not converge to zero. The converter begins to exhibit period-doubling bifurcation at the line frequency [24]. Therefore, the criterion of the stability of the converter is the eigenvalues of matrix M.

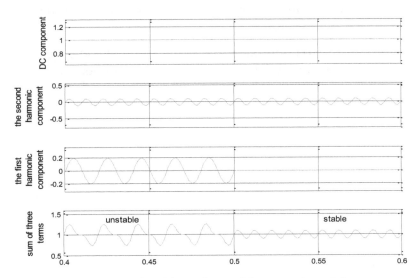

Figure 2. Illustration of stable and unstable operation of the converter.

4. Non-Linear Phenomena of the OCC Boost PFC Converter

To verify the above-mentioned theory, simulations and experiments are conducted. The same circuit topology is adopted (Figure 1). The parameters in the converter are shown in Table 1, unless otherwise specified.

Table 1. Parameters in the OCC Boost PFC converter.

Symbol	Quantity	Unit
T_s	15	μs
ω_m	100π	rad/s
L	2	mH
C	100	μF
R_{f1}	849	$k\Omega$
R_{f2}	37.3	$k\Omega$
R_{gm}	10.25	$k\Omega$
C_z	32	nF

Table 1. *Cont.*

Symbol	Quantity	Unit
C_p	32	pF
V_{ref}	7	V
R_s	0.645	Ω
$R_{(load)}$	1600	Ω
g_m	40	μS

In the converter, the input voltage and the output capacitance are two important parameters, which are selected in the design process. In this paper, the effect of these parameters on the non-linear phenomena of the converter is studied. Figure 3 shows the stability boundaries obtained from theoretical calculation based on Equation (36) and simulation experiment. From Figure 3, we have the following conclusions.

1.   The effect of the input voltage on the stability of the converter. Figure 3 shows that when the capacitance is fixed and the input voltage increases, the converter may lose stability.

2.   The effect of the output capacitance on the stability of the converter. Figure 3 shows that when the input voltage is increased, a larger output capacitance is needed in order to assure stable operation of the converter. This result is important because a larger output capacitance affects the dynamic performance of the converter.

3.   The difference between the two boundaries lies in the fact that some approximations are taken in the analysis, and only the first and the second harmonic components are taken into consideration.

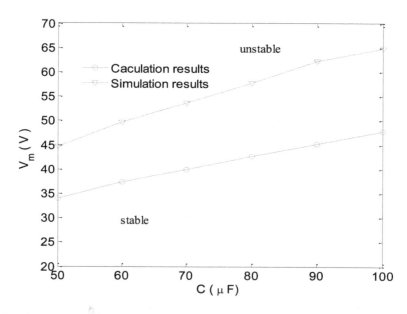

Figure 3. Stability boundaries of the input voltage obtained from the theoretical calculation and simulation experiment.

The simulation waveforms of the output voltage and the inductor current are shown in Figures 4 and 5 when the input voltage $V_m = 40$ V and $V_m = 66.5$ V, respectively. (For the MATLAB model file, please contact the corresponding author by e-mail: zhangrui@cqust.eud.cn.) In Figure 4, the converter operate stably. In Figure 5, the converter exhibits line frequency instability as a result of the period-doubling bifurcation at the line frequency. The instability reduces the power factor of the converter to be considerably lower than 1. If the input voltage increases further, the converter may exhibit chaotic phenomena.

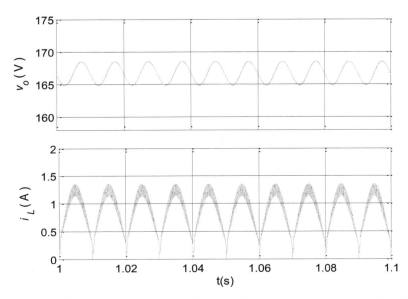

Figure 4. Simulation waveforms of the output voltage and the inductor current when the input voltage $V_m = 40$ V, the load $R = 1600\ \Omega$ and the output capacitance $C = 100\ \mu$F.

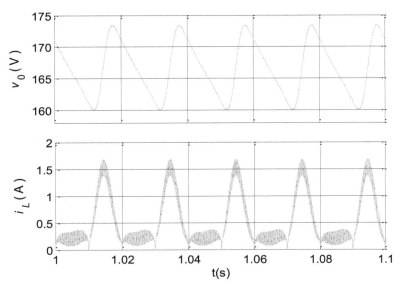

Figure 5. Simulation waveforms of the output voltage and the inductor current when the input voltage $V_m = 66.5$ V, the load $R = 1600\ \Omega$ and the output capacitance $C = 100\ \mu$F.

5. Experimental Verifications

To verify the line frequency instability from the theoretical analysis, an experimental circuit prototype was implemented using IR1150 (Infineon Technologies AG, Neubiberg, Germany), which is a classical OCC Boost PFC IC. In our experiment, the circuit parameters are identical to the above theoretical analysis. The current probe (FLUKE i5s, Everett, WA, USA) AC (400 mV/A) current clamp is used to detect the line current. The input voltage, the inductor current and the output voltage (AC coupling) are shown in Figures 6 and 7 when the input voltage $V_m = 40$ V and $V_m = 68$ V, respectively. In Figure 6, the converter operates in a stable manner and the frequencies of all waveforms are 100 Hz. In Figure 7, the converter exhibits period-doubling phenomena. Furthermore, the first harmonic frequency of the inductor current and the output voltage ripple are 50 Hz, which is half of the rectified AC voltage. As shown in Figures 6 and 7, the experimental results are consistent with the analytical results. It is important to note that when the PFC converter exhibits period-doubling bifurcation, the voltage ripple of the output voltage is larger compared with the normal operation. This value is important for the performance and lifetime of a electrolytic capacitor.

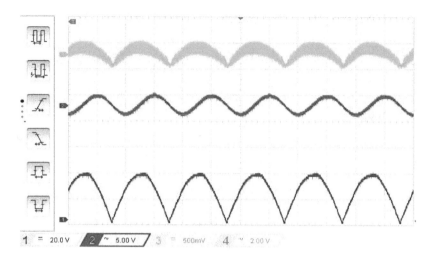

Figure 6. Experimental waveforms of the converter when the input voltage $V_m = 40$ V, the load $R = 1600\ \Omega$ and the output capacitance $C = 100\ \mu$F. CH1: the input voltage (20 V/div), CH3: the inductor current (500 mV/div) and CH2: the output voltage (5 V/div) (AC coupling).

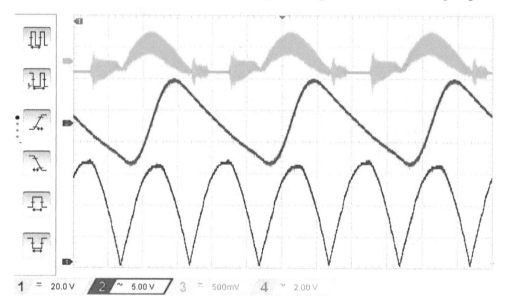

Figure 7. Experimental waveforms of the converter when the input voltage $V_m = 68$ V, the load $R = 1600\ \Omega$ and the output capacitance $C = 100\ \mu$F. CH1: the input voltage (20 V/div), CH3: the inductor current (500 mV/div) and CH2: the output voltage (5 V/div) (AC coupling).

6. Conclusions

The OCC PFC converters have a simpler control circuit compared to the traditional averaged current mode PFC converters. In this paper, the method of double averaging was adopted to study the dynamics of an OCC Boost PFC converter. The first averaging is applied to the switching period, and the second averaging is applied to the line period. We derived the round-trip signal transfer function of the first harmonic component in the converter, and the stability of the converter is decided by the eigenvalues of the round-trip signal transfer function. By calculating the eigenvalues, we gave theoretical prediction of the stability of the converter under different output capacitors. Simulation and experimental results verified theoretical prediction. The method of double averaging can predict nonlinear phenomena which traditional method cannot predict. It is important to note that when the OCC PFC converter exhibits line frequency instabilities, the power factor decreases dramatically. Therefore, theoretical analysis in this paper is of great importance in designing the OCC PFC converter.

Author Contributions: R.Z. and W.M. designed the experiment, drafted the manuscript; L.W. and M.H. conducted the experiment; L.C., H.Z. and Y.Z. reviewed and refined the paper.

Acknowledgments: This research is supported by the National Natural Science Foundation of China under Grant 51377188.

Appendix A

Some important properties:

$$\left(\frac{du(t)}{dt}\right)_k = \frac{du_k(t)}{dt} + jk\omega_m u_k. \tag{A1}$$

$$(u(t)\cos 2\omega_m t)_k = \frac{1}{2}(u_{k-2} + u_{k+2}). \tag{A2}$$

$$\left(u^2(t)\right)_0 = u_0^2 + 2|u_1|^2 + 2|u_2|^2. \tag{A3}$$

$$\left(u^2(t)\right)_1 = 2u_0 u_1 + 2(u_{1r}u_{2r} + u_{1i}u_{2i} + j(u_{1r}u_{2i} - u_{1i}u_{2r})). \tag{A4}$$

$$\left(u^2(t)\right)_2 = (u_1^*)^2 + 2u_0 u_2. \tag{A5}$$

where $u_1 = u_{1r} + ju_{1i}$, $u_2 = u_{2r} + ju_{2i}$.

References

1. García, O.; Cobos, J.A.; Prieto, R.; Alou, P.; Uceda, J. Single phase power factor correction: A survey. *IEEE Trans. Power Electron.* **2003**, *18*, 749–755. [CrossRef]
2. Giaouris, D.S.; Banerjee, B.; Zahawi, V. Pickert, Control of fast scale bifurcations in power-factor correction converters. *IEEE Trans. Circuits Syst. II* **2007**, *54*, 805–809. [CrossRef]
3. Lai, Z.; Smedley, K.M. A family of continuous-conduction-mode power-factor-correction controllers based on the general pulse-width modulator. *IEEE Trans. Power Electr.* **1998**, *13*, 501–510.
4. Orabi, M.; Nimoniya, T. Non-linear dynamic of power factor correction converter. *IEEE Trans. Ind. Electron.* **2003**, *50*, 1116. [CrossRef]
5. Iu, H.H.C.; Zhou, Y.F.; Tse, C.K. Fast-scale instability in a PFC boost converter under average current-mode control. *Int. J. Circ. Theor. App.* **2003**, *31*, 611–624. [CrossRef]
6. Orabi, M.; Ninomiya, T. Stability investigation of the cascade twostage PFC converter. *IEICE Trans. Commun.* **2004**, *E87-B*, 3506–3514.
7. Chu, G.; Tse, C.K.; Wong, S.C. Line-frequency instability of PFC power supplies. *IEEE Trans. Power Electron.* **2009**, *24*, 469–482. [CrossRef]
8. El Aroudi, A.; Orabi, M.; Haroun, R.; Martinez-Salamero, L. Asymptotic slow-scale stability boundary of PFC AC-DC power converters: Theoretical prediction and experimental validation. *IEEE Trans. Ind. Electron.* **2011**, *58*, 3448–3460. [CrossRef]
9. Wang, F.; Zhang, H.; Ma, X. Analysis of slow-scale instability in boost PFC converter using the method of harmonic balance and floquet theory. *IEEE Trans. Circuits Syst. Regul. Pap.* **2010**, *57*, 405–414. [CrossRef]
10. El Aroudi, A.; Orabi, M. Stabilizing technique for AC-DC boost PFC converter based on time delay feedback. *IEEE Trans. Circuits Syst. Express Briefs* **2010**, *57*, 56–60. [CrossRef]
11. Ma, W.; Wang, M.; Liu, S.; Li, S.; Yu, P. Stabilizing the average current-mode-controlled boost PFC converter via washout-filter-aided method. *IEEE Trans. Circuits Syst. Express Briefs* **2011**, *58*, 595–599. [CrossRef]
12. Zou, J.; Ma, X.; Tse, C.K.; Dai, D. Fast-scale bifurcation in power-factor-correction buck-boost converters and effects of incompatible periodicities. *Int. J. Circuit Theory Appl.* **2006**, *34*, 251–264. [CrossRef]
13. Wu, X.; Tse, C.K.; Dranga, O.; Lu, J. Fast-scale instability of single-stage power-factor-correction power supplies. *IEEE Trans. Circuits Syst. Regul. Pap.* **2006**, *53*, 204–213.

14. Wong, S.-C.; Tse, C.K.; Orabi, M.; Ninomiya, T. The Method of Double Averaging: An Approach for Modeling Power-Factor-Correction Switching Converters. *IEEE Trans. Circuits Syst. Regul. Pap.* **2006**, *53*, 454–464. [CrossRef]

15. Orabi, M.; Haron, R.; Youssef, M.Z. Stability analysis of PFC converters with one-cycle control. In Proceedings of the 31st International Telecommunications Energy Conference, Incheon, Korea, 18–22 October 2009.

16. Smedley, K.M.; Cuk, S. One-Cycle Control of Switching Converters. *IEEE Trans. Power Electr.* **1995**, *10*, 625–633. [CrossRef]

17. Smedley, K.M.; Cuk, S. Dynamics of One-Cycle controlled cuk converters. *IEEE Trans. Power Electr.* **1995**, *10*, 634–639. [CrossRef]

18. Fang, C.C.; Abed, E.H. Robust feedback stabilization of limit cycles in PWM DC-DC converters. *Nonlinear Dyn.* **2002**, *27*, 295–309. [CrossRef]

19. Fang, C.-C. Sampled-Data modeling and analysis of One-Cycle control and charge control. *IEEE Trans. Power Electr.* **2001**, *16*, 345–350. [CrossRef]

20. Lee, F.; Iwens, R.; Yu, Y.; Triner, J. Generalized computer-aided discrete time domain modeling and analysis of dc-dc converters. In Proceedings of the 1977 IEEE Power Electronics Specialists Conference, Palo Alto, CA, USA, 14–16 June 1977.

21. Sanders, S.R.; Verghese, G.C. Synthesis of averaged circuit models for switched power converters. *IEEE Trans. Circuits Syst.* **1991**, *8*, 905–915. [CrossRef]

22. Maksimovic, D.; Zane, R.; Erickson, R. Impact of digital control in power electronics. In Proceedings of the 16th International Symposium on Power Semiconductor Devices & ICs, Kitakyushu, Japan, 24–27 May 2004.

23. Caliskan, V.A.; Verghese, O.C.; Stankovic, A.M. Multifrequency averaging of DC/DC converters. *IEEE Trans. Power Electron.* **1999**, *1*, 124–133. [CrossRef]

24. Dorf, R.C.; Bishop, R.H. *Modern Control Systems*, 11th ed.; Pearson Education, Inc.: Upper Saddle River, NJ, USA, 2008.

5

Extended Kalman Filter Based Sliding Mode Control of Parallel-Connected Two Five-Phase PMSM Drive System

Tounsi Kamel [1], **Djahbar Abdelkader** [1], **Barkat Said** [2], **Sanjeevikumar Padmanaban** [3] and **Atif Iqbal** [4,*]

[1] Department of Electrical Engineering, LGEER laboratory, U.H.B.B-Chlef University, Chlef 02000, Algeria; t_kamel@outlook.com (T.K.); a_djahbar@yahoo.fr (D.A.)

[2] Laboratoire de Génie Électrique, Faculté de Technologie, Université de M'Sila, M'Sila 28000, Algeria; sa_barkati@yahoo.fr

[3] Department of Energy Technology, Aalborg University, 6700 Esberg, Denmark; sanjeevi_12@yahoo.co.in

[4] Department of Electrical Engineering Qatar University, Doha, Qatar

* Correspondence: atif.iqbal@qu.edu.qa

Abstract: This paper presents sliding mode control of sensor-less parallel-connected two five-phase permanent magnet synchronous machines (PMSMs) fed by a single five-leg inverter. For both machines, the rotor speeds and rotor positions as well as load torques are estimated by using Extended Kalman Filter (EKF) scheme. Fully decoupled control of both machines is possible via an appropriate phase transposition while connecting the stator windings parallel and employing proposed speed sensor-less method. In the resulting parallel-connected two-machine drive, the independent control of each machine in the group is achieved by controlling the stator currents and speed of each machine under vector control consideration. The effectiveness of the proposed Extended Kalman Filter in conjunction with the sliding mode control is confirmed through application of different load torques for wide speed range operation. Comparison between sliding mode control and PI control of the proposed two-motor drive is provided. The speed response shows a short rise time, an overshoot during reverse operation and settling times is 0.075 s when PI control is used. The speed response obtained by SMC is without overshoot and follows its reference and settling time is 0.028 s. Simulation results confirm that, in transient periods, sliding mode controller remarkably outperforms its counterpart PI controller.

Keywords: five-phase permanent magnet synchronous machine; five-leg voltage source inverter; multiphase space vector modulation; sliding mode control; extended Kalman filter

1. Introduction

Recently, five-phase AC machine drives have gained an increasing attention for a wide variety of industrial applications such as electric vehicles, aerospace applications, naval propulsion systems and paper mills. Major advantages of using a five-phase machine over three-phase machine are better fault tolerant, higher torque density, reduced torque pulsations, improvement of the drive noise characteristic and decrease in the required rating per inverter leg [1–3]. In addition, there are three possible connections for the windings, which is able to enlarge the speed operation range compared with three-phase machines. [4].

Five-phase machines include either induction or synchronous machines. However, compared with induction machine, under the synchronous machines category, the permanent magnet synchronous machine possesses many advantages such as high-power density, better torque generating capability and high conversion efficiency [2]. The rotor excitation of the PMSM is provided by PMs. The PMSM do not need extra DC power supply or field windings in order to provide rotor excitations. So, the power losses related to the filed windings are eliminated in the PMSM. In addition, the magnets and redundant teeth in stators allow magnetic decoupling from the different groups of windings [5,6]. Therefore, more and more multiphase permanent magnet synchronous machines are addressed in a variety of specialized literatures [7–10]. Fortunately, with the increasing development of the technology in traction and industrial applications such as for electrical railways and steel processing, the parallel/series-connected multi-machine systems fed by a single supply become strongly suggested. The reasons for that are: low cost drive, compactness and lightness [11,12]. However, the series-connected system suffers from some serious drawbacks compared with parallel-connected system. In such connection, both beginnings and ending of each phase should be brought out to the terminal box of each multiphase machine. Connecting the phase endings into the star point within the machine can eliminate this disadvantage, as it is the case for parallel connection [13]. Further, the series-connected machines suffer from drawback of poor efficiency because of higher losses.

This paper therefore exploits the maturity of the control ideas proposed for series-connected multiphase multi-machine drives [14–16], as the starting point and extends them to parallel-connected multiphase multi-machine drives.

Actually, to control the torque and flux of any multiphase only direct and quadrature current components are used. The remaining components can be used to control the other machines which are fed by a single multi-leg inverter. This constitutes the main idea behind the concept of parallel-connected multiphase multi-machine drive system fed by a single multi-leg inverter supply. This idea has been developed for all induction machines with even and odd supply phase numbers as pointed in [1,17], respectively.

Usually, in order to control the multiphase drive, the standard controllers have been widely used. However, neglected dynamics, parameter variations, friction forces and load disturbances are the main disturbances and uncertainties that can affect the effective functioning of the drive system. So, it will be very difficult to limit these disturbances effectively if linear control methods like PI controllers are adopted [18]. To overcome the aforementioned problems, other advanced methods have been proposed [1,19–21]. These approaches include, among others, the sliding mode control (SMC). The SMC is a nonlinear control method known to have robust control characteristics under restricted disturbance conditions or when there are limited internal parameter modeling errors as well as when a there are some nonlinear behaviors [22]. The robustness of the SMC is guaranteed usually by using a switching control law. Unfortunately, this switching strategy often leads to a chattering phenomenon. In order to mitigate the chattering phenomenon, a common method is to use the smooth function instead of the switching function [18,23,24].

The five-phase PMSM is invariably supplied by a five-phase voltage source inverter (VSI). There are many techniques to control the five-phase inverter such as carrier based PWM (CBPWM) and space vector modulation (SVM). However, SVM has become the most popular due to its ease of digital implementation and higher DC bus utilization [25–27]. To develop the SVM technique for the parallel-connected two five-phase PMSMs configuration, the concept of multiple 2-D subspaces is used. The idea is to select in each of the two planes, completely independently, a set of four active space vectors neighboring the corresponding reference. So, it can be possible to create two voltage space vector references independently, by using the same approach and the same analytical expressions as for the case of purely sinusoidal output voltage generation. However, in the first switching period, the space vector modulator will apply α-β voltage reference. In the next switching period the space vector modulator will apply x-y voltage reference [27].

In the most electric drives, an accurate knowledge on rotor position is crucial for feedback control. It can be achieved from some types of shaft sensors such as an optical encoder or resolver connected to the rotor shaft [11,28]. However, the use of these sensors will increase the cost and reduce the reliability of the drive and may suffer from some restrictions such as temperature, humidity and vibration. In order to overcome these shortcomings, a number of researchers have developed the well-known sensorless control technology. Various sensorless algorithms have been investigated and reported in many publications [29–31]. The main idea of sensorless control of parallel-connected two five-phase machines is to estimate the rotor positions and their corresponding speeds through an appropriate way using measurable quantities such as five-phase currents and voltages. However, few applications deal with the sensorless control of multiphase machines such as, model reference system [32], Kalman filtering technique [33] and sliding mode observer of five-phase induction motor [10]. Unlike the other approaches, EKF is more attractive because it delivers rapid, precise and accurate estimation. The feedback gain used in EKF achieves quick convergence and provides stability for the observer [34]. For stochastic systems, the extended Kalman filter is the preferable solution capable to provide states estimation or of both the states and parameters estimation.

The purpose of this paper is to study a sensorless sliding mode control scheme using the extended Kalman filter of parallel connected two five-phase PMSMs fed by a single five-leg inverter. To meet this end, the SMC is implemented for speeds and currents control and EKF is used for sensorless operation purposes. The resulting control scheme combines the features of the robust control and the stochastic observer to enhance the performances of the proposed two-machine drive. The performance of the estimation and control scheme is tested with challenging variations of the load torque and velocity reference. The obtained results prove that the two machines are totally decoupled under large speeds and loads variations, although they are connected in parallel and supplied by a single inverter. In addition to that, a comparison between SMC and the traditional PI for sensorless operation is also considered.

2. Modeling of Multiphase AC Drive System

The two-machine drive system under consideration is shown in Figure 1. It consists of a five-leg inverter feeding two five-phase PMSMs. The five-phase PMSM has five-phase windings spatially shifted by 72 electrical degrees. In Figure 1 each stator is star-connected with isolated neutral point which eliminates the zero sequence voltages. It can be seen from Figure 1 that the phase transposition rules of parallel-connected two five-phase PMSMs system are as follows [17]: as_1-as_2, bs_1-cs_2, cs_1-es_2, ds_1-bs_2, es_1-ds_2. Where indices 1 and 2 identify the two machines as indicated in Figure 1. So, the relationships between voltages and currents are given as:

$$v_{abcde} = \begin{bmatrix} v_a^{inv} \\ v_b^{inv} \\ v_c^{inv} \\ v_d^{inv} \\ v_e^{inv} \end{bmatrix} = \begin{bmatrix} v_{as1} = v_{as2} \\ v_{bs1} = v_{cs2} \\ v_{cs1} = v_{es2} \\ v_{ds1} = v_{bs2} \\ v_{es1} = v_{ds2} \end{bmatrix} \quad i_{abcde} = \begin{bmatrix} i_a^{inv} \\ i_b^{inv} \\ i_c^{inv} \\ i_d^{inv} \\ i_e^{inv} \end{bmatrix} = \begin{bmatrix} i_{as1} + i_{as2} \\ i_{bs1} + i_{cs2} \\ i_{cs1} + i_{co2} \\ i_{ds1} + i_{bs2} \\ i_{es1} + i_{ds2} \end{bmatrix} \quad (1)$$

The main five dimensional systems can be decomposed into five dimensional uncoupled subsystems (d-q-x-y-0). Let the correlation between the original phase variables and the new (d-q-x-y-0) variables are given by $f_{dqxy} = Cf_{abcde}$, where C is the following invariant transformation matrix:

$$[C] = \frac{2}{5} \begin{bmatrix} \cos(\theta) & \cos(\theta - 2\pi/5) & \cos(\theta - 4\pi/5) & \cos(\theta + 4\pi/5) & \cos(\theta + 2\pi/5) \\ \sin(\theta) & \sin(\theta - 2\pi/5) & \sin(\theta - 4\pi/5) & \sin(\theta + 4\pi/5) & \sin(\theta + 2\pi/5) \\ \cos(\theta) & \cos(\theta + 4\pi/5) & \cos(\theta - 2\pi/5) & \cos(\theta + 2\pi/5) & \cos(\theta - 4\pi/5) \\ \sin(\theta) & \sin(\theta + 4\pi/5) & \sin(\theta - 2\pi/5) & \sin(\theta + 2\pi/5) & \sin(\theta - 4\pi/5) \\ 1/2 & 1/2 & 1/2 & 1/2 & 1/2 \end{bmatrix} \quad (2)$$

By applying the transformation matrix (2) on Equation (1), the voltage and current components of the five-phase VSI become:

$$
\begin{bmatrix} v_d^{inv} \\ v_q^{inv} \\ v_x^{inv} \\ v_y^{inv} \\ v_0^{inv} \end{bmatrix} = [C]v_{abcde} = \begin{bmatrix} v_{ds1} = v_{xs2} \\ v_{qs1} = -v_{ys2} \\ v_{xs1} = v_{ds2} \\ v_{ys1} = v_{qs2} \\ 0 \end{bmatrix}, \quad \begin{bmatrix} i_d^{inv} \\ i_q^{inv} \\ i_x^{inv} \\ i_y^{inv} \\ i_0^{inv} \end{bmatrix} = [C]i_{abcde} = \begin{bmatrix} i_{ds1} + i_{xs2} \\ i_{qs1} - i_{ys2} \\ i_{xs1} + i_{ds2} \\ i_{ys1} + i_{qs2} \\ 0 \end{bmatrix} \tag{3}
$$

From (3), it is evident that the inverter voltage d-q components can control the first machine (PMSM1), while the second machine (PMSM2) can be controlled separately using the inverter voltage x-y components.

The model of each five-phase PMSM is presented in a rotating d-q-x-y frame as:

$$
\begin{aligned}
v_{dsj} &= r_{sj}i_{dsj} + L_{dj}\frac{di_{dsj}}{dt} - \omega_j L_{qsj}i_{qsj} \\
v_{qsj} &= r_{sj}i_{qsj} + L_{qj}\frac{di_{qsj}}{dt} + \omega_j L_{dsj}i_{dsj} + \omega_j \Phi_{fj} \\
v_{xsj} &= r_{sj}i_{xsj} + L_{lsj}\frac{di_{xsj}}{dt} \\
v_{ysj} &= r_{sj}i_{ysj} + L_{lsj}\frac{di_{ysj}}{dt}
\end{aligned} \tag{4}
$$

where $j = 1,2$. $v_{dj}, v_{qj}, v_{xj}, v_{yj}$ are the stator voltages in the d, q, x, y axes, respectively. $i_{dj}, i_{qj}, i_{xj}, i_{yj}$ are the stator currents in d, q, x, y axes, respectively. L_{dj}, L_{qj}, L_{lsj} are inductances in the rotating frames. r_{sj} is the stator resistance.

The torques equations for the first and the second machines are given by:

$$
\begin{aligned}
T_{em1} &= \tfrac{5p_1}{2}(\Phi_{f1}i_{qs} + (L_{d1} - L_{q1})i_{qs}i_{ds}) \\
T_{em2} &= \tfrac{5p_2}{2}(\Phi_{f2}i_{ys} + (L_{d2} - L_{q2})i_{ys}i_{xs})
\end{aligned} \tag{5}
$$

where p_j are pole pairs, Φ_{fj} are the permanent magnet fluxes.

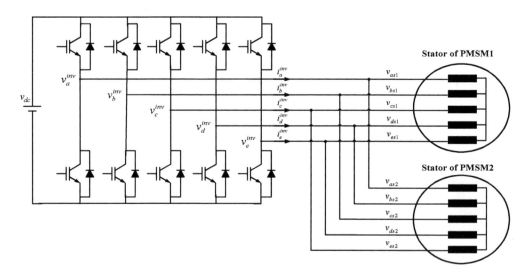

Figure 1. A parallel connected five-phase two-motor drive.

The proposed sensor-less control of the parallel-connected two five-phase permanent magnet synchronous machines is presented in Figure 2, where the two main parts EKF and SMC are considered. The EKF is designed to estimate the rotor position, speed and load torque of each machine by using a current observer. The feedback actual speed, estimated speed and load torques are the inputs of the speeds SMCs to determine the q_1-y_2 axes reference current components. The other current components are maintained to zero. The measured currents are processed in the current SMCs to obtain as outputs the *dqxy* axes reference voltages components. These reference voltages are transformed into the *abcde* frame and transformed again to *αβxy* frame to become input signals to the SVM blocks. The SVM transmits the signals to the inverter to drive the two five-phase PMSMs connected in parallel.

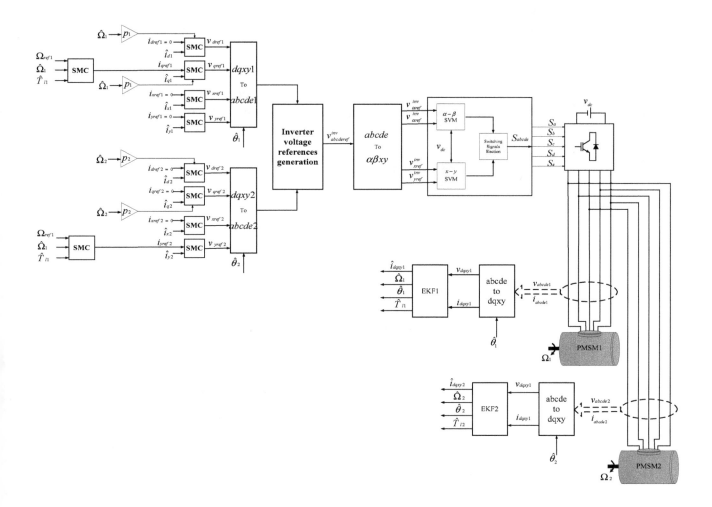

Figure 2. Sensor-less SMC of parallel-connected two five-phase PMSMs drive system.

3. Sliding Mode Controller (SMC)

The design of a sliding mode controller requires mainly two stages. The first stage is choosing an appropriate sliding surface. The second stage is designing a control law, which will drive the state variables to the sliding surface and will keep them there.

3.1. Sliding Surfaces Choice

In order to prescribe the desired dynamic characteristics of the controlled system, the following general form of sliding surface can be adopted [35].

$$S(x) = (\frac{d}{dt} + \lambda)^{r-1} e(x) \tag{6}$$

With: $e(x) = x_{ref} - x$. λ: is a positive coefficient. r: is the relative degree, which is the number of times required to differentiate the surface before the input u {\display style u} appears explicitly.

3.2. Controller Design

In order to drive the state variables to the sliding surface, the following control law is defined as:

$$u = u_{eqc} + u_{dic} \tag{7}$$

The equivalent control u_{eqc} is capable to keep the state variables on the switching surface, once they reach it and to achieve the desired performance under nominal model. It is derived as the solution of the following equation:

$$S(x) = \dot{S}(x) = 0 \tag{8}$$

The discontinuous control u_{dic} is needed to assure the convergence of the system states to sliding surfaces in finite time and it should be designed to eliminate the effect of any unpredictable perturbation. The discontinuous control input can be determined with the help of the following Lyapunov function candidate:

$$V = \frac{1}{2}S(x)^2 \tag{9}$$

The stability is shown under two conditions as:

- The Lyapunov function V is positive definite.
- The derivative of the sliding function should be negative $\dot{V} = \dot{S}(x)S(x) < 0 \ (\forall S)$.

The so-called reaching stability condition ($\dot{V} = S\dot{S} < 0$) is fulfilled using the following discontinuous control:

$$u_{dic} = G\,sign(S(x)) \tag{10}$$

where G is a control gain.

In order to reduce the chattering phenomenon, a saturation function instead of the switching one can be used. The saturation function depicted in Figure 3 is expressed as follows:

$$sat(S(x)) = \begin{cases} \text{sgn}(S(x)) & if \quad |S(x)| > \delta \\ \frac{S(x)}{\delta} & if \quad |S(x)| < \delta \end{cases} \tag{11}$$

With δ is the boundary layer width.

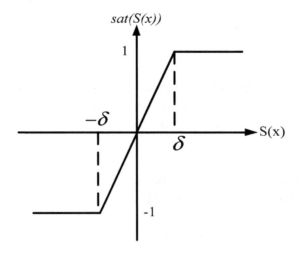

Figure 3. Saturation function.

4. Sliding Mode Control of the Five-Phase Two-Machine AC Drive System

4.1. Speeds SMC Design

The first task in the speeds SMC design process is to select suitable sliding surfaces $S(\Omega_j)$. Since the relative degree is one, the following sliding surfaces are adopted:

$$S(\Omega_j) = \Omega_{refj} - \Omega_j \tag{12}$$

By taking the derivative of sliding surfaces (12) with respect to time and using the machines motion equations, it yields:

$$\dot{S}(\Omega_1) = \dot{\Omega}_{ref1} - \frac{5p_1(L_{d1}-L_{q1})i_{ds1}+5p_1\Phi_{f1}}{2J_1}i_{qs1} + \frac{T_{l1}}{J_1} + \frac{f_1\Omega_1}{J_1}$$
$$\dot{S}(\Omega_2) = \dot{\Omega}_{ref2} - \frac{5p_2(L_{d2}-L_{q2})i_{xs2}+5p_2\Phi_{f2}}{2J_2}i_{ys2} + \frac{T_{l2}}{J_2} + \frac{f_2\Omega_2}{J_2} \tag{13}$$

where J_j, f_j and T_{lj} are moment of inertia, damping coefficient and load torque of each machine. The currents controls i_{qsref1} and i_{ysref2} are defined by:

$$i_{qs1} = i_{qseqc1} + i_{qsdic1}$$
$$i_{ys2} = i_{yseqc2} + i_{ysdic2} \tag{14}$$

where:

$$i_{qseqc1} = -\frac{J_1\dot{\Omega}_{ref1}+T_{l1}+f_1\Omega_1}{\frac{5}{2}p_1(L_{d1}-L_{q1})i_{ds1}+\frac{5}{2}p_1\Phi_{f1}}; i_{yseqc2} = -\frac{J_2\dot{\Omega}_{ref1}+T_{l2}+f_2\Omega_2}{\frac{5}{2}p_2(L_{d2}-L_{q2})i_{xs2}+\frac{5}{2}p_2\Phi_{f2}}$$
$$i_{qsdic1} = G_{\Omega1}sat(S(\Omega_1)); i_{ysdic2} = G_{\Omega2}sat(S(\Omega_2))$$

During the convergence mode, the condition $\dot{V} = \dot{S}(x)S(x) < 0 \, (\forall S)$ must be verified. By replacing (14) into (13), we get:

$$\dot{S}(\Omega_1) = -\frac{5p_1(L_{d1}-L_{q1})i_{ds1}+5p_1\Phi_{f1}}{2J_1}i_{qsdic1} < 0$$
$$\dot{S}(\Omega_2) = -\frac{5p_2(L_{d2}-L_{q2})i_{xs2}+5p_2\Phi_{f2}}{2J_2}i_{ysdic2} < 0 \tag{15}$$

4.2. Currents SMC Design

The control objectives are to track the desired currents trajectories. So, the sliding surfaces can be calculated as follows:

$$S(i_{dsj}) = i_{dsrefj} - i_{dsj}$$
$$S(i_{qsj}) = i_{qsrefj} - i_{qsj}$$
$$S(i_{xsj}) = i_{xsrefj} - i_{xsj}$$
$$S(i_{ysj}) = i_{ysrefj} - i_{ysj} \tag{16}$$

The time derivative of (16) is:

$$\dot{S}(i_{dsj}) = \dot{i}_{ds\,refj} - \dot{i}_{dsj}$$
$$\dot{S}(i_{qsj}) = \dot{i}_{qs\,refj} - \dot{i}_{qsj}$$
$$\dot{S}(i_{xsj}) = \dot{i}_{xs\,refj} - \dot{i}_{xsj}$$
$$\dot{S}(i_{ysj}) = \dot{i}_{ys\,refj} - \dot{i}_{ysj} \tag{17}$$

Using (4), the Equation (17) can be rewritten as:

$$\dot{S}(i_{dsj}) = \dot{i}_{dsrefj} + \frac{r_{sj}}{L_{dj}}i_{dsj} - \frac{L_{qj}}{L_{dj}}\omega_j i_{qsj} - \frac{1}{L_{dj}}v_{dsj}$$
$$\dot{S}(i_{qsj}) = \dot{i}_{qsrefj} + \frac{r_{sj}}{L_{qj}}i_{qsj} + \frac{L_{dj}}{L_{qj}}\omega_j i_{dsj} + \frac{\omega_j \Phi_{fj}}{L_{qj}} - \frac{1}{L_{qj}}v_{qsj}$$
$$\dot{S}(i_{xsj}) = \dot{i}_{xsrefj} + \frac{r_{sj}}{L_{lsj}}i_{xsj} - \frac{1}{L_{lsj}}v_{xsj}$$
$$\dot{S}(i_{ysj}) = \dot{i}_{ysrefj} + \frac{r_{sj}}{L_{lsj}}i_{ysj} - \frac{1}{L_{lsj}}v_{ysj}$$

$$(18)$$

So, it is possible to choose the control laws for stator voltages as follows:

$$v_{ds\,refj} = v_{dseqcj} + v_{dsdicj}$$
$$v_{qs\,refj} = v_{qseqcj} + v_{qsdicj}$$
$$v_{xs\,refj} = v_{xseqcj} + v_{xsdicj}$$
$$v_{ys\,refj} = v_{yseqcj} + v_{ysdicj}$$

$$(19)$$

where:

$$v_{dseqcj} = (\dot{i}_{dsrefj} + \frac{r_{sj}}{L_{dj}}i_{dsj} - \frac{L_{qj}}{L_{dj}}\omega_j i_{qsj})L_{dj}; v_{qseqcj} = (\dot{i}_{qsrefj} + \frac{r_{sj}}{L_{qj}}i_{qsj} + \frac{L_{dj}}{L_{qj}}\omega_j i_{dsj} + \frac{\omega_j \Phi_{fj}}{L_{qj}})L_{qj}$$
$$v_{xseqcj} = (\dot{i}_{xsrefj} + \frac{r_{sj}}{L_{lsj}}i_{xsj})L_{lsj}; v_{yseqcj} = (\dot{i}_{ysrefj} + \frac{r_{sj}}{L_{lsj}}i_{ysj})L_{lsj}$$

$$v_{dsdicj} = G_{dsj}sat(S(i_{dsj})); v_{qsdicj} = G_{qsj}sat(S(i_{qsj})); v_{xsdicj} = G_{xsj}sat(S(i_{xsj})); v_{ysdicj} = G_{ysj}sat(S(i_{ysj}))$$

During the convergence mode, the condition $\dot{V} = \dot{S}(x)S(x) < 0 \ (\forall \ S)$ must be verified. By replacing (19) into (18), we get:

$$\dot{S}(i_{dsj}) = -\frac{1}{L_{dj}}v_{dsdicj}$$
$$\dot{S}(i_{qsj}) = -\frac{1}{L_{qj}}v_{qdicsj}$$
$$\dot{S}(i_{xsj}) = -\frac{1}{L_{lsj}}v_{xsdicj}$$
$$\dot{S}(i_{ysj}) = -\frac{1}{L_{lsj}}v_{ysdicj}$$

$$(20)$$

The control voltages given by (19) are transformed in *abcde* frame and then the inverter phase voltage references are calculated according to the following expression:

$$v_a^{inv} = v_{as\,ref1} + v_{as\,ref2}$$
$$v_b^{inv} = v_{bs\,ref1} + v_{cs\,ref2}$$
$$v_c^{inv} = v_{cs\,ref1} + v_{es\,ref2}$$
$$v_d^{inv} = v_{ds\,ref1} + v_{bs\,ref2}$$
$$v_e^{inv} = v_{es\,ref1} + v_{ds\,ref2}$$

$$(21)$$

5. Space Vector Modulation Technique for Parallel Connected Multiphase AC Drive System

5.1. Five-Leg VSI Modeling

The five-leg inverter output phase-to-neutral voltages can be expressed as:

$$v_a^{inv} = \frac{v_{dc}}{5}(4S_a - S_b - S_c - S_d - S_e)$$
$$v_b^{inv} = \frac{v_{dc}}{5}(-S_a + 4S_b - S_c - S_d - S_e)$$
$$v_c^{inv} = \frac{v_{dc}}{5}(-S_a - S_b + 4S_c - S_d - S_e)$$
$$v_d^{inv} = \frac{v_{dc}}{5}(-S_a - S_b - S_c + 4S_d - S_e)$$
$$v_e^{inv} = \frac{v_{dc}}{5}(-S_a - S_b - S_c - S_d + 4S_e)$$

$$(22)$$

where v_{dc} denotes DC-link voltage and S_i, $i = a, b, c, d, e$ refer to switching functions.

The five-phase inverter has totally thirty-two space voltage vectors, thirty non-zero voltage vectors and two zero voltage vectors. These space vectors can be projected on α-β subspace as well as on x-y subspace as shown in Figure 4. Every plane is divided in ten sectors, each occupying a $36°$ angle around the origin by means of the following two space vectors [27]:

$$v_{\alpha\beta}^{inv} = \tfrac{2}{5}(v_a^{inv} + v_b^{inv}e^{j\alpha} + v_c^{inv}e^{j2\alpha} + v_d^{inv}e^{j3\alpha} + v_e^{inv}e^{j4\alpha})$$
$$v_{xy}^{inv} = \tfrac{2}{5}(v_a^{inv} + v_b^{inv}e^{j2\alpha} + v_c^{inv}e^{j4\alpha} + v_d^{inv}e^{j6\alpha} + v_e^{inv}e^{j8\alpha})$$

(23)

where $\alpha = \tfrac{2\pi}{5}$.

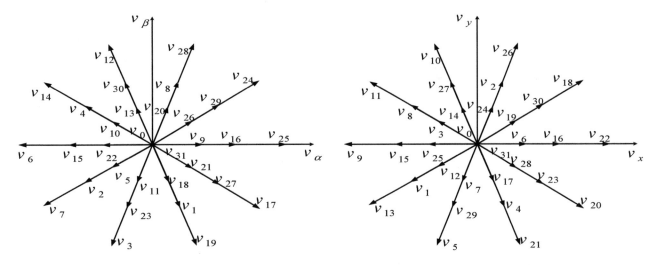

Figure 4. Space vectors of a five-phase inverter in two 2-D subspaces.

From Figure 3 the space vectors are divided into three groups in accordance with their magnitudes: small, medium and large space vector groups. The magnitudes are identified with indices s, m and l and are given as: $|V_s| = 4/5\cos(2\pi/5)v_{dc}$, $|V_m| = 2/5v_{dc}$ and $|V_l| = 4/5\cos(\pi/5)v_{dc}$, respectively [27,35–38]. It can be observed from Figure 3 that medium length space vectors of the α-β plane are mapped into medium length vectors in the x-y plane and large vectors of the α-β plane are mapped into small vectors in the x-y plane and vice-versa.

5.2. SVM Method for Five-Leg VSI

The reference voltage can be obtained by averaging a certain number of active space vectors for adequate time intervals, without saturating the VSI. Four active space vectors are required to reconstruct the reference voltage vector [27,36–38].

The dwell times for active space vectors T_{1m}, T_{1l}, T_{2m}, T_{2l} are:

$$T_{1l} = \frac{|v_{ref}|\sin(s\pi/5-\vartheta)}{(|V_l|+|V_s|)\sin(\pi/5)}T_s$$
$$T_{2m} = \frac{|V_s||v_{ref}|\sin(s\pi/5-\vartheta)}{|V_m|(|V_l|+|V_s|)\sin(\pi/5)}T_s$$
$$T_{1l} = \frac{|v_{ref}|\sin(s\pi/5-\vartheta)}{(|V_l|+|V_s|)\sin(\pi/5)}T_s$$
$$T_{2m} = \frac{|V_s||v_{ref}|\sin(s\pi/5-\vartheta)}{|V_m|(|V_l|+|V_s|)\sin(\pi/5)}T_s$$
$$T_o = T_s - (T_{1l} + T_{1m} + T_{2l} + T_{2m})$$

(24)

where: T_s is the switching period and ϑ is the voltage reference vector position.

The control strategy adopted herein is based on the approach proposed in [27,36]. Indeed, in the first switching period, the space vector modulator will apply α-β voltage reference. In the next switching period the space vector modulator will apply x-y voltage reference as shown in Figure 5. The selection of switching signals is depicted in Figure 6. So, two independent space vector modulators are further utilized to realize the required two voltage space vector references, with dwell times calculated independently in the two planes using (24).

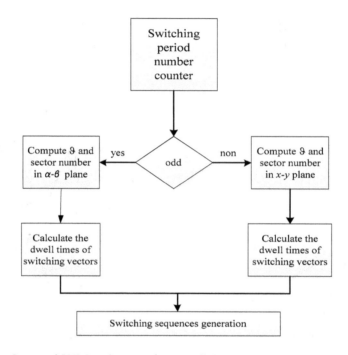

Figure 5. Steps of SVM technique for parallel connected two-machine drive.

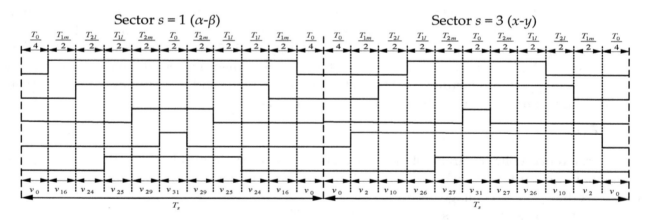

Figure 6. Switching pattern obtained with SVM.

6. Extended Kalman Filter Based Speed Estimator for Parallel Connected Two Motor Drivel

Normally, speed observers used for three-phase machines can be easily extended to multi-phase multi-machine drives. For each machine, the speed estimator requires only stator voltages and currents components. The SMC block diagram based on extended kalman filter of parallel-connected two five-phase machines drive system is shown in Figure 2. The main task of EKF is to find the best estimate of state variables and the unknown load torques since the knowledge of the load torque is necessary for each speed SMC implementation.

In the five-phase PMSM control case, d, q, x, y currents and voltages are measured and the Equation (4) is sampled to obtain a discrete state space representation to be used in the observer synthesis. Assuming that the sampling interval T_e is very short compared to the system dynamics, the augmented discrete-time of each five-phase PMSM model is given as follows:

$$x_{j(k+1)} = A_{jk}x_{jk} + B_{jk}u_{kj} + w_{jk}$$
$$y_{jk} = C_{jk}x_{jk} + v_{jk}$$

(25)

With:

$$x_{jk} = [i_{dsjk}\ i_{qsjk}\ i_{xsjk}\ i_{ysjk}\ \Omega_{jk}\ \theta_{jk}\ T_{ljk}]^t$$

$$u_{jk} = \begin{bmatrix} 0 & v_{qk}^{inv} & 0 & v_{yk}^{inv} \\ v_{dk}^{inv} & 0 & v_{xk}^{inv} & 0 \end{bmatrix}^t$$

$$y_{jk} = [i_{dsjk}\ i_{qsjk}\ i_{xsjk}\ i_{ysjk}]^t$$

$$A_{jk} = \begin{bmatrix} 1 - T_e\left(\frac{r_{sj}}{L_{dj}}\right) & p_j\Omega_j\frac{L_{qj}}{L_{dj}} & 0 & 0 & 0 & 0 & 0 \\ -p_j\Omega_j\frac{L_{qj}}{L_{dkj}} & 1 - T_e\left(\frac{r_{sj}}{L_{qj}}\right) & 0 & 0 & p_j\Phi_{ff}\frac{T_e}{L_{qj}} & 0 & 0 \\ 0 & 0 & 1 - T_e\left(\frac{r_{sj}}{L_{lsj}}\right) & 0 & 0 & 0 & 0 \\ 0 & 0 & 0 & 1 - T_e\left(\frac{r_{sj}}{L_{lsj}}\right) & 0 & 0 & 0 \\ T_e\frac{5P_j}{2J_j}\left(L_{dj}-L_{qj}\right)i_{qsj} & T_e\frac{5P_j}{2J_j}\Phi_{ff} & 0 & 0 & 1 - T_e\left(\frac{f_j}{J_j}\right) & 0 & 0 \\ 0 & 0 & 0 & 0 & T_eP_j & 1 & 0 \\ 0 & 0 & 0 & 0 & 0 & 0 & 1 \end{bmatrix}$$

$$B_{jk} = \begin{bmatrix} T_e\frac{1}{L_{dj}} & 0 \\ 0 & T_e\frac{1}{L_{dj}} \\ T_e\frac{1}{L_{lsj}} & 0 \\ 0 & T_e\frac{1}{L_{lsj}} \\ 0 & 0 \\ 0 & 0 \\ 0 & 0 \end{bmatrix} \quad C_{jk} = \begin{bmatrix} 1 & 0 & 0 & 0 & 0 & 0 & 0 \\ 0 & 1 & 0 & 0 & 0 & 0 & 0 \\ 0 & 0 & 1 & 0 & 0 & 0 & 0 \\ 0 & 0 & 0 & 1 & 0 & 0 & 0 \end{bmatrix}$$

where x_{jk}, u_{jk} and y_{jk} are the augmented state vector and input vector and output vector at the sampling instant k of machine j, respectively. A_{jk} and B_{jk} are discrete system matrix and discrete input matrix for each machine, respectively. w_{jk} and v_{jk} are the system noise and measurement noise, respectively.

The added white-noise vectors are Gaussian and uncorrelated from each other with zero mean and covariance Q_j and R_j, respectively. The covariance matrices Q_j and R_j of these noises are defined as:

$$Q_j = E\left\{w_{jk}w_{jk'}^T\right\}, \quad \text{for}\quad k \neq k'$$
$$R_j = E\left\{v_{jk}v_{jk'}^T\right\}, \quad \text{for}\quad k \neq k'$$

(26)

In a first main stage the state $x_{j(k+1)}$ is predicted using discrete matrices and previous state. In a second main stage, the feedback correction weight matrix K_j (filter coefficients) is used to have an accurate prediction of the state $x_{j(k+1/k)}$. This is obtained by computing K_j depending not only on the error made but also with an adjustment using weight P_j (covariance state matrix). This allows estimating accurately x_j with respect to Q_j and R_j covariance matrices corresponding respectively to state noise and measurement noise levels [30]. Using Equation (25), the rotor speeds and load torques can be estimated by the extended Kalman filter algorithm described as follows:

- Sate prediction:

$$\hat{x}_{j(k+1/k)} = A_{jk}x_{j(k/k)} + B_{jk}u_{jk}$$

(27)

- Estimation of the matrix of the covariance error:

$$\hat{P}_{j(k+1/k)} = A_{jk}P_{j(k/k)}A_{jk}^T + Q_j$$

(28)

- Kalman coefficient update:

$$K_{j(k+1)} = \hat{P}_{j(k+1/k)}C_{jk}^T[C_{jk}\hat{P}_{j(k+1/k)}C_{jk}^T + R_j]^{-1}$$

(29)

- State estimation:

$$\hat{x}_{j(k+1/k+1)} = \hat{x}_{j(k+1/k)} + K_{j(k+1)}(y_{j(k+1)} - C_{jk}\hat{x}_{j(k+1/k)})$$

(30)

- Covariance error matrix update:

$$\hat{P}_{j(k+1/k+1)} = \hat{P}_{j(k+1/k)} - K_{j(k+1)}C_{jk}\hat{P}_{j(k+1/k)} \tag{31}$$

where $\hat{x}$ is the system state, u_{jk} is the system input vector, y is the system output vector, P, Q and R are the covariance matrices, C is the transformation matrix.

7. Numerical Simulation Results

In order to verify the applicability of the proposed control scheme for the two-machine drive system of Figure 2, the following simulations are performed using two identical 2-pole, 50 Hz five-phase PMSM. The parameters of each machine are listed in Table 1. The performance of the SMC controller is compared with that of the conventional controller. The tuning parameters for the PI controllers and SMC controllers are also given in Table 2. Many simulation tests are performed in order to verify the independence of the control of the two machines in sensor-less mode.

Table 1. Five-phase PMSM parameters.

p_j	L_{dj}	L_{qj}	L_{lsj}	Φ_{fj}	J_j	r_{sj}	f_j
2	8.5 mH	8 mH	0.2 mH	0.175 Wb	0.004 kg m^2	1 Ω	0

Table 2. PI and SMC parameters.

	Speed Controller	i_{sd} Controller	i_{sq} Controller	i_{sx} Controller	i_{sy} Controller
PI	$k_p = 0.8$ $k_i = 40$	$k_p = 33$ $k_i = 32{,}000$	$k_p = 33$ $k_i = 32{,}000$	$k_p = 33$ $k_i = 32{,}000$	$k_p = 33$ $k_i = 32{,}000$
SMC	$G_{\Omega j} = 5$	$G_{idj} = 4000$	$G_{iqj} = 7000$	$G_{ixj} = 4000$	$G_{iyj} = 7000$

The behavior of the overall drive system is presented in Figures 7–11 at different test conditions. Figure 7 shows then estimated speeds, currents and torques of the unloaded two machines for many different speeds references. At the beginning, the first machine is running at 100 rad/s, at $t = 0.7$ s it decelerated to -10 rad/s, after that, it is accelerated again to the speed 60 rad/sat $t = 1.4$ s. For the second machine the speed reference is set at 50 rad/s, 25 rad/s, 100 rad/s, -100 rad/s and 80 rad/s at $t = 0$ s, 0.4 s, 0.9 s, 1.2 s, 1.7 s, respectively. Effect of the speed rotation reversion of one machines on the other machine performance is investigated Figure 8. In this test, most of the time when one machine is rotating at $+100$ rad/s the other is running at the opposite speed.

Some additional reversing tests are conducted next to further verify decoupling of the control of the two machines. Figure 9 displays results for the case when the speed Ω_2 is kept at standstill, while Ω_1 is reversed from: $+100$ to -100 rad/s at $t = 0.5$ s and returns to zero at $t = 1$ s. At the subsequent test, the speed Ω_1 is held at zero, while Ω_2 is reversed from 100 to -100 rad/s at 1.5 s.

Figure 10 shows the speeds, torques and currents of the two-machine drive controlled by both PI and SMC controller in the presence of load torques variations. The reference of the first speed is fixed at 100 rad/s, while the speed reference of the second machine is fixed at 50 rad/s. Load torques are applied on the two machines at $t = 0.5$ s and $t = 0.7$ s, respectively.

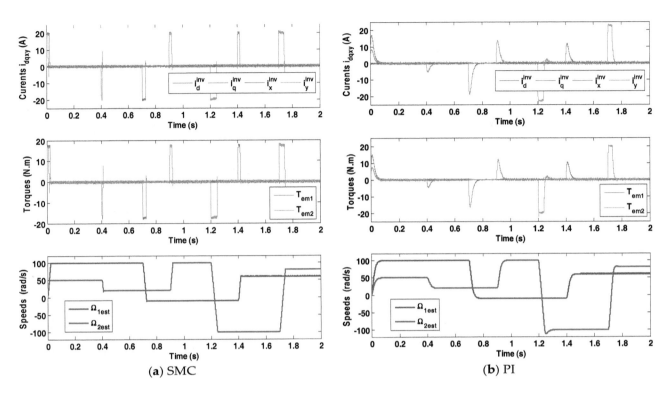

Figure 7. Dynamic responses of parallel-connected two five-phase PMSMs system at different reference speeds values.

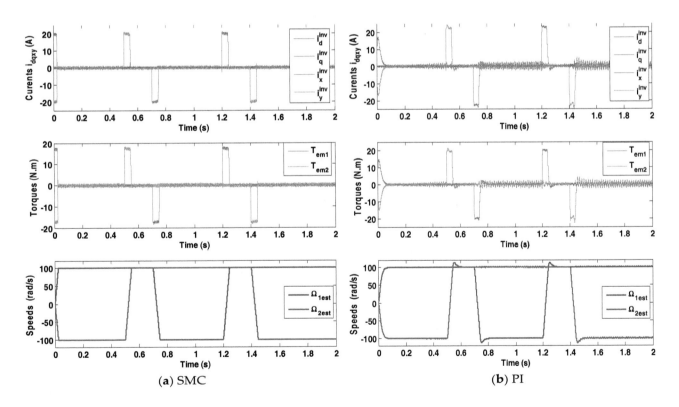

Figure 8. Dynamic responses of parallel-connected two five-phase PMSMs system: when the two motors are operating in the opposite directions.

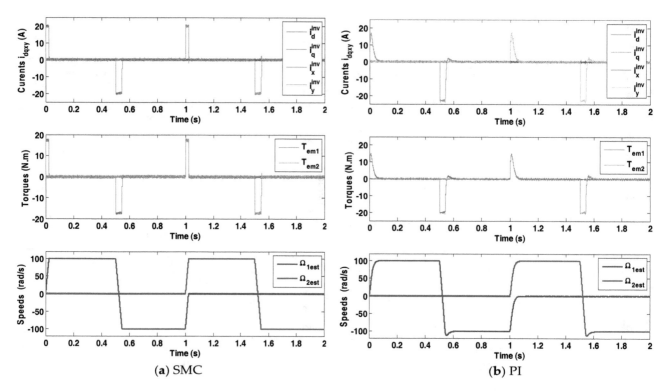

(a) SMC   (b) PI

Figure 9. Dynamic responses of parallel-connected two five-phase PMSMs system: when one machine is at standstill and the other is still running.

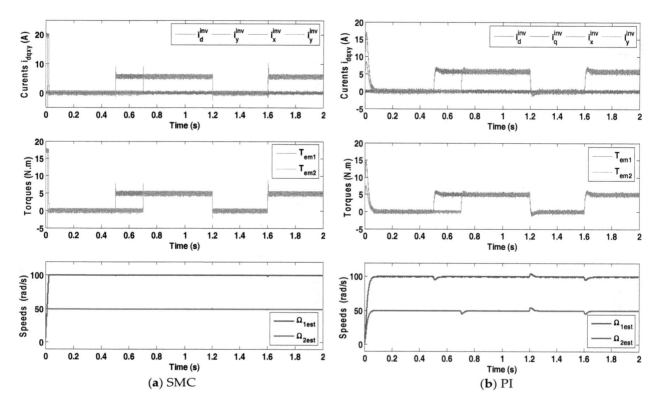

(a) SMC   (b) PI

Figure 10. Dynamic responses of parallel-connected two five-phase PMSMs system at different loading conditions.

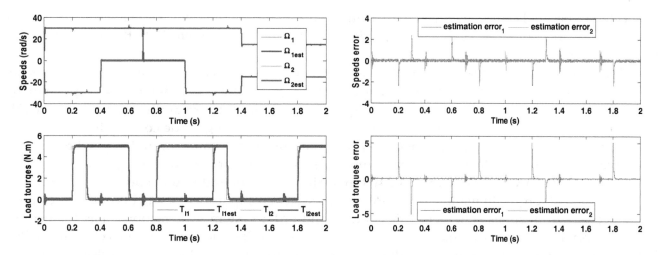

Figure 11. Actual and estimated speeds and load torques and their corresponding estimation errors (SMC case).

It is clear from all estimated speeds characteristics that in every test the speed estimators provide accurate speed estimations. These results also prove that both speeds machines are independently controlled even in sensor-less mode. Indeed, the speed variation of the first machine in the two-machine drive system does not affect the behavior of the speed of the second machine even in reversal conditions.

The electromagnetic torque generated by each machine during the simulated speed step response is shown in Figures 7–10. Note that the generated torques are directly proportional to the q-x axes currents and fully decoupled from d-y axes currents.

Comparison of results in Figure 7 shows once more that the control of the two machines is completely decoupled. There is hardly any evidence of torque disturbance of one machine during the reversal of the other one. Furthermore, the direct axis currents responses remain completely unaffected during these transients.

As shown from Figure 8, the starting and reversing transients of one machine do not have any tangible consequence on the operation of the second machine. The decoupled control is preserved and the characteristics of both machines are unaffected.

Figure 9 illustrates results for the case when the speed of one machine is kept at zero, while the second is reversed. Speed of one machine and its electromagnetic torque remain completely undisturbed during the reversion of the other machine, indicating a complete decoupling of the control.

Figure 10 shows inverter current characteristic, motor torques and estimated speed of motors at different loading conditions for parallel-connected two five-phase machines drive system. It is clear from Figure 10 that when one machine is loaded or unloaded, the second machine performance is unaffected; which proves once again that both motors connected in parallel are totally decoupled. In case of sliding mode control, no variation whatsoever can be observed in the speeds responses of the both machines during these transients.

The estimated and actual values of speeds and load torques as well as their estimations errors are reported in Figure 11. The EKF algorithms give accurate and fast speeds estimations over entire speed range including low speed and standstill operations with low speed errors, even in transients. Furthermore, the estimated values of loads torques are very close to their applied ones. Consequently, the load torque estimation errors are almost zeros; this reflects the stability of EKF during load torques variations. These results confirm that the extended Kalman filter is very suitable for two-machine drive system.

It is worth to notice that there is no impact on the speed and electromagnetic torque of one machine when the speed or the load of the other machine in parallel-connected system changes. Thus, through proper phase transformation rules, the decoupled control of two five-phase PMSMs connected in parallel can be achieved with a single supply from a five-phase voltage source inverter. Furthermore, measured and estimated speeds are in excellent agreement in both steady state and transient operations.

Figures 7–10 illustrate the behavior of the dq-axes and xy-axes inverter currents for both controllers. In case of the PI controller, the stator currents i_q^{inv} and i_y^{inv} peak above 17 A, then decay exponentially to the steady-state while the currents i_d^{inv} and i_x^{inv} are maintained at zero as illustrated in Figures 7b, 8b, 9b and 10b. Figures 7a, 8a and 9a show currents in case of sliding mode control. In contrast, the i_q^{inv} and i_y^{inv} currents peak slightly above 20 A and continue on this value, until the speed reaches its reference value, this leads to a short settling time, as shown in Figures 7a, 8a, 9a and 10a.

Figures 7–9 show the behavior of the two-machine drive under different speeds step variations and without load torque. In Figures 7b, 8b and 9b, the system comportment using PI controller exhibits the expected step response characteristics of a second order system. The response has a short rise time, an overshoot of approximately 18% during reverse modes and settling times close to 0.075 s. Figures 7a, 8a and 9a show comparable dynamic behavior using SMC. However, it is clear from these figures that the system reaches steady-state at 0.028 s without overshoot.

From Figure 10b and by analyzing the transient of two-machine drive controlled by PI controller, it is easy to observe speeds drops taken place at the moments of loads changes. These speed drops are compensated by the PI controller after a necessary recovery time. Figure 10a shows the drive responses in the same load conditions with PI control. At the moments of load variations, the SMCs keep the speeds close to their references without overshoots and without drops. Therefore, the SMC can be considered as more robust under loads variations.

A general comparison between SMC and PI is given in Table 3. Compared to PI controller, SMC shows superiority in terms of settling time and overshoot. However, it needs more energy in starting transient then that needed by the conventional controller.

Table 3. Comparison between SMC and PI.

Comparison Criterion	SMC	PI
Settling time (s)	0.028	0.075
Recovery time (at abrupt load) (s)	0.0045	0.05
Overshoot in reversal mode (%)	0	18%
Starting current (A)	20	17
Starting torque (Nm)	18	15
Speeds drops (%)	0	5

8. Conclusions

In this paper, sensor-less non-linear sliding mode control based on the Lyapunov theory of parallel-connected two five-phase PMSMs drive fed by a single inverter has been developed in order to make the system asymptotically stable. In the proposed control scheme, the Extended Kalman Filter is used for rotor speeds, positions and load torques estimations, while a sliding mode controller is used for speeds and currents control. The sliding mode control has several advantages such as, robustness, high precision, stability and simplicity, very low Settling time. The added value of EKF based sensor-less control is the improvement in system dynamics through the accuracy in speeds, rotor positions and load torques estimations. The effectiveness of the control approach has also been verified through extensive computer simulations and compared with PI controller. The response has a short rise time, an overshoot during reverse modes in PI controller and settling times close to 0.075 s. However, the speed response obtained by SMC is without overshoot and follows its reference and settling times close to 0.028 s. The results also show that the torque obtained by the PI control decreases progressively, while the torque obtained by the SMC is maintained longer at its maximum value, until the speed reaches its reference value. Speeds drops taken place at the moments of loads changes in PI controller. The SMC keep the speeds close to their references without overshoots and without drops.

Therefore, the SMC can be considered as more robust under loads variations. SMC shows better speed tracking performance at both dynamic and steady state than conventional PI controller in the situation reverse modes and load torque variations. Thus, simulation results have verified the proposed whole system has great robust to external disturbances. The simulation of the two-machine drive under various test conditions confirmed that the control of the parallel-connected two five-phase machines is truly decoupled even in sensor-less mode. These results affirm also the ability of the observer to guarantee good estimations in steady state and transients as well. Simulation results point out also that using sliding mode control the dynamic performance of the two-machine drive is further improved compared with the conventional PI controller.

Acknowledgments: No source of funding for this research investigation.

Author Contributions: All authors contributed equally for the final decimation of the research investigation as a full article.

References

1. Navid, R.A. Sliding-mode control of a six-phase series/parallel connected two induction motors drive. *ISA Trans.* **2014**, *53*, 1847–1856.
2. Chen, H.H.; Chong, X.S. Current control for single-inverter-fed series-connected five-phase PMSMS. In Proceedings of the IEEE International Symposium on Industrial Electronics, Taipei, Taiwan, 28–31 May 2013; pp. 1–6.
3. Liliang, G.; John, E.F. A space vector switching strategy for three-level five-phase inverter drives. *IEEE Trans. Ind. Electron.* **2010**, *57*, 2332–2343. [CrossRef]
4. Sneessens, C.; Labbe, T.; Baudart, F.; Matagne, E. Position sensorless control for five-phase permanent-magnet synchronous motors. In Proceedings of the International Conference on Advanced Intelligent Mechatronics, Besançon, France, 8–11 July 2014; pp. 794–799.
5. Wang, Z.; Wang, X.J.C.; Cheng, M.; Hu, Y. Direct torque control of T-NPC inverters fed double-stator-winding PMSM drives with SVM. *IEEE Trans. Power Electron.* **2018**, *33*, 1541–1553. [CrossRef]
6. Wang, X.; Wang, Z.; Cheng, M.; Hu, Y. Remedial strategies of T-NPC three-level asymmetric six-phase PMSM drives based on SVM-DTC. *IEEE Trans. Ind. Electron.* **2017**, *64*, 6841–6853. [CrossRef]
7. Lei, Y.; Ming-liang, C.; Jian-qing, S.; Fei, X. Current Harmonics Elimination Control Method for Six-Phase PM Synchronous Motor Drives. *ISA Trans.* **2015**, *59*, 443–449.
8. Leila, P.; Hamid, A.T. Sensorless Direct Torque Control of Five-Phase Interior Permanent-Magnet Motor Drives. *IEEE Trans. Ind. Appl.* **2007**, *43*, 952–959.
9. Siavash, S.; Lusu, G.; Hamid, A.T.; Leila, P. Wide Operational Speed Range of Five-Phase Permanent Magnet Machines by Using Different Stator Winding Configurations. *IEEE Trans. Ind. Electron.* **2012**, *59*, 2621–2631.
10. Hosseyni, A.; Ramzi, T.; Faouzi, M.M.; Atif, I.; Rashid, A. Sensorless Sliding Mode Observer for a Five-Phase Permanent Magnet Synchronous Motor Drive. *ISA Trans.* **2015**, *58*, 462–473. [CrossRef] [PubMed]
11. Ahmad, A.A.; Dahaman, I.; Pais, S.; Shahid, I. Speed-Sensorless Control of Parallel-Connected PMSM Fed By A Single Inverter Using MRAS. In Proceedings of the IEEE International Power Engineering and Optimization Conference, Melaka, Malaysia, 6–7 June 2012; pp. 35–39.
12. Zhang, H.; Luo, S.; Yu, Y.; Liu, L. Study On Series Control Method For Dual Three-Phase PMSM Based On Space Vector Pulse Width Modulation. *Int. J. Control Autom.* **2015**, *8*, 197–210. [CrossRef]
13. Martin, J.; Emil, L.; Slobodan, N.V. Independent Control of Two Five-Phase Induction Machines Connected In Parallel To A Single Inverter Supply. In Proceedings of the IEEE Industrial Electronics Conference, Paris, France, 6–10 November 2006; pp. 1257–1262.
14. Levi, E.; Jones, M.; Vukosavic, S.N.; Iqbal, A.; Toliyat, H.A. Modeling, control, and experimental investigation of a five-phase series-connected two-motor drive with single inverter supply. *IEEE Trans. Ind. Electron.* **2007**, *54*, 1504–1516. [CrossRef]
15. Levi, E.; Jones, M.; Slobodan, N.V.; Hamid, A.T. Steady-State Modeling of Series-Connected Five-Phase And Six-Phase Two-Motor Drives. *IEEE Trans. Ind. Appl.* **2008**, *44*, 1559–1568. [CrossRef]

16. Mekri, F.; Charpentier, J.F.; Semail, E. An Efficient Control of A Series Connected Two-Synchronous Motor 5-Phase with Non-Sinusoidal EMF Supplied By A Single 5-Leg VSI: Experimental And Theoretical Investigations. *Electr. Power Syst. Res.* **2012**, *92*, 11–19. [CrossRef]

17. Jones, M.; Vukosavic, S.N.; Levi, E. Parallel-Connected Multiphase Multidrive Systems with Single Inverter Supply. *IEEE Trans. Ind. Electron.* **2019**, *56*, 2047–2057. [CrossRef]

18. Zhang, X.; Lizhi, S.; Zhao, K.; Sun, L. Nonlinear Speed Control for PMSM System Using Sliding-Mode Control and Disturbance Compensation Techniques. *IEEE Trans. Power Electron.* **2013**, *28*, 1358–1365. [CrossRef]

19. Fatemi, F.S.M.J.R.; Navid, R.A.; Jafar, S.; Saeed, A. Speed Sensorless Control of a Six-Phase Induction Motor Drive Using Backstepping Control. *IET Power Electron.* **2014**, *7*, 114–123. [CrossRef]

20. Anissa, H.; Ramzi, T.; Atif, I.; Med, F.M. Backstepping Control for a Five-Phase Permanent Magnet Synchronous Motor Drive. *Int. J. Power Electron. Drive Syst.* **2015**, *6*, 842–852.

21. Ahmed, M.; Karim, F.M.; Abdelkader, M.; Abdelber, B. Input Output Linearization And Sliding Mode Control of a Permanent Magnet Synchronous Machine Fed by a Three Levels Inverter. *J. Electr. Eng.* **2006**, *57*, 205–210.

22. Le-Bao, L.; Ling-Ling, S.; Sheng-Zhou, Z.; Qing-Quan, Y. PMSM Speed Tracking And Synchronization of Multiple Motors Using Ring Coupling Control and Adaptive Sliding Mode Control. *ISA Trans.* **2015**, *58*, 635–649.

23. Lin, S.F.J.; Hung, Y.C.; Tsai, M.T. Fault-Tolerant Control for Six-Phase PMSM Drive System via Intelligent Complementary Sliding Mode Control Using TSKFNN-AMF. *IEEE Trans. Ind. Electron.* **2013**, *60*, 5747–5762. [CrossRef]

24. Chen, S.; Luo, Y.; Pi, Y.G. PMSM Sensorless Control with Separate Control Strategies and Smooth Switch from Low Speed to High Speed. *ISA Trans.* **2015**, *58*, 650–658. [CrossRef] [PubMed]

25. Dujic, D.; Jones, M.; Levi, E.; Prieto, J.; Barrero, F. Switching Ripple Characteristics of Space Vector PWM Schemes for Five-Phase Two-Level Voltage Source Inverters—Part 1: Flux Harmonic Distortion Factors. *IEEE Trans. Ind. Electron.* **2011**, *58*, 2789–2798. [CrossRef]

26. Iqbal, A.; Levi, E. Space Vector Modulation Schemes for a Five-Phase Voltage Source Inverter. In Proceedings of the European Conference on Power Electronics and Applications, Dresden, Germany, 11–14 September 2005; pp. 1–12.

27. Dujic, D.; Grandi, G.; Jones, M.; Levi, E. A Space Vector PWM Scheme for Multi Frequency Output Voltage Generation with Multiphase Voltage-Source Inverters. *IEEE Trans. Ind. Electron.* **2008**, *55*, 1943–1955. [CrossRef]

28. Quang, N.K.; Hieu, N.T.; Ha, Q.P. FPGA-Based Sensorless PMSM Speed Control Using Reduced-Order Extended Kalman Filters. *IEEE Trans. Ind. Electron.* **2014**, *12*, 6574–6582. [CrossRef]

29. Zhuang, X.; Rahman, M.F. Comparison of a Sliding Observer and a Kalman Filter for Direct-Torque-Controlled IPM Synchronous Motor Drives. *IEEE Trans. Ind. Electron.* **2012**, *59*, 4179–4188.

30. Dong, X.; Zhang, S.; Liu, J. Very-Low Speed Control of PMSM Based on EKF Estimation with Closed Loop Optimized Parameters. *ISA Trans.* **2013**, *52*, 835–843.

31. Abbas, N.K.; Jafar, S. MTPA Control of Mechanical Sensorless IPMSM Based on Adaptive Nonlinear Control. *ISA Trans.* **2016**, *61*, 348–356.

32. Khan, M.R.; Iqbal, A. MRAS Based Sensorless Control of a Series-Connected Five-Phase Two-Motor Drive System. *J. Electr. Eng. Technol.* **2008**, *3*, 224–234. [CrossRef]

33. Khan, M.R.; Iqbal, A. Extended Kalman Filter Based Speeds Estimation of Series-Connected Five-Phase Two-Motor Drive System. *Simul. Model. Pract. Theory* **2009**, *17*, 1346–1360. [CrossRef]

34. Ali, W.H.; Gowda, M.; Cofie, P.; Fuller, J. Design of a Speed Controller Using Extended Kalman Filter for PMSM. In Proceedings of the IEEE 57th International Midwest Symposium on Circuits and Systems, College Station, TX, USA, 8–11 July 2014; pp. 1101–1104.

35. Slotine, J.J. *Applied Nonlinear Control*; Tice Hall: Englewood Cliffs, NJ, USA, 1991.

36. Iqbal, A.; Levi, E. Space Vector PWM for a Five-Phase VSI Supplying Two Five-Phase Series-Connected Machines. In Proceedings of the 12th International Power Electronics and Motion Control Conference, Portoroz, Slovenia, 30 August–1 September 2006; pp. 222–227.

37. Dujic, D.; Jones, M.; Levi, E. Generalised Space Vector PWM for Sinusoidal Output Voltage Generation with Multiphase Voltage Source Inverters. *Int. J. Ind. Electron. Drives* **2009**, *1*, 1–13. [CrossRef]
38. Jones, M.; Dordevic, O.; Bodo, N.; Levi, E. PWM Algorithms for Multilevel Inverter Supplied Multiphase Variable-Speed Drives. *Electronics* **2012**, *16*, 22–31. [CrossRef]

All SiC Grid-Connected PV Supply with HF Link MPPT Converter: System Design Methodology and Development of a 20 kHz, 25 kVA Prototype

Serkan Öztürk [1], Mehmet Canver [2], Işik Çadirci [1] and Muammer Ermiş [2,*

[1]   Department of Electrical and Electronics Engineering, Hacettepe University, Beytepe, Ankara 06800, Turkey; ozturk@ee.hacettepe.edu.tr (S.Ö.); cadirci@ee.hacettepe.edu.tr (I.Ç.)

[2]   Department of Electrical and Electronics Engineering, Middle East Technical University, Ankara 06800, Turkey; mehmet.canver@artielektronik.com.tr

*   Correspondence: ermis@metu.edu.tr

Abstract: Design methodology and implementation of an all SiC power semiconductor-based, grid-connected multi-string photovoltaic (PV) supply with an isolated high frequency (HF) link maximum power point tracker (MPPT) have been described. This system configuration makes possible the use of a simple and reliable two-level voltage source inverter (VSI) topology for grid connection, owing to the galvanic isolation provided by the HF transformer. This topology provides a viable alternative to the commonly used non-isolated PV supplies equipped with Si-based boost MPPT converters cascaded with relatively more complex inverter topologies, at competitive efficiency figures and a higher power density. A 20 kHz, 25 kVA prototype system was designed based on the dynamic model of the multi-string PV panels obtained from field tests. Design parameters such as input DC link capacitance, switching frequencies of MPPT converter and voltage source inverter, size and performance of HF transformer with nanocrystalline core, DC link voltage, and LCL filter of the VSI were optimized in view of the site dependent parameters such as the variation ranges of solar insolation, module surface temperature, and grid voltage. A modified synchronous reference frame control was implemented in the VSI by applying the grid voltage feedforward to the reference voltages in abc axes directly, so that zero-sequence components of grid voltages are taken into account in the case of an unbalanced grid. The system was implemented and the proposed design methodology verified satisfactorily in the field on a roof-mounted 23.7 kW multi-string PV system.

Keywords: dynamic PV model; grid-connected VSI; HF-link MPPT converter; nanocrystalline core; SiC PV Supply

1. Introduction

Various grid-connected photovoltaic system concepts and topologies have been summarized previously [1–4]: (i) micro inverters [5,6]; (ii) residential systems supplied from a PV string [7,8]; (iii) commercial/residential systems supplied from multiple PV strings having their own maximum power point tracking (MPPT) converters and a central inverter [9,10]; and (iv) commercial/utility-scale PV plants supplied from a common DC link with a central inverter [11]. The central inverter in [11] performs direct conversion of PV power to AC from multiple PV strings. MPPT and reliable and efficient conversion of PV power to AC at grid frequency are the major design issues for two stage grid-connected PV systems [12]. In the PV systems performing direct conversion of PV power to AC, the voltage of the common DC link is varied by the central inverter against the changes in solar insolation and panel temperature [13]. An MPPT algorithm integrated into a two-layer controller is recommended in [14] for direct conversion of PV power through a three-phase grid-connected VSI.

A neural network based MPPT algorithm is proposed in [15] to improve the dynamic performance of DC capacitor voltage at the input of VSI and to maintain faster tracking response against sudden changes in solar irradiance. Grid-connected PV supplies can be classified into two groups: (i) those having transformerless inverter topologies [16]; and (ii) those having magnetic transformers usually on the grid side [17]. Transformerless PV supplies require more complex and expensive inverter topologies as compared to PV supplies having transformers for galvanic isolation to prevent the flow of common-mode currents [18–22].

In the vast majority of commercially available PV supplies, new generation hybrid IGBTs are currently employed to synthesize grid frequency voltages and/or currents. In the next generation PV inverters, simpler circuit designs such as two-level instead of multi-level inverters are expected to be used for fewer components and hence higher reliability [23]. Wide bandgap (WBG) devices such as SiC power MOSFETs will be employed for higher switching frequency to reduce the size of passive components, heatsinks and hence the volume of the PV supply, as foreseen in [23]. WBG semiconductors show superior material properties, enabling potential power device operation at higher temperatures, voltages, and switching speeds than current Si technology [24].

In recent years, performances of all SiC PV supplies consisting of DC/DC converters and three-phase grid-connected inverters are assessed, as reported in [25–27]. These use either three-level T-type inverter topology [26] or cascaded multi-level inverter topology [27]. All SiC inverters such as single-phase H-bridge converter [28], neutral point clamped (NPC) T-type three-phase inverter [29], and three-phase five-level T-type inverter [30] are used to perform direct conversion of PV power to the grid. Design and performances of some non-application-specific two-level three-phase SiC inverters are presented in [31–34].

Since the power converters in PV supplies are presently switched at relatively higher frequencies, ranging from a few kHz to a few tens of kHz, sizing and design of these converters require the dynamic model of the PV cells, modules or arrays instead of the well-known static model. Some attempts have been made to obtain dynamic model parameters of various types of PV cells or modules by theoretical calculations and/or measurements [35–40]. The use of internal solar cell diffusion capacitance to replace the input capacitor in boost derived DC-DC converters has been proposed in [41] for energy harvesting applications. In almost all of the two-stage PV supplies, boost type DC-DC converters have been used for MPPT. Recently, in this area, small size SiC MOSFET [42–45] and SiC JFET based [46] MPPT converters are reported in the literature. A 1-MW solar power inverter with boost converter using all SiC power modules has been presented in [47]. A 3.5 kW, 100 kHz all SiC buck/boost interleaved MPPT converter is proposed in [48] for higher conversion efficiency. Design, construction, and testing of a general purpose 750-V, 100-kW, 20-kHz bidirectional dual active bridge DC–DC converter using all SiC dual modules with HF transformer is described in [49] to provide maximum conversion efficiency from the DC-input to the DC-output terminals.

Design of the AC filter on the grid side of a VSI is a compromise between performance and size. A step-by-step design procedure of an LCL filter for a grid-connected three-phase PWM voltage source converter is proposed in [50]. A new design method is recommended in [51] for LCL and LLCL filters with passive damping to be used in single-phase grid-connected VSIs. Peña-Alzola et al. [52] examined passive damping losses in LCL filters. Design methodology, analysis and performance of LCL filters for three-phase grid-connected inverters are also described in [53,54]. Stability regions of active damping control for LCL filters are discussed in [55]. A generalized LCL filter design algorithm for grid-connected VSIs is presented in [56].

The common control methods for three-phase grid-connected inverters are: (i) synchronous reference frame control (in dq-axes) [57–59]; (ii) stationary reference frame control (in $\alpha\beta$-axes) [60]; and (iii) natural reference frame control (in abc-axes) [61]. The main drawback of the synchronous reference frame control is the poor compensation capability of the low-order harmonics in the case of PI based controllers [60]. To cope with this drawback, the control system can be equipped with proportional resonant harmonic compensators, one pair for each prominent low order sidekick harmonics produced by the system itself [62]. In stationary reference frame control method, two proportional resonant controllers can be used in $\alpha\beta$-axes for the fundamental component to eliminate possible steady-state error which may arise in PI control. However, the number of PR controllers for low-order harmonic compensation are doubled. In *abc* control strategy, an individual controller is required for each line current on the grid side. In this control method, all types of controllers such as PI controller, PR controller, hysteresis controller, dead-beat controller and repetitive controller can be adopted [61].

This paper recommends a grid-connected, all SiC Power MOSFET based multi-string PV supply with a HF link MPPT converter in each string and a two-level central inverter. The design methodology described in the paper utilizes a custom dynamic model of the roof-mounted multi-string PV system, and site dependent parameters. These parameters are variation ranges of solar insolation, module surface temperature, and grid voltage. In the design procedure the following are optimized: (i) input DC link capacitance; (ii) switching frequencies of MPPT converter and VSI; (iii) the size and performance of HF transformer with nanocrystalline core; (iv) the DC link voltage; and (v) the LCL filter of VSI. Corresponding field test results of the implemented 20 kHz, 25 kVA prototype are presented. It has been shown that the HF link converter in cascade with a two-level VSI provides a viable alternative to multi-string PV supplies with a competitive efficiency, lower harmonic distortion and a much higher power density. This PV supply topology with a HF link MPPT converter makes possible the use of a simple and reliable, two-level, voltage source PV inverter for grid connection. In the case of PV supplies with non-isolated MPPT converters or those performing direct conversion of PV power to the grid, either complex inverter topologies or bulky grid-side common-mode filters or coupling transformers would be required to minimize the flow of undesirable common-mode currents.

2. System Description

2.1. General

Block diagram representation of all SiC grid-connected PV supply with HF link is as shown in Figure 1a. With the advents in SiC power MOSFETs, the kVA rating of three-phase, two-level voltage source all SiC central inverter can reach 175–400 kVA, by employing commercially available three half-bridge all SiC modules in Table 1 (by February 2018) for direct connection to 400 V l-to-l, 50 Hz utility grid. Each MPPT converter with DC/AC/DC link can be designed to process 15–25 kW PV power by employing SiC half-bridge modules, TO-package SiC Schottky diodes, and a nanocrystalline transformer core. The MPPT converter rating in the application described in this paper is limited to 25 kW by considering the partial shading risk of multi-string PV panels. In the experimental system, five mono-crystalline PV strings of 23.7 kW peak under standard test conditions, with adjustable tilt angles occupy nearly 375 m^2 in the roof area as shown in Figure 2. In this study, the experimental work was carried out on the prototype system given in Figure 1b.

Table 1. Commercially available all SiC power MOSFET half-bridge modules that can be employed in central inverter in Figure 1a/by February 2018.

Manufacturer	Half-Bridge Module	V_{DS}, V	I_D, A
ROHM [63]	BSM300D12P2E001	1200	300
CREE [64]	CAS325M12HM2	1200	325
SEMIKRON [65]	SKM500MB120SC	1200	541
Mitsubishi [66]	FMF800DX-24A	1200	800

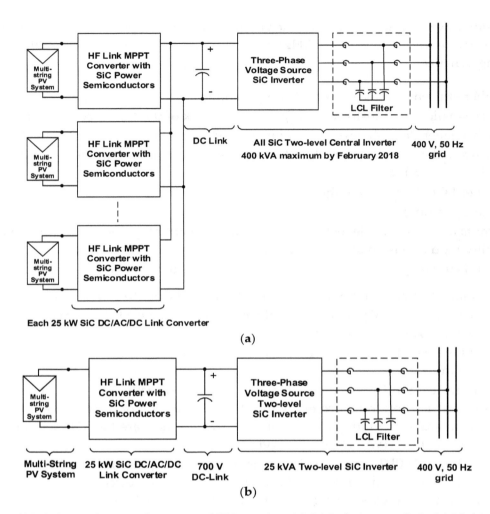

Figure 1. All SiC three-phase grid-connected PV supply with high-frequency link: (**a**) Multi-string PV inverter system; and (**b**) Experimental set-up.

Figure 2. Roof-mounted 23.7 kW at 1000 W/m^2 and 25 $^\circ$C multi-string PV system containing 95 CSUN250-72M modules (five parallel connected strings of 19 series modules; with adjustable tilt angles of 15°, 30°, 45° and 60°); coordinates 39.890917, 32.782278; five modules in the photograph are idle; terminal box of the system inside the laboratory is shown at the right bottom corner of photograph while the pyranometer is at left bottom corner.

Advantages and disadvantages of the multi-string PV supply system in Figure 1a can be summarized as follows:

(i)   In medium power applications, as a wide bandgap device, SiC power MOSFETs can be switched at higher frequencies (a few tens of kHz) in comparison with Si IGBTs. This ability brings us the following benefits:

- Higher efficiency at the same switching frequency;
- Much smaller HF transformer in DC/AC/DC link of the MPPT converter;
- Much smaller electrolytic or metal film DC link capacitor at the output of the MPPT converter;
- Much smaller distributed metal film DC link capacitor bank in the laminated bus/PCB of the inverter circuit;
- Smaller LCL filter bank at the output of the PV inverter;
- Silent operation;
- Permits the use of two-level three-phase bridge inverter topology to conform with the power quality regulations on the grid side; and
- Extends the range of high order current harmonics that can be eliminated.

(ii)   They eliminate the need for a bulky, grid-side PV transformer to provide electrical isolation.

(iii) Separation of installed PV panels into multiple strings having MPPT converters minimizes the undesirable effects of partial shading.

(iv)   The central inverter has higher efficiency and lower cost in comparison with the usage of several smaller scale inverters.

(v)   In most countries, the PV system in Figure 1a can be directly connected to the low-voltage (LV) side of distribution transformers without the permission of distribution system operators. Although standard power ratings for distribution transformers are 100, 250, 400, and 630 kVA [67], higher ratings up to 2 MVA are also in service.

(vi)   Since the HF link MPPT converter is cascaded with the grid-connected PV inverter, overall efficiency is 1–2% lower than that of PV converters performing direct conversion of PV power to the grid. However, the PV systems having non-isolated MPPT converters or performing direct conversion of PV power to the grid should have more complex converter topologies to minimize common-mode currents.

2.2. MPPT Converter with HF Link

A prototype of the MPPT converter with HF link shown in Figure 1b is designed and implemented. MPPT converter is composed of a SiC Power MOSFET based single-phase H-bridge converter, a HF transformer, and a SiC Schottky diode rectifier. The circuit diagram of the power stage, the control circuitry, and top and side views of the implemented MPPT converter prototype are as shown in Figure 3.

2.3. Three-Phase Two-Level Voltage Source Inverter

A prototype of the three-phase two-level voltage source inverter shown in Figure 1b was designed and implemented. This converter topology was chosen for the following reasons:

- The ability of SiC power MOSFETs at high switching frequencies for lower harmonic current content;
- Minimum power semiconductor count gives higher reliability in comparison with multi-level converters;
- Common-mode current is not a design concern because of the HF transformer in the MPPT converter; and
- PV inverter delivers power to one of the most common low voltage utility grid.

The circuit diagram of the power stage and its control circuitry, and top and side views of the implemented inverter prototype are shown in Figure 4.

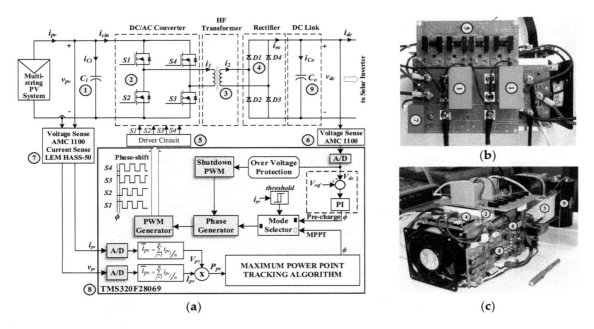

Figure 3. All SiC MPPT converter with high-frequency link: (**a**) power stage and its control circuitry; (**b**) yop view of the developed hardware; and (**c**) side view. (1) Metal Film 30//30 µF Input Capacitor Bank (EPCOS B32778G0306K); (2) All-SiC Half Bridge Module (CREE CAS120M12BM2); (3) Nanocrystalline Core (SU-102b) Based HF Transformer; (4) SiC Schottky Diode (CREE C4D40120D); (5) Dual Channel SiC MOSFET Driver (CREE CGD15HB62P); (6) Fully-Differential Isolation Amplifier (TI AMC-1100); (7) Hall-Effect Current Transducer (LEM HASS 50-S); (8) Microcontroller (TI TMS320F28069); (9) Electrolytic 3400 µF Output Capacitors (EPCOS B43456-A9688-M).

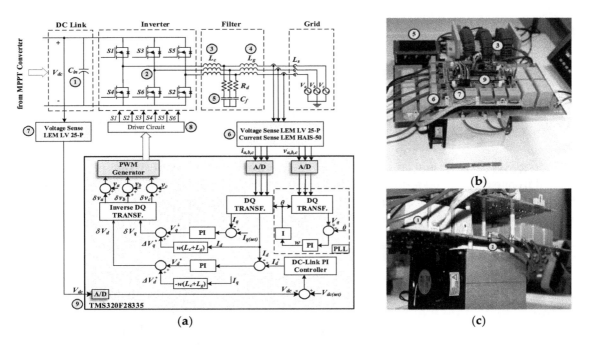

Figure 4. All SiC three-phase grid-connected two-level inverter: (**a**) power stage and its control circuitry; (**b**) top view of the developed hardware; and (**c**) side view. (1) Metal Film 12 × 40 µF Input Capacitor Bank (VISHAY MKP1848640094Y); (2) All-SiC Six-Pack Three-Phase Module (CREE CCS050M12CM2); (3,4) X-Flux L_c = 250 µH and L_g = 50 µH Filter Inductor (Magnetics 0078337A7); (5) Metal Film 15 µF Filter Capacitor (VISHAY MKP1848S61510JY5F); (6) Hall-Effect Current Transducer (LEM HAIS 50-P); (7) Voltage Transducer (LEM LV25-P); (8) Single Channel SiC MOSFET Driver (CREE CRD-001); (9) Experimenter Kit (TI TMDSDOCK28335).

3. Modeling and System Design

The recommended modeling and system design methodology is described in this section for the existing multi-string PV system shown in Figure 2. For another multi-string PV system configuration, the same design principles can be applied.

3.1. Dynamic Model of Multi-String PV System

Even in the steady-state operation of PV supplies, their power converters such as the MPPT converter and the inverter operate in the periodic transient state, instead of pure DC operation. Therefore, in the analysis and design of such systems, a proper dynamic model of the multi-string PV system is to be used. Several attempts have been made to obtain dynamic models of a PV module [35–41]. However, in this research work, the dynamic model parameters of the roof-mounted multi-string PV system shown in Figure 2 were obtained from a set of experimental data. It includes also all cabling and wirings up to the input terminals of the MPPT converter. These parameters were then combined with the static model of the CSUN250-72M modules available in MATLAB/Simscape/Power Systems R2016b for use in the design work.

Equivalent circuits of the multi-string PV system in Figure 2 consisting of 95 pieces of CSUN250-72M modules (5 × 19 modules) are as shown in Figure 5. Static model in Figure 5a is determined for the PV array in Figure 2 by using the PV array block developed by the National Renewable Energy Laboratory (NREL) System Advisor Model and available in MATLAB/Simscape/Power Systems R2016b. However, the design of the PV supply presented in this paper is based on the dynamic model of the PV system shown in Figure 5b.

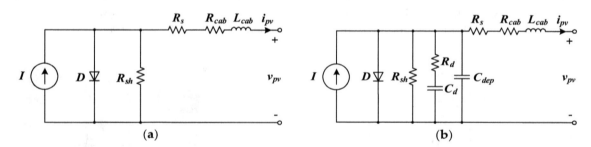

(a)   (b)

Figure 5. Equivalent circuit of the multi-string PV system in Figure 2 consisting of 95 pieces of CSUN250-72M modules (5 × 19 modules). (a) Static model: R_{sh} = 490 Ω and R_s = 1.33 Ω are, respectively, the equivalent shunt and series resistances calculated by MATLAB/Simscape/Power Systems; and R_{cab} = 10 mΩ and L_{cab} = 45 µH are the equivalent parameters of all cabling and wiring and estimated from experimental results given in Figure 6. (b) Dynamic model: C_d = 4 µF is the equivalent diffusion capacitance and R_d = 3 Ω is its series resistance, and C_{dep} = 600 nF is the equivalent depletion layer capacitance, which were estimated from experimental results given in Figure 6.

The topology of this dynamic model is nearly the same as the one given in [41]. The dynamic model parameters are estimated from the results of two tests conducted on the multi-string PV system. In the first test, multi-string PV system is solidly short-circuited at the terminal box and the short-circuit current is recorded as shown in Figure 6a. In the second test, a slightly inductive resistive bank is suddenly connected to the open-circuited multi-string PV system terminals and the terminal voltage and the current are recorded as shown in Figure 6b. Parameters of the multi-string PV system estimated from the current and voltage records shown in Figure 6 are as given in the caption of Figure 5.

These experimental records are compared with the simulation results obtained by MATLAB/Simscape/Power Systems, for the static and dynamic models separately under the same test conditions, as given in Figure 6. It is clear from this figure that the dynamic model gives much better results than those of the static model, and it can therefore be satisfactorily used in the design of HF link MPPT converter.

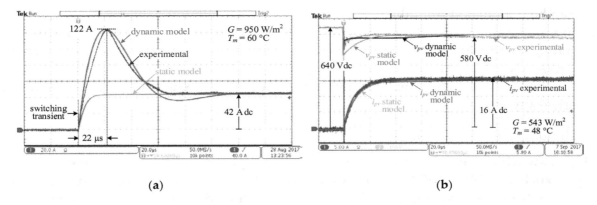

(a)   **(b)**

Figure 6. Transient response of the multi-string PV system at the input of the MPPT converter (Experimental: Tektronix MSO3034 oscilloscope, Tektronix TCPA300 current probe amplifier, Tektronix TCP404XL current probe; static model as given in Figure 5a; dynamic model as given in Figure 5b; Pyranometer EKO MS-410; Fluke Ti29 infrared camera): **(a)** from open-circuit to short-circuit; and **(b)** from open-circuit to partial resistive load (R_L = 42.5 Ω, L_L = 278 μH measured by GW INSTEK LCR-817 LCR meter at the resistor temperature of 72 °C).

3.2. Three-Phase Two-Level VSI

In this subsection, optimum DC link voltage and switching frequency are determined and the design of the inverter circuit, its control circuitry and LCL filter are described. The design of the HF link converter is directly affected by the optimum DC link voltage determined in this subsection.

3.2.1. Optimum DC Link Voltage

DC link voltage, V_{dc}, is kept constant by the inverter against the variations in PV power over the entire operating range of the PV supply. At the optimum value of V_{dc}, modulation index, M in Equation (1) of the inverter circuitry should vary in a range as close as possible to unity in order to minimize harmonic distortion of the output line currents.

$$M = (2\sqrt{2}/\sqrt{3})(V/V_{dc}) \qquad (1)$$

where V is the l-to-l value of either grid voltage, V_s or consumers load voltage, V_t. Furthermore, in calculating the variation range of M, permissible changes in 400 V l-to-l, 50 Hz grid voltage are as specified in IEC 60038 2002-07 Standard Voltages [68]. This standard specifies the maximum changes in grid voltage, V_s as +6% to −0% while a further ±4% at the consumers load voltage, V_t. In view of these considerations, the variation ranges of M are calculated and plotted (Figure 7).

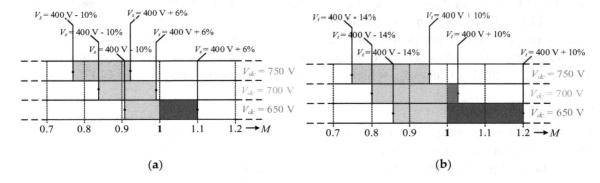

(a)   **(b)**

Figure 7. The variation range of modulation index, M as a function of chosen DC link voltage, V_{dc} and operating voltage (acceptable low voltage variation ranges for 230/400 V, 50 Hz systems are as specified in IEC 60038 2002-07 Standard Voltages): **(a)** for the supply side voltage range of V_s = 400 V +6% to −10%; and **(b)** for the utilization voltage range of V_t = 400 V +10% to −14%.

As can be understood from the results in Figure 7, the optimum value of V_{dc} is around 700 V. This choice may result in operation slightly in over-modulation region as given in Figure 7b. Since inverter rating is 25 kVA/22.3 kW for the available multi-string PV system, the inverter can deliver nearly 11 kVAr inductive or more to bring V_t back to V_s = 400 V + 6%. Application of third harmonic injection method [69,70] might be an alternative design approach in which the optimum value of V_{dc} is to be nearly 600 V for V_s = 400 V + 6%.

3.2.2. Optimum Switching Frequency

Higher switching frequencies for the two-level three-phase inverter with different modulation techniques such as SPWM, SVPWM, etc., excluding SHEM, result in low harmonic distortion for the line currents injected into the grid [71]. In this research work, sinusoidal PWM is chosen as the modulation technique because of its simplicity, and its ability to illustrate basic design guidelines. Since SiC Power MOSFETs can be switched at higher frequencies in comparison with Si IGBTs for the same power dissipation and solar inverter rating, power loss components of solar inverter with SPWM modulation are calculated for a reasonable operating frequency range, e.g., at f_{sw} = 10, 20, and 30 kHz, by using the expressions and manufacturers' design tools given in [72].

The associated pie-charts are shown in Figure 8. All wiring and cabling losses between discrete components, inverter, LCL filter, and grid are ignored in the preparation of Figure 8. In addition, extra power losses due to the switching ripple current through the power MOSFETs are not considered by the power loss calculation tools mentioned above in the calculation of conduction and switching losses. In summary, slightly higher loss content and a lower efficiency are expected for the solar inverter in the field tests.

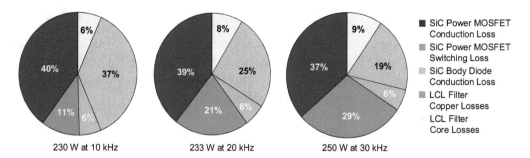

Figure 8. Power loss components of solar inverter for different switching frequencies at P_o = 22.3 kW. Numerical values are rounded.

In the design of LCL filter at different switching frequencies, only the converter side inductance, L_c, is optimized to keep its peak-to-peak current ripple constant at 25%. Power loss components in the associated LCL filters are then used in the preparation of pie charts in Figure 8. Although 10 kHz switching frequency reduces the power dissipation marginally in comparison with that of the 20 kHz, a considerably larger LCL filter size is to be used. Therefore, in the design and implementation of the solar inverter, f_{sw} = 20 kHz is chosen, which is a compromise between losses and LCL filter size.

3.2.3. LCL Filter Design

An LCL filter consisting of inverter side inductors L_c, shunt capacitors C_f, passive damping resistors R_d, and grid side inductors L_g are considered in design, as shown in Figure 4a. The LCL filter should be designed to have not more than 10 A peak-to-peak ripple superimposed on 36 A rms fundamental current in L_c at 25 kVA, and 400 V l-to-l. Peak-to-peak ripple remains nearly constant over the entire operating range of the all SiC three-phase grid-connected two-level inverter and is 25% at 25 kVA, 400 V l-to-l. The corner frequency of the LCL filter is chosen around 1/3rd of the 20 kHz switching frequency. These choices are consistent with the recommendations given in various papers [50–56]. The transfer function Bode plots of undamped LCL filter for three different L_c, C_f, and L_g parameter sets are given in Figure 9a. All of them provide nearly 100 dB attenuation at switching frequency. Among these, L_c = 250 µH, C_f =15 µF, and L_g = 50 µH parameter set is chosen for the implementation. Red colored parameter set (L_c =350 µH, C_f =20 µF, L_g = 50 µH) is not chosen because its L_c is nearly 40% greater than that of the optimum design and provides unnecessarily high attenuation. Although the green colored parameter set (L_c = 150 µH, C_f = 10 µF, and L_g = 50 µH) gives minimum LCL size, it is also not chosen in the implementation because it makes narrower the control range of the phase shift angle, and hence may cause undesirable oscillations in the output power and possible instability. On the other hand, field experience has shown that larger LCL filter size reduces harmonic content of the line currents and maintains stability of the inverter.

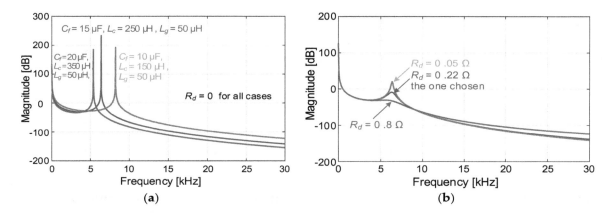

Figure 9. Transfer characteristics of various LCL filters shown in: (**a**) undamped filters, R_d = 0; and (**b**) effects of R_d on the transfer characteristic on the chosen LCL filter (L_c = 250 µH, C_f = 15 µF, and L_g = 50 µH).

Very high amplification of the current component at resonance frequency can be entirely eliminated by either passive or active damping technique. In this research work, passive damping is preferred and the damping resistance R_d is connected in series with C_f. In choosing the optimum value of R_d, a compromise is needed between copper losses and damping effect.

The effects of various damping resistors on the current transfer function bode plots of the chosen parameter set are as given in Figure 9b. Although lower R_d values are less dissipative, their damping effect is inadequate. On the other hand, higher R_d values provide strong damping at resonance frequency at the expense of higher losses and reduced attenuation at high frequencies. R_d = 0.22 Ω is therefore chosen for the implementation.

3.2.4. Controller Design

In this paper, active power delivered to the grid is controlled by using the rotating reference frame synchronized with the grid frequency by implementing a modified version of the control technique presented in [50]. The block diagram of the designed and implemented DSP (TMS320F28335) based controller is shown in Figure 4a. Line to neutral voltages $v_{a,b,c}$ and line currents $i_{a,b,c}$ on the grid side, and DC link voltage V_{dc} are the inputs to the controller. These quantities are sampled at 10 kHz/channel. Set values of the DC link voltage, $V_{dc(set)}$ and $I_{q(set)}$ are adjusted, respectively, to 700 V and 0 A in the control software. PWM signals applied to the driver circuits are the outputs of the control system. Control actions are achieved firstly in rotating DQ reference frame which is synchronized with the grid frequency by the PLL circuit, and then in ABC reference frame.

To be able to lock to the grid, reference value of V_q should be 0. PI controller in the PLL circuit calculates synchronous speed, ω which is equal to angular frequency of the grid voltage. ω is then integrated to give space angle θ, where θ defines relative position of synchronously rotating reference frame with respect to the stationary ABC reference frame, i.e., relative position of rotating d-axis with respect to stationary a-axis. I_q is compared with $I_{q(set)} = 0$ for zero reactive power and then processed in the PI controller to generate reference signal $V_q{}^*$. Cross-coupling term $\Delta V_q{}^*$ is then superimposed on $V_q{}^*$ to compensate for potential drop on the total filter inductance and also for better transient response in the feed-forward form of cross-coupling terms. Actual value of DC link voltage V_{dc} is compared with its set value $V_{dc(set)}$ and resulting error signal is then processed in a PI controller to generate reference signal $I_d{}^*$. $I_d{}^*$ is then compared with actual current I_d and processed in a PI controller to generate reference value, $V_d{}^*$. Cross-coupling term $\Delta V_d{}^*$ is superimposed on $V_d{}^*$ to compensate for potential drop on the total filter inductance and also for better transient response.

The above operations yield main control signals δV_d and δV_q in synchronously rotating reference frame. Instead of superimposing δV_d and δV_q on d- and q-axis components, V_d and V_q of the grid voltages, δV_d and δV_q are transformed back to abc-axes, resulting in δ_{va}, δ_{vb}, and δ_{vc}, and these control signals are then superimposed on the actual grid voltage waveforms v_a, v_b, and v_c. This modification results in lower harmonic distortion in the line current waveforms because when the actual grid voltages are transformed to dq reference frame (instead of $dq0$) and then used in the control together with δV_d and δV_q, odd multiples of third harmonic (zero sequence component) would not be taken into account.

3.3. MPPT Converter with HF Link

In this subsection, switching frequency of the H-bridge converter and the transformer turns-ratio are optimized in view of the following design constraints:

(i)   Multi-string PV system and its dynamic model should be known. For this purpose, the system in Figure 2 and its dynamic model in Figure 5b are prespecified for the experimental set-up in Figure 1b.

(ii)   The variation range of global solar insolation, G, should be known for the geographical location at which the PV system is going to be installed, i.e., $G \leq 1000 \text{ W/m}^2$ for the experimental set-up.

(iii)   The variation range of module surface temperature, T_m, should be estimated, i.e., $10 \leq T_m \leq 70 \,°\text{C}$ for the experimental set-up.

(iv)   DC link voltage, V_{dc}, is kept constant at 700 V by the solar inverter in the experimental set-up.

(v)   DC link capacitance, C_o in Figure 3a is taken to be 3400 μF.

All calculations are carried out on MATLAB/Simscape/Power Systems by using the equivalent circuit in Figure 3a in which leakage inductances of the HF transformer are assumed to be $L_p = 3.2$ μH and $L_s = 7.4$ μH, respectively, on the primary and secondary sides, and switching losses of all SiC power MOSFETs and Schottky diodes are neglected.

3.3.1. Optimum Transformer Turns-Ratio, n

The lowest and the highest maximum power point voltages for the multi-string PV system in Figure 2 are, respectively, $V_{mpp(min)} = 434$ V at $G = 50 \text{ W/m}^2$, $T_m = 70 \,°\text{C}$, and $V_{mpp(max)} = 600$ V at $G = 1000 \text{ W/m}^2$, $T_m = 10 \,°\text{C}$. Transformer turns-ratio is defined as $n = N_s/N_p$, where N_s and N_p are the number of series turns of the secondary and primary windings, respectively. The duty cycle, D, of the phase-shifted H-bridge converter is as defined in Equation (2).

$$D = t_{on}/(t_{on} + t_{off}) \tag{2}$$

where t_{on} is the power transfer period of phase-shifted H-bridge, and t_{off} is the sum of freewheeling and no-conduction periods. It is desirable to maintain the operation of H-bridge converter at relatively high D values over the entire operating range to keep corresponding peak values of SiC power semiconductor and transformer currents relatively low. The variation ranges of D for two different n values and extreme operating conditions are shown in Figure 10.

On the other hand, for an ideal MPPT converter, the lowest MPPT voltage, $V_{mpp(min)}$, over the entire operating range can be related to the chosen DC link voltage, $V_{dc} = 700$ V in terms of n and D as given in Equation (3). As an example, for $D = 1.0$ and $V_{mpp(min)} = 434$ V at $G = 50$ W/m^2 and $T_m = 70\,°$C, n is found to be 1.61 from Equation (3). However, an n value lower than 1.61 can be chosen, since the total leakage inductance of the transformer provides boosting action in a practical MPPT converter.

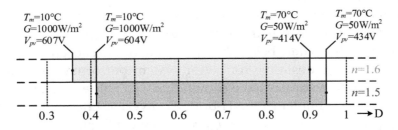

Figure 10. Variation range of duty cycle, D, for the practical MPPT converter as a function of n in between extreme operating conditions.

In the implemented MPPT converter, $n = 1.52$ is chosen for which $D = 0.94$ and $V_{mpp(min)} = 434$ V at $G = 50$ W/m^2 and $T_m = 70\,°$C. This choice ensures power transfer even at the minimum G and maximum T_m conditions and provides a margin for better transient response.

$$V_{dc} = 700 \leq V_{mpp(min)}\, n\, D \qquad (3)$$

3.3.2. Choice of DC Link Capacitor of H-Bridge Converter

Operation modes of the MPPT converter in Figure 3 are as defined in Figure 11. The controllable section of the MPPT converter is the phase-shifted H-bridge converter. The stray inductance of the implemented DC-bus on PCB is estimated as $L_{stray} = 15.2$ nH. Two discrete metallized film capacitors are connected across the DC link, one for each leg of the H-bridge converter. Total DC link capacitance is denoted by C_i in Figures 3 and 11. Suppose now that initially S_1 and S_3 are conducting in power transfer mode in the positive half-cycle as shown in Figure 11a.

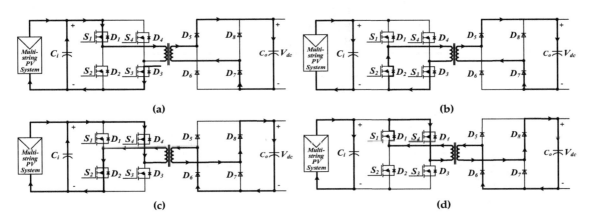

Figure 11. Operation modes of the MPPT converter for one complete cycle: (**a**) Mode 1: Power transfer in positive half-cycle; (**b**) Mode 2: Freewheeling in positive half-cycle; (**c**) Mode 3: Power transfer in negative half-cycle; and (**d**) Mode 4: Freewheeling in negative half-cycle.

When S_1 is turned off, current is commutated to D_2 which starts the freewheeling mode through S_3 and D_2 as shown in Figure 10b. The freewheeling mode is then followed by the OFF mode after the current decays to zero. For the negative half cycle, other diagonal switches S_2 and S_4 are turned on for power transfer mode which is followed by freewheeling mode through S_4 and D_1, respectively, shown in Figure 11c,d. Note that the converter is operated in the discontinuous conduction mode, owing to the absence of a lossy and bulky output filter inductor in the design.

Fall time of the SiC power MOSFET used in this research work is nearly 50 ns with increased gate resistance, R_g. During the commutation period, current closes its path mainly through C_i. The potential drop on L_{stray} is therefore $V_{stray} = L_{stray} \cdot (\Delta I / \Delta t) = 51$ V for maximum possible device current of $I_D = 200$ A at $f_{sw} = 20$ kHz when $G = 1000$ W/m^2, $T_m = 10$ °C on the predefined geographical site. V_{stray} will then be superimposed on drain-to-source voltage, v_{DS}, of outgoing SiC power MOSFET. Since open-circuit voltage of the multi-string PV system is 800 V, peak value of v_{DS} never exceeds 850 V in the worst case which is safely below the v_{DS} rating of the chosen SiC power MOSFETs.

A high C_i value is always desirable for better system performance at the expense of higher size and hence cost. Simulation studies have shown that a C_i value in the range from 20//20 μF to 40//40 μF can be chosen in the implemented H-bridge converter. Effects of C_i on peak-to-peak ripple content of i_{pv} and v_{pv}, $i_{ci(rms)}$, and form factor of i_{cin} are shown in Figure 12 for standard test conditions ($G = 1000$ W/m^2, $T_m = 25$ °C) and $f_{sw} = 20$ kHz. These curves show that:

- $i_{ci(rms)}$ and form factor of icin are not affected by C_i; and
- Peak-to-peak ripple content of i_{pv} and v_{pv} reduces as C_i is increased. Lower peak-to-peak content on the PV side is always desirable, not only for potential drop on all series inductances but also for MPPT efficiency.

In view of these characteristics, C_i greater than or equal to 30//30 μF seems to be suitable for the implemented H-bridge converter, provided that the commercially available metallized film capacitors can carry this rms current. In the implemented system $C_i = 30//30$ μF is chosen which is bigger than the DC link capacitor recommended by the SiC power MOSFET manufacturer [73]. To justify that 30//30 μF meets the entire operating range of the MPPT converter for the predefined geographical location and their commercial availability, characteristics in Figure 13 and manufacturer's data in Table 2 are given.

Table 2. Technical characteristics of some metallized film capacitors (1100 V DC at 70 °C) [74].

Capacity, μF	Code	I, A rms	ESR, mΩ
20	B32778T0206	13	11.9
30	B32778G0306	17.5	8.2
40	B32778G0406	21.5	6.2

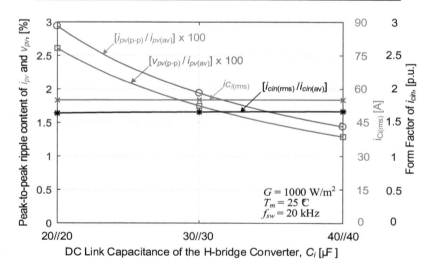

Figure 12. The variations in peak-to-peak ripple contents of i_{pv} and v_{pv}, $i_{ci(rms)}$, and form factor of i_{cin} against DC link capacitance of H-bridge converter for standard test conditions (these theoretical results were obtained using the dynamic model in Figure 5b; $f_{sw} = 20$ kHz assumed).

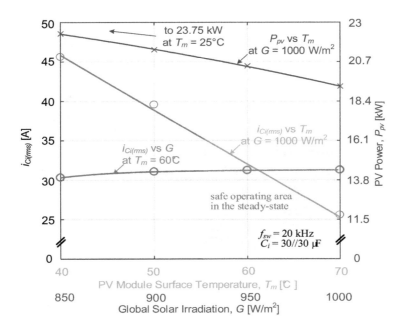

Figure 13. Effects of PV module surface temperature and global solar irradiation on $i_{ci(rms)}$ for 20 kHz switching frequency (these theoretical results were obtained from the dynamic model in Figure 5b; i_{ci} is defined in Figure 3a; $C_i = 30//30$ µF is assumed; variations in power delivered by the multi-string PV system in Figure 2 against PV module surface temperature are also given on this figure).

3.3.3. Optimum Switching Frequency of H-bridge Converter

Variations in ripple contents of v_{pv}, i_{pv}, i_{Ci}, and i_{cin} against switching frequency, f_{sw}, obtained by simulation studies at standard test conditions for the chosen $C_i = 30//30$ µF are as given in Figure 14.

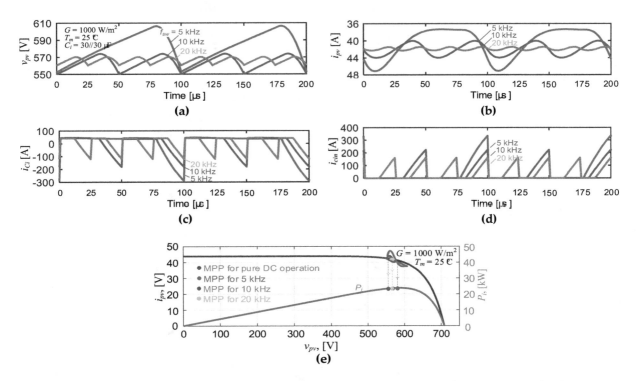

Figure 14. Variations in ripple content for different switching frequencies, f_{sw}: (**a**) PV panel voltage, v_{pv}; (**b**) PV panel current, i_{pv}; (**c**) input capacitor current, i_{Ci}; (**d**) total current transferred to MPPT converter; i_{cin}, and (**e**) effects of f_{sw} on maximum power point of multi-string PV system in steady-state.

As can be understood from these waveforms, operation at higher switching frequency reduces ripple contents and hence rms values of all currents. Lower rms values for i_{cin} and i_{Ci} are better in the selection not only of SiC power MOSFET but also of DC link capacitor, C_i. These variations are quantified and presented in graphical form as a function of f_{sw} in Figure 15. It can be concluded from these characteristics that the phase-shifted MPPT converter should be switched at a frequency greater than or equal to 15 kHz. Figure 14e justifies this statement. Dynamic variations in MPPs arising from ripple content of i_{pv} are marked on i_{pv}/v_{pv} characteristic given in Figure 14e. Calculated mean MPPs are also marked on the same figure. The ideal case is pure DC operation. As f_{sw} is increased dynamic MPP curve converges to that of pure DC operation, e.g., MPP power for f_{sw} = 20 kHz is nearly the same with that of pure DC operation.

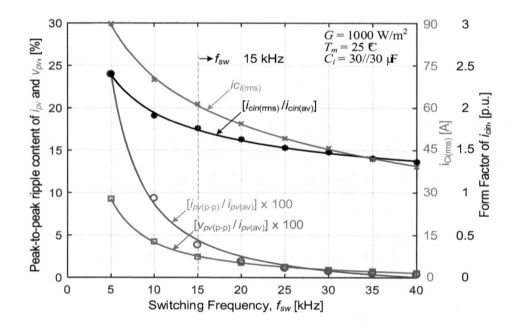

Figure 15. The variations in peak-to-peak ripple contents of i_{pv} and v_{pv}, $i_{Ci(rms)}$, and form factor of i_{cin} as a function of switching frequency (i_{pv}, v_{pv}, i_{Ci}, and i_{cin} are as defined in Figure 3a; these results were obtained by using the dynamic model in Figure 5b for the standard test conditions marked on the figure; DC link capacitance of the H-bridge converter in the MPPT converter is assumed to be $C_i = 30//30$ μF).

Thus far, in this subsection, only the factors for standard test conditions affecting the selection of f_{sw} have been considered. However, the entire operating range of the MPPT converter installed at the predefined geographical site gives more valuable information about the selection of f_{sw}. For this purpose, the variation range of duty ratio, D, for the H-bridge converter is calculated for different switching frequencies, and given in Figure 16. D should vary in a narrow range and at relatively high values to keep rms value of the semiconductor current, and hence its form factor at relatively low values. In view of these discussions, f_{sw} of the H-bridge converter should be at least 20 kHz.

In the selection of f_{sw}, the size of the HF transformer and the switching and conduction losses of H-bridge converter and conduction losses of Schottky diode rectifier should also be considered. Total semiconductor losses in the MPPT converter as a function of f_{sw} are given in Table 3. These losses exclude all wiring and cabling losses. It is seen in Table 3 that total power semiconductor losses in MPPT converter becomes minimum at f_{sw} = 20 kHz.

Table 3. Power semiconductor losses in MPPT converter against switching frequency (Operating Conditions: $P_{pv} = 25.1$ kW, $G = 1000$ W/m^2, $T_m = 10\,°C$, $T_j = 80\,°C$, $R_g = 10\,\Omega$).

H-Bridge Converter					Diode Bridge Rectifier		Total SiC MOSFET and SiC Diode Losses
f_{sw} (kHz)	I_{peak} (A)	I_{rms} (A)	P_{cond} (W)	P_{sw} (W)	I_{ave} (A)	P_{cond} (W)	P_{total} (W)
5	380	84	452	82	17.3	111	645
10	280	72	332	140	17.3	111	583
20	200	60	230	232	17.3	111	573
30	160	54	187	313	17.3	111	611
40	140	50	160	388	17.4	111	659

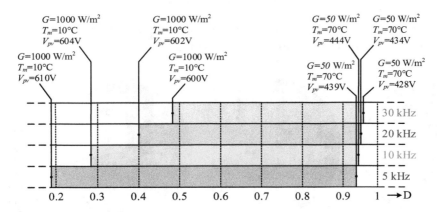

Figure 16. The variations in duty ratio, D of the H-bridge circuit in MPPT converter over the entire operating range for $f_{sw} = 5$, 10, 20, and 30 kHz, $V_{dc} = 700$ V, and $n = N_s/N_p = 1.52$.

3.3.4. HF Transformer Design

In this application, advanced types of ferrite, amorphous metal cobalt-base, amorphous metal iron-base, and nanocrystalline core materials can be used. Their recommended peak flux densities for operating frequency around 20 kHz are 0.3, 0.5, 1.3, and 1.0 T, respectively. The nanocrystalline core material has lower core loss than ferrite and much lower core loss than high flux density, amorphous metal iron-base material operating at the same peak flux density and frequency. On the other hand, although core loss density of the amorphous metal cobalt-base material is comparable to that of nanocrystalline material, it requires higher core volume resulting in higher total core loss and higher cost because of lower operating flux density. Nanocrystalline core material is therefore chosen in the design of the HF transformer with natural air cooling.

Rated values of the target transformer are specified as 23 kW, 20 kHz rated frequency, 700 V peak secondary voltage, and $n = N_s/N_p = 1.52$. SU102b nanocrystalline core is then used in the implementation of the HF transformer. Core loss, AC copper loss and total power loss of HF transformer against peak flux density are as given in Figure 17. Total power loss variation curve in Figure 17 shows that optimum design point can be chosen in between 0.25 T and 0.35 T. Design of the HF transformer is completed by choosing peak flux density as 0.3 T for minimum power dissipation at 20 kHz.

To test whether the design of MPPT converter is optimum, all power loss components excluding wiring and cabling losses are calculated and given in Figure 18 for three different operating frequencies and maximum PV power in the entire operating range of the MPPT converter installed at the predefined geographical site. Since $f_{sw} = 20$ kHz causes minimum power loss in the MPPT converter, the optimum design principles given in this subsection are justified.

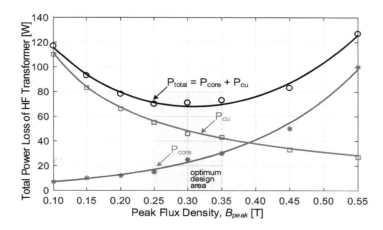

Figure 17. HF transformer power loss against peak flux density (23 kW, 700 V peak secondary voltage, f_{sw} = 20 kHz and n = 1.52 for SU102b nanocrystalline core).

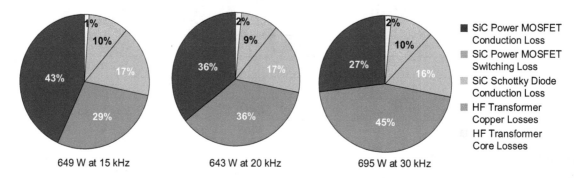

Figure 18. Power loss components of MPPT converter for different switching frequencies when P_{pv} = 25.1 kW. Numerical values are rounded.

3.3.5. Controller Design

The block diagram of the implemented DSP (TMS 320F28069) based controller is shown in Figure 3a. The controller is designed to fulfill the following tasks: (i) precharging the DC link capacitors, C_o, of the HF link MPPT converter; (ii) MPPT operation; and (iii) overvoltage protection and drain-to-source voltage monitoring for shoot-through protection. For this purpose, DC link voltage, V_{dc}, and PV panel voltage and current, V_{pv} and I_{pv}, are input to the controller, at a sampling rate of 40 kHz/channel. Voltage transducers used are of Fully-Differential Isolation Amplifier (TI AMC-1100) type for noise immunity, and the current transducer is Hall-Effect type (LEM HASS 50-S).

When the sun rises, the PI controller in Figure 3a starts to operate by applying narrow pulses to limit the charging current of C_o to a safe value. Precharging period is less than 30 s, during which D does not exceed 0.15. Set value of the DC link voltage, $V_{dc(set)}$, is specified to be 700 V in the control software. Whenever V_{dc} reaches 700 V, the inverter control system is activated to start the transfer of power to the grid. Just after the inverter operation, the controller enables MPPT algorithm based on an adaptive version of Perturb and Observe method [75], which runs only once every 300 ms, and stops whenever the error in active power between any two consecutive iterations is less than 0.1%. The algorithm then starts afresh after 1 s. In each iteration, the magnitude of duty ratio perturbation, Δd, is only ±0.25% of the previous duty ratio, D. Sampled I_{pv} and V_{pv} data are averaged over a period of 25 ms (256 samples), not only to filter out the measurement noise but also for the use in overvoltage protection software. The inverter tends to keep V_{dc} constant, and overvoltage protection facility in the controller of MPPT converter does not allow a rise in V_{dc} more than 10% in the case of loss of control. Furthermore, v_{DS} monitoring is carried out by an analog chip within the SiC driver [76].

4. Field Test Results

4.1. HF Link MPPT Converter

Variations in ac components of v_{pv} and i_{pv} while the MPPT converter is supplied from the multi-string PV system in Figure 2 are given in Figure 19, for operation at two different switching frequencies. The following observations can be made about these waveforms:

(i)　Since the experimental results are the same with the theoretical ones based on the dynamic model of the PV array, the mathematical model and system design methodology can be successfully used in the design of the SiC power MOSFET-based HF link MPPT converter.

(ii)　v_{pv} and i_{pv} are nearly pure DC at f_{sw} = 20 kHz, i.e., only 6 V p-p ripple is superimposed on $V_{pv(av)}$ = 533 V and 0.3 A p-p ripple is superimposed on $I_{pv(av)}$ = 15.9 A. This justifies the optimum switching frequency of around 20 kHz, for the SiC HF link converter.

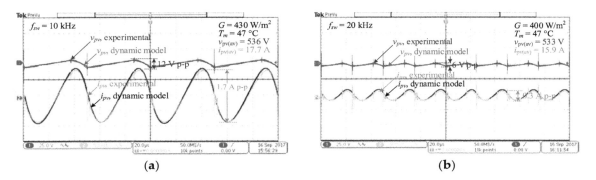

(a) **(b)**

Figure 19. Variations in ac components of v_{pv} and i_{pv} while the MPPT converter supplied from multi-string PV system in Figure 2 is operating at two different switching frequencies in the steady-state: (a) f_{sw} = 10 kHz; and (b) f_{sw} = 20 kHz. (v_{pv} and i_{pv} are as defined in Figure 3a. Theoretical results have been obtained for the dynamic model in Figure 5b; v_{pv} and i_{pv} are recorded by using Tektronix MSO3034 oscilloscope, Tektronix P5205A high voltage differential probe, Tektronix TCP404XL current probe and Tektronix TCPA300 current probe amplifier.)

Drain-to-source voltage, v_{DS}, of SiC power MOSFET S_3 and the line current i_1 waveform of the H-bridge in Figure 3 are also recorded as shown in Figure 20. Positive half cycles of i_1 correspond to the drain current i_d waveform of S_3. Note that all SiC power MOSFETs turn-on at zero-current owing to the ramp current waveform of the discontinuous conduction mode. At the turn-off, however, only S_3 and S_4 are switched at zero current at the end of the OFF period as illustrated in Figure 11b,d, due to the phase shifted operation. S_1 and S_2, however, are switched off at the peak of the transformer primary current, i_1. The glitches superimposed on v_{DS} waveform of S_3 in Figure 20 are attributed to the noise coupled to the oscilloscope voltage probe during switching-off of the other SiC power MOSFETs.

HF transformer voltage and current waveforms on both the primary and the secondary sides are as shown in Figure 21, at nearly full-load. Operation modes of the HF link converter, as defined in Figure 11 are marked on various segments of the recorded voltage and current waveforms in Figure 21. The voltage spikes at the turn-off of S_1 and S_2 (just at the beginning of freewheeling modes of S_3-D_2 and S_4-D_1) in Figure 21a are caused by the ringing between switches' output capacitances (C_{oss} = 880 pF) and the primary stray inductance of the current path between C_i and S_1 or S_2 (calculated from layout as $L_{stray1} \approx 27$ nH). The corresponding oscillation frequency is measured as 33 MHz, as expected. This effect is more pronounced at the secondary side waveforms in Figure 21b, owing to the resonance between the two outgoing Schottky diode output capacitances ($C_j \approx 1000$ pF each) and the secondary leakage inductance (L_s = 7.4 µH), in series with the stray inductances ($L_{stray2} \approx 2.4$ µH) between C_o and the outgoing diodes, resulting in an oscillation frequency of 2.3 MHz. The slight voltage drops during freewheeling modes in Figure 21b are caused by L_{stray2}.

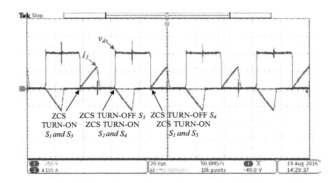

Figure 20. Drain-to-source voltage, v_{DS}, of SiC power MOSFET (S_3) and line current, i_1, waveforms in the H-bridge circuit recorded by Tektronix MSO3034 oscilloscope, Tektronix P5205A high voltage differential probe and Rogowski CWTUM/3/B current probe (f_{sw} = 20 kHz).

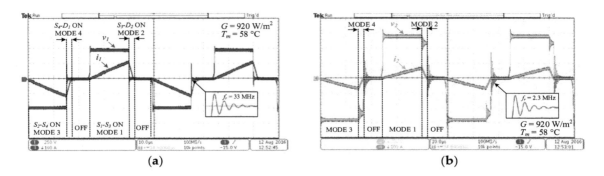

Figure 21. HF transformer voltage and current waveforms recorded by Tektronix MSO3034 oscilloscope, Tektronix P5205A high voltage differential probe and Rogowski CWTUM/3/B current probe (f_{sw} = 20 kHz, Test conditions are marked on the figure): (**a**) primary side; and (**b**) secondary side.

AC components of rectifier output current, i_{ro}, and the converter output current waveform, i_{dc}, in Figure 22 are recorded by a Rogowski current probe. Note that 72 A p-p rectifier output current ripple at full-load is filtered down to 12 A p-p by the low ESR DC link capacitor C_o.

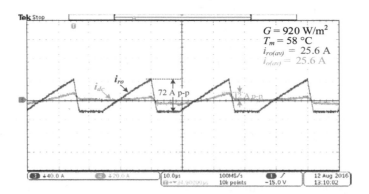

Figure 22. AC components of rectifier output current, i_{ro}, and the converter output current waveform, i_{dc}, recorded by Rogowski CWTUM/3/B current probe (average values of i_{ro} and i_{dc} and test conditions are marked on the figure; f_{sw} = 20 kHz).

4.2. Voltage Source Inverter

Since the three-phase two-level VSI is built by using a full SiC six-pack module, only the drain-to-source voltages v_{DS} and unfiltered line currents i_a, i_b, and i_c can be recorded. Figure 23 shows the circuit diagram of first leg of the inverter and the associated unfiltered line current, i_a.

To investigate effects of dead band on the turn-off performance of SiC power MOSFETs, v_{DS4} and i_a are recorded around the zero-crossing point of the unfiltered line-a current waveform for various dead times, as shown in Figure 24. These waveforms are as shown in Figure 24a when dead band is adjusted to 400 ns. Since current is very low, C_{oss4} and C_{oss1} are, respectively, charging and discharging slowly. When M_1 is turned on at the end of the dead band period, C_{oss4} is not charged yet to $V_{dc} = 700$ V and the C_{oss1} is not discharged entirely. The residual voltage on C_{oss1} will then be superimposed on DC link voltage which appears across the drain-source terminals of S_4 (v_{DS4}). A shorter dead band causes a larger overshoot on v_{DS4} waveform. This phenomenon does not occur around the peak value of unfiltered line-a current. This is because C_{oss4} and C_{oss1}, respectively, charges and discharges more rapidly since the current is high.

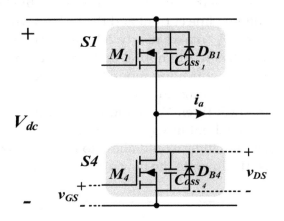

Figure 23. First leg of PV inverter and supply line-a (S:SiC power MOSFET, M:MOSFET part, C_{oss}: output capacitance, DB: body diode, and i_a: unfiltered line-a current).

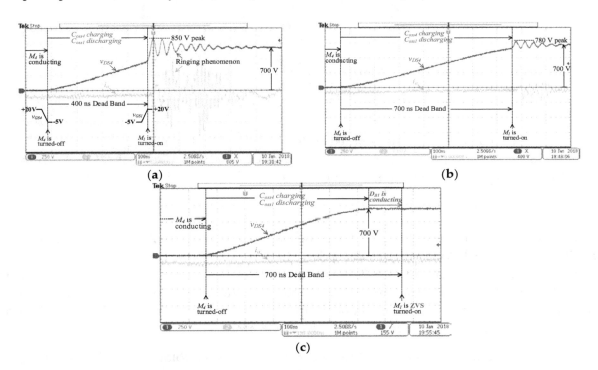

Figure 24. Effects of dead band on turn-off behavior of S_4 around zero crossing of the unfiltered line current (symbols are as defined in Figure 22 and v_{DS} and i_a are recorded by Tektronix MSO3034 oscilloscope, Tektronix P5205A high voltage differential probe, Tektronix TCP404XL current probe, Tektronix TCPA300 current probe amplifier): (**a**) 400 ns dead band; (**b**) 700 ns dead band; and (**c**) 700 ns dead band with ZVS turn-on of M_1.

On the other hand, ringing phenomenon is observed on v_{DS4} when S_1 is turned on due to the damped high frequency oscillation between the stray inductance of DC bus and C_{oss1}. In general, shorter dead time reduces low order harmonic distortion in line current waveforms at the expense of higher switching loss at turn-on and high frequency harmonic component. As can be seen in Figure 24b, a longer dead time (700 ns) reduces peak value of v_{DS4} and alleviates ringing phenomenon. At a current level slightly higher than that of Figure 24a,b, 700 ns dead time eliminates entirely ringing phenomenon and leads to ZVS turn-on of M_1 as shown in Figure 24c at the expense of higher low order harmonic distortion. In view of these considerations, 400 ns dead time is used in the implementation.

To investigate the transient performance of the VSI control system, multi-string PV system is suddenly disconnected from the input of the MPPT converter while the PV supply is delivering nearly 11 kW to the grid. The recorded filtered line current and the DC link voltage waveforms are as shown in Figure 25. Just after disconnection V_{dc} makes nearly 20% undershoot and then settles down to 98% of its rated value in nearly 480 ms. It is worth noting that, after reaching minimum V_{dc}, the voltage source converter starts to operate in rectification mode to allow power transfer from the grid to the DC link, thus maintaining V_{dc} at 700 V. Transient response is affected primarily by the size of the DC link capacitor and secondarily the LCL filter. V_{dc} in Figure 25 would decay more rapidly in the case of a smaller C_o, thus increasing undershoot in V_{dc}. To compensate for this phenomenon a larger size LCL filter could be used. A larger LCL filter would allow to increase the control range and hence the voltage source converter could settle down to the new operation state much more rapidly.

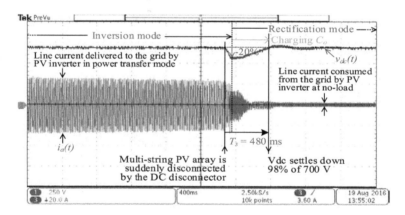

Figure 25. $v_{dc(t)}$ and $i_{a(t)}$ waveforms recorded by Tektronix MSO3034 oscilloscope, Tektronix P5205A high voltage differential probe Tektronix TCP404XL current probe and Tektronix TCPA300 current probe amplifier when the multi-string PV array is suddenly disconnected from MPPT converter.

4.3. Harmonic Distortion

Snapshots of the grid-side electrical quantities are given in Figure 26 for operation at full-load (Figure 26, left) and half-load (Figure 26, side). Figure 26a gives the three-phase voltages and line current waveforms, and Figure 26b the associated harmonic current contents according IEC61000-4-7:2002 harmonic measurement method [77], including both the line harmonics and the interharmonics. The rms quantities and output powers are recorded as shown in Figure 26c. The following observations can be made about these waveforms:

- The inverter operates connected to the 50-Hz AC grid with a total harmonic distortion, THDv $\approx$ 1.4% for the line-to-line voltages, and dominant 5th and 7th harmonics. The current THDi is recorded to be 3.8% at full-load, and 4.3% at half-load, with dominant 5th and 7th harmonics as in the AC grid. These current THD values correspond to current TDD values, respectively, of 3.4% and 1.9% by taking 25 kVA as the apparent power rating of the VSI.

- The resulting individual line current harmonics obtained experimentally are found to be within the recommended limits by the IEEE Std.-519-2014 [78] for all supply conditions, as can be seen from Figures 26b and 27 for a harmonic spectrum up to the 50th.

- The inverter operates successfully at unity pf, (pf $\approx$ 0.999 recorded), under both the full-load and the half-load conditions, according to the preset value, $I_{q(set)} = 0$.

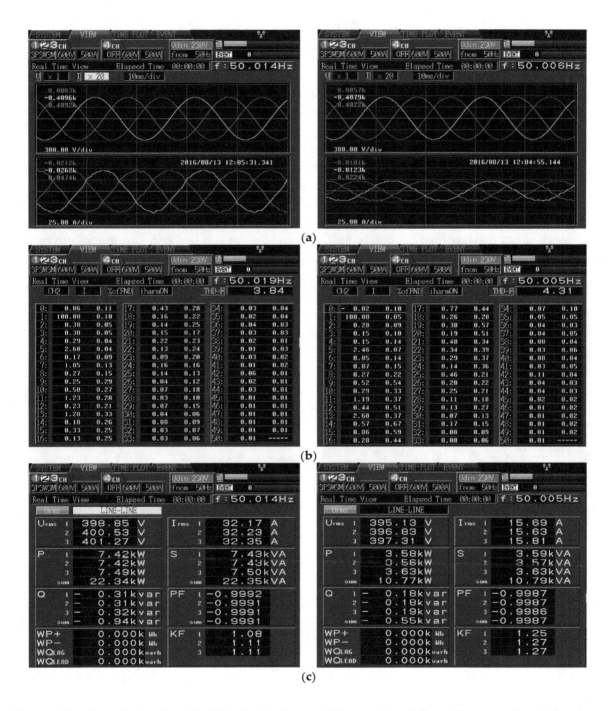

Figure 26. Snapshots of grid-side electrical quantities captured from the experimental set-up in Figure 1b by Hioki Power Analyzer PW3198 at P_o = 22 kW (left) and P_o = 11 kW (right): (**a**) line-to-line voltage and line current waveforms; (**b**) harmonic content of line current waveforms; and (**c**) output powers.

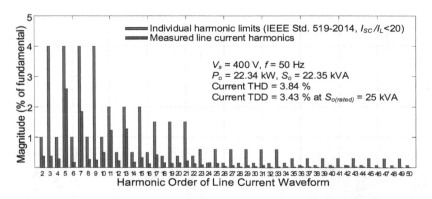

Figure 27. Sample harmonic spectra for the line current waveform injected by the inverter to the utility grid (deduced from the records of Hioki Power Analyzer PW3198 in Figure 26b).

In this research work, the LCL filter is optimized to yield minimum filter size and hence cost. To illustrate the effects of LCL filter size on the harmonic distortion of line current waveforms, L_g is increased from 50 μH to 1.5 mH, as preferred by several researchers in their implementations, and then the harmonic distortion record is repeated at 10 kVA. These records are given in Figure 28. THD and TDD values of the line current waveforms are 2.26% and 0.9%, respectively. It is seen that a larger size and hence more dissipative and costly LCL filter yields much lower harmonic distortion in line current waveforms.

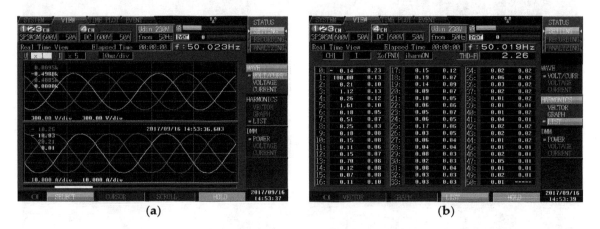

Figure 28. Snapshots of grid-side electrical quantities captured from the experimental set-up in Figure 1b by Hioki Power Analyzer PW3198 at P_o = 10 kW for L_g = 1.5 mH: (**a**) line-to-line voltage and line current waveforms; and (**b**) harmonic content of line current waveforms.

4.4. Efficiency

Efficiencies of HF link MPPT converter, VSI, and the overall grid-connected PV supply are obtained separately by field measurements for different operating conditions. Experimental results are given in comparison with theoretical values. For efficiency calculations, power components P_i, $P_{o(MPPT)}$, and P_o are as defined in Figure 29. For different operating conditions, $P_i = v_{pv(av)} \cdot i_{pv(av)}$ and $P_{o(MPPT)} = v_{dc(av)} \cdot i_{dc(av)}$ are calculated from measured $v_{pv(av)}$, $i_{pv(av)}$, $v_{dc(av)}$, and $i_{dc(av)}$ data as described in the caption of Figure 29. A sample set of $v_{pv(av)}$, $i_{pv(av)}$, $v_{dc(av)}$, and $i_{dc(av)}$ waveforms is given in Figure 30. Experimental efficiency values for the MPPT converter calculated from field data for different operating conditions are given in Table 4. For the corresponding insolation levels and module temperatures, the theoretical efficiency values are also calculated by using the MPPT converter model including the dynamic model of the multi-string PV system and running it on MATLAB Simulink at f_{sw} = 20 kHz. These theoretical values are also marked in Table 4. The following conclusions can be drawn from these results:

(i)   Maximum efficiency occurs at nearly half-load.

(ii)   Full-load efficiency (97.3%) is only 0.5% lower than the maximum efficiency (97.7%).

(iii) Experimental values are slightly lower than corresponding theoretical values (discrepancies, $\delta\eta \leq 4\%$ at low power levels and $\delta\eta < 1\%$ at high powers).

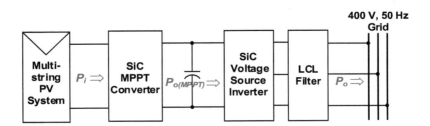

Figure 29. Definition of power components for efficiency calculations. P_i is calculated from the measured $v_{pv(av)}$ and $i_{pv(av)}$ data. $P_{o(MPPT)}$ is calculated from the measured $v_{dc(av)}$ and $i_{dc(av)}$ data. Measuring instruments are Tektronix MSO3034 oscilloscope together with its moving average filters, Tektronix P5205A high voltage differential probe, Tektronix TCP404XL current probe, Tektronix TCPA300 current probe amplifier, and Fluke 80i-110s current probe for i_{dc}. P_o is measured by Hioki Power Analyzer PW3198. Instantaneous values v_{pv}, i_{pv}, v_{dc} and i_{dc} are as defined in Figure 3 before averaging for steady-state operation.

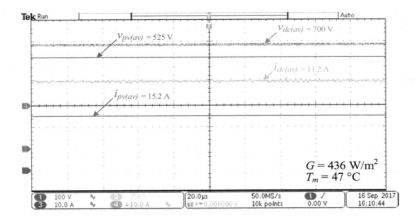

Figure 30. A sample set of $v_{pv(av)}$, $i_{pv(av)}$, $v_{dc(av)}$ and $i_{dc(av)}$ waveforms recorded at $G = 436$ W/m^2, and $T_m = 47\,°$C.

Experimental efficiency values for the SiC VSI calculated from field data for different $P_{o(MPPT)}$ values are given in Table 5. Theoretical values of P_o are calculated by subtracting all inverter losses from experimental values of $P_{o(MPPT)}$. For the SiC VSI, computer simulations are carried out using the Wolfspeed SpeedFit design simulation software [72] to calculate SiC MOSFET losses, and Magnetics Inductor Design Tool [79] to determine the LCL filter losses. The following conclusions can be drawn from the results in Table 5:

(i)   Maximum efficiency (98.6%) occurs nearly at 40% of full 22.3 kW-load.

(ii)   Efficiency is 98.1% at 88% of full kW-load.

(iii) Experimental values are slightly lower than corresponding theoretical values (discrepancies, $\delta\eta \leq 2\%$ at low power levels and $\delta\eta < 1\%$ at high powers).

The variations in efficiency of the all-SiC PV supply are calculated from field data and given in Figure 31 as a function of P_o. Maximum efficiency is observed to be 97%. Full-load efficiency is estimated to be slightly higher than 96%. At very low power levels such as 10% of the full-load, the overall efficiency is around 92%. These efficiency values are comparable with those of new generation PV supplies containing boost type MPPT converters and hybrid IGBT based inverters of the same power ratings and supplied from the existing multi-string PV system in Figure 2.

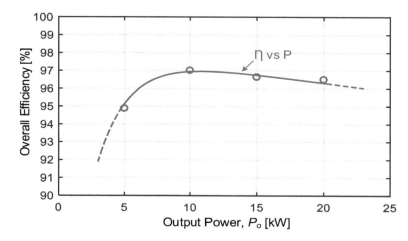

Figure 31. Efficiency of all SiC grid-connected PV supply against output power, $\eta = [P_o/P_i] \times 100$ (experimental values of P_o and P_i are obtained as described in the caption of Figure 29).

In the case where the multi-string PV system is initially not available and the types of the modules are not prespecified, the optimum multi-string PV system configuration and its technical characteristics will be a design issue for the overall grid-connected PV supply. The structure and technical characteristics of the multi-string PV system mainly affect the efficiency of the MPPT converter and hence efficiency of the overall system. To illustrate this fact, simulation studies for two different cases which employ 100 CSUN250-72M modules were carried out. In Case 1, 100 modules are connected to give 4×25 multi-string PV system to illustrate the effects of high operating voltage and low current on the efficiency of MPPT converter. In Case 2, the same modules are connected to give 5×20 multi-string PV system with lower operating voltage and higher current. The simulation results are as given in Figure 32. As can be understood from Figure 32, under standard test conditions, Case 1 gives 99% converter efficiency at nearly half-load and, when it is combined with inverter efficiency, the maximum efficiency of the overall system may reach 98.1%. This is because the operation of the MPPT converter with HF link in Figure 3 at a lower PV current, i_{pv}, reduces conduction loss components remarkably, thus improving the converter efficiency.

Table 4. Comparison of experimental and theoretical efficiency values of MPPT converter.

Test Condition *	P_i, kW		$P_{o(MPPT)}$, kW		Efficiency, $\eta = [P_{o(MPPT)}/P_i]$ 100, %	
G and T_m	Experimental *	Theoretical †	Experimental *	Theoretical °	Experimental	Theoretical
$G = 70$ W/m² $T_m = 32\ °C$	1.66	1.60	1.49	1.55	89.76	93.47
$G = 150$ W/m² $T_m = 35\ °C$	3.28	3.14	3.08	3.13	93.90	95.55
$G = 250$ W/m² $T_m = 38\ °C$	4.51	5.19	4.32	5.01	95.78	96.69
$G = 370$ W/m² $T_m = 40\ °C$	8.91	8.23	8.70	8.04	97.64	97.69
$G = 490$ W/m² $T_m = 49\ °C$	10.61	10.21	10.37	10.02	97.73	98.20
$G = 730$ W/m² $T_m = 57\ °C$	15.52	15.42	15.14	15.04	97.55	97.56
$G = 980$ W/m² $T_m = 60\ °C$	20.14	20.61	19.59	20.05	97.26	97.28

* Experimental values of P_i and $P_{o(MPPT)}$ are determined as defined in Figure 29 for different test conditions.
† Theoretical values of P_i are calculated for the same test conditions by using the dynamic model in Figure 5b.
° Theoretical values of $P_{o(MPPT)}$ are calculated by adding all MPPT converter losses to theoretical values of P_i.

Table 5. Comparison of experimental and theoretical efficiency values of SiC inverter.

$P_{o(MPPT)}$, kW	P_o, kW		Efficiency, $\eta = [P_o/P_{o(MPPT)}]$ 100, %	
Experimental *	Experimental *	Theoretical †	Experimental	Theoretical
3.08	2.99	3.05	97.05	99.00
4.31	4.21	4.27	97.54	99.05
8.69	8.57	8.61	98.55	99.10
10.37	10.21	10.27	98.46	99.10
15.13	14.88	14.98	98.32	99.05
19.58	19.21	19.38	98.09	99.00

* Experimental values of $P_{o(MPPT)}$ and P_o are obtained as defined in Figure 29 for different operating conditions.
† Theoretical values of P_o are calculated by subtracting all inverter losses from experimental values of $P_{o(MPPT)}$.

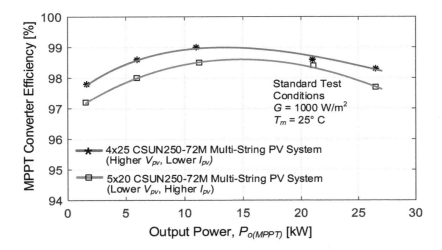

Figure 32. Effects of the configuration of multi-string PV system on the efficiency of MPPT converter with HF link in Figure 3 (Simulation results).

5. Conclusions

A system design methodology for an all SiC grid-connected PV supply with HF link MPPT converter has been proposed and a prototype of 25 kVA converter operating at 20 kHz has been implemented for verification. Owing to the very high dv/dt (>10 kV/μs) ratings of SiC power semiconductors, common-mode EMI is more pronounced in SiC based non-isolated converters. In this work, galvanic isolation of the proposed MPPT converter overcomes the common-mode EMI problem, thus enabling the grid connection using a simple and reliable three-phase, two-level inverter. In the design of the HF link MPPT converter operating at 20 kHz, a dynamic model of the multi-string PV system, parameters of which are obtained from the field test results, is used. More realistic MPPT converter parameters are shown to be obtained in the paper by using the dynamic PV model in the design procedure in comparison with the well-known static PV models.

The optimum switching frequency of the 25 kVA three-phase two-level inverter is determined as 20 kHz in the design procedure in view of inverter losses. The resulting 25 kVA, 20 kHz SiC VSI has 98.5% maximum efficiency which is slightly higher than or comparable with those of new-generation IGBT (Si IGBT + antiparallel SiC Schottky Diode) based counterparts for the existing multi-string PV system in Figure 2. This relatively high switching frequency not only reduces the size of the passive components, such as the LCL filter and the HF transformer but also the size of the cooling aggregates. LCL filter of the VSI, which is optimized by considering stability concerns of the controller in the design, provides nearly 100 dB attenuation at 20 kHz and its size is at least ten times smaller than those of LCL filter designs reported in the literature, even for lower size converters. TDD of the grid-connected VSI is measured to be 3.9% at nearly full-load and its individual current harmonics up to 50th conform with IEC Std. 61000-4-7:2002 even for the weakest grid. A higher grid-side inductance (L_g = 1.5 mH) of the LCL filter lowers the current THDs considerably, which is measured to be 2.3% at nearly half-load.

The resulting SiC MPPT converter operating at 20 kHz and supplied from the existing 5×19 multi-string PV system in Figure 2 has 98% measured maximum efficiency, which is comparable with IGBT based and lower than SiC based boost type MPPT converters. Power densities are calculated as nearly 1.8 kW/lt and 1.6 kW/lt for forced air-cooled SiC MPPT converter and SiC grid-connected VSI, respectively. These figures are higher than those of forced air-cooled new-generation IGBT based converters and much higher than those of natural air-cooled new-generation IGBT based converters.

Author Contributions: M.E. and I.Ç. conceived the presented idea. M.C. and M.E. developed and implemented the SiC inverter. S.Ö. and I.Ç. developed and implemented the SiC MPPT converter. All authors discussed the results and contributed to the final manuscript.

Acknowledgments: The authors would like to acknowledge Energy Institute, TUBITAK Marmara Research Center, Ankara branch for the scholarship provided to M.C. for the research on SiC converters.

References

1. Kjaer, S.B.; Pedersen, J.K.; Blaabjerg, F. A review of single-phase grid-connected inverters for photovoltaic modules. *IEEE Trans. Ind. Appl.* **2005**, *41*, 1292–1306. [CrossRef]

2. Mechouma, R.; Azoui, B.; Chaabane, M. Three-phase grid connected inverter for photovoltaic systems, a review. In Proceedings of the 2012 First International Conference on Renewable Energies and Vehicular Technology, Hammamet, Tunisia, 26–28 March 2012; pp. 37–42.

3. Orłowska-Kowalska, T.; Frede, B.; José, R. *Advanced and Intelligent Control in Power Electronics and Drives*; Springer: Berlin, Germany, 2014.

4. Barater, D.; Lorenzani, E.; Concari, C.; Franceschini, G.; Buticchi, G. Recent advances in single-phase transformerless photovoltaic inverters. *IET Renew. Power Gener.* **2016**, *10*, 260–273. [CrossRef]

5. Chen, S.M.; Liang, T.J.; Yang, L.S.; Chen, J.F. A Boost Converter with Capacitor Multiplier and Coupled Inductor for AC Module Applications. *IEEE Trans. Ind. Electron.* **2013**, *60*, 1503–1511. [CrossRef]

6. Liao, C.Y.; Lin, W.S.; Chen, Y.M.; Chou, C.Y. A PV Micro-inverter With PV Current Decoupling Strategy. *IEEE Trans. Power Electron.* **2017**, *32*, 6544–6557. [CrossRef]

7. Villanueva, E.; Correa, P.; Rodriguez, J.; Pacas, M. Control of a Single-Phase Cascaded H-Bridge Multilevel Inverter for Grid-Connected Photovoltaic Systems. *IEEE Trans. Ind. Electron.* **2009**, *56*, 4399–4406. [CrossRef]

8. Lorenzani, E.; Immovilli, F.; Migliazza, G.; Frigieri, M.; Bianchini, C.; Davoli, M. CSI7: A Modified Three-Phase Current-Source Inverter for Modular Photovoltaic Applications. *IEEE Trans. Ind. Electron.* **2017**, *64*, 5449–5459. [CrossRef]

9. Chen, Y.M.; Lo, K.Y.; Chang, Y.R. Multi-string single-stage grid-connected inverter for PV system. In Proceedings of the 2011 IEEE Energy Conversion Congress and Exposition, Phoenix, AZ, USA, 17–22 September 2011; pp. 2751–2756.

10. Edrington, C.S.; Balathandayuthapani, S.; Cao, J. Analysis and control of a multi-string photovoltaic (PV) system interfaced with a utility grid. In Proceedings of the IEEE PES General Meeting, Minneapolis, MN, USA, 25–29 July 2010; pp. 1–6.

11. SMA, SUNNY CENTRAL—High Tech Solution for Solar Power Stations. Products Category Brochure. Available online: http://www.sma-america.com/ (accessed on 21 February 2018).

12. Subudhi, B.; Pradhan, R. A Comparative Study on Maximum Power Point Tracking Techniques for Photovoltaic Power Systems. *IEEE Trans. Sustain. Energy* **2013**, *4*, 89–98. [CrossRef]

13. Müller, N.; Renaudineau, H.; Flores-Bahamonde, F.; Kouro, S.; Wheeler, P. Ultracapacitor storage enabled global MPPT for photovoltaic central inverters. In Proceedings of the 2017 IEEE 26th International Symposium on Industrial Electronics (ISIE), Edinburgh, UK, 19–21 June 2017; pp. 1046–1051.

14. Miguel, C.; Fernando, M.; Riganti, F.F.; Antonino, L.; Alessandro, S. A neural networks-based maximum power point tracker with improved dynamics for variable dc-link grid-connected photovoltaic power plants. *Int. J. Appl. Electromagn. Mech.* **2013**, *43*, 127–135.

15. Mancilla-David, F.; Arancibia, A.; Riganti-Fulginei, F.; Muljadi, E.; Cerroni, M. A maximum power point tracker variable-dc-link three-phase inverter for grid-connected PV panels. In Proceedings of the 2012 3rd IEEE PES Innovative Smart Grid Technologies Europe (ISGT Europe), Berlin, Germany, 14–17 October 2012; pp. 1–7.

16. Kerekes, T.; Teodorescu, R.; Liserre, M.; Klumpner, C.; Sumner, M. Evaluation of Three-Phase Transformerless Photovoltaic Inverter Topologies. *IEEE Trans. Power Electron.* **2009**, *24*, 2202–2211. [CrossRef]

17. Walker, G.R.; Sernia, P.C. Cascaded DC-DC converter connection of photovoltaic modules. *IEEE Trans. Power Electron.* **2004**, *19*, 1130–1139. [CrossRef]

18. Koutroulis, E.; Blaabjerg, F. Design Optimization of Transformerless Grid-Connected PV Inverters Including Reliability. *IEEE Trans. Power Electron.* **2013**, *28*, 325–335. [CrossRef]

19. Meneses, D.; Blaabjerg, F.; García, Ó.; Cobos, J.A. Review and Comparison of Step-Up Transformerless Topologies for Photovoltaic AC-Module Application. *IEEE Trans. Power Electron.* **2013**, *28*, 2649–2663. [CrossRef]

20. Araujo, S.V.; Zacharias, P.; Mallwitz, R. Highly Efficient Single-Phase Transformerless Inverters for Grid-Connected Photovoltaic Systems. *IEEE Trans. Ind. Electron.* **2010**, *57*, 3118–3128. [CrossRef]

21. Zhang, L.; Sun, K.; Xing, Y.; Xing, M. H6 Transformerless Full-Bridge PV Grid-Tied Inverters. *IEEE Trans. Power Electron.* **2014**, *29*, 1229–1238. [CrossRef]

22. Rodriguez, J.; Bernet, S.; Steimer, P.K.; Lizama, I.E. A Survey on Neutral-Point-Clamped Inverters. *IEEE Trans. Ind. Electron.* **2010**, *57*, 2219–2230. [CrossRef]

23. Bala, S. Next Gen PV Inverter Systems Using WBG Devices. In *High Pen PV Through Next-Gen PE Technologies Workshop*; NREL: Golden, CO, USA, 2016.

24. Millán, J.; Godignon, P.; Perpiñà, X.; Pérez-Tomás, A.; Rebollo, J. A Survey of Wide Bandgap Power Semiconductor Devices. *IEEE Trans. Power Electron.* **2014**, *29*, 2155–2163. [CrossRef]

25. Mookken, J.; Agrawal, B.; Liu, J. Efficient and Compact 50 kW Gen2 SiC Device Based PV String Inverter. In Proceedings of the International Exhibition and Conference for Power Electronics, Intelligent Motion, Renewable Energy and Energy Management, Nuremberg, Germany, 20–22 May 2014; pp. 1–7.

26. Wei, S.; He, F.; Yuan, L.; Zhao, Z.; Lu, T.; Ma, J. Design and implementation of high efficient two-stage three-phase/level isolated PV converter. In Proceedings of the 18th International Conference on Electrical Machines and Systems (ICEMS), Pattaya, Thailand, 25–28 October 2015; pp. 1649–1654.

27. Shi, Y.; Li, R.; Xue, Y.; Li, H. High-Frequency-Link-Based Grid-Tied PV System With Small DC-Link Capacitor and Low-Frequency Ripple-Free Maximum Power Point Tracking. *IEEE Trans. Power Electron.* **2016**, *31*, 328–339. [CrossRef]

28. Islam, M.; Mekhilef, S. Efficient Transformerless MOSFET Inverter for a Grid-Tied Photovoltaic System. *IEEE Trans. Power Electron.* **2016**, *31*, 6305–6316. [CrossRef]

29. Sintamarean, N.C.; Blaabjerg, F.; Wang, H.; Yang, Y. Real Field Mission Profile Oriented Design of a SiC-Based PV-Inverter Application. *IEEE Trans. Ind. Appl.* **2014**, *50*, 4082–4089. [CrossRef]

30. Shi, Y.; Wang, L.; Xie, R.; Shi, Y.; Li, H. A 60-kW 3-kW/kg Five-Level T-Type SiC PV Inverter With 99.2% Peak Efficiency. *IEEE Trans. Ind. Electron.* **2017**, *64*, 9144–9154. [CrossRef]

31. Colmenares, J.; Peftitsis, D.; Rabkowski, J.; Sadik, D.P.; Tolstoy, G.; Nee, H.P. High-Efficiency 312-kVA Three-Phase Inverter Using Parallel Connection of Silicon Carbide MOSFET Power Modules. *IEEE Trans. Ind. Appl.* **2015**, *51*, 4664–4676. [CrossRef]

32. Rabkowski, J.; Peftitsis, D.; Nee, H.P. Design steps towards a 40-kVA SiC inverter with an efficiency exceeding 99.5%. In Proceedings of the Twenty-Seventh Annual IEEE Applied Power Electronics Conference and Exposition (APEC), Orlando, FL, USA, 5–9 February 2012; pp. 1536–1543.

33. Yin, S.; Tseng, K.J.; Tong, C.F.; Simanjorang, R.; Gajanayake, C.J.; Gupta, A.K. A 99% efficiency SiC three-phase inverter using synchronous rectification. In Proceedings of the IEEE Applied Power Electronics Conference and Exposition (APEC), Long Beach, CA, USA, 20–24 March 2016; pp. 2942–2949.

34. Laird, Y.Y.I.; Yuan, X.; Scoltock, J.; Forsyth, A.J. A Design Optimization Tool for Maximizing the Power Density of 3-Phase DC–AC Converters Using Silicon Carbide (SiC) Devices. *IEEE Trans. Power Electron.* **2018**, *33*, 2913–2932. [CrossRef]

35. Payan, D.; Catani, J.P.; Schwander, D. Solar Array Dynamic Simulator Prevention of Solar Array Short-Circuits due to Electrostatic Discharge. In Proceedings of the Space Power, Sixth European Conference, Porto, Portugal, 6–10 May 2002; pp. 609–615.

36. Schwander, D. Dynamic Solar Cell Measurement Techniques: New Small Signal Measurement Techniques. In Proceedings of the Space Power, Sixth European Conference, Porto, Portugal, 6–10 May 2002; pp. 603–608.

37. Blok, R.; van den Berg, E.; Slootweg, D. Solar Cell Capacitance Measurement. In Proceedings of the Space Power, Sixth European Conference, Porto, Portugal, 6–10 May 2002; pp. 597–602.

38. Herman, M.; Jankovec, M.; Topic, M. Optimisation of the I-V measurement scan time through dynamic modelling of solar cells. *IET Renew. Power Gener.* **2013**, *7*, 63–70. [CrossRef]

39. Qin, L.; Xie, S.; Yang, C.; Cao, J. Dynamic model and dynamic characteristics of solar cell. In Proceedings of the IEEE ECCE Asia Downunder, Melbourne, Australia, 3–6 June 2013; pp. 659–663.

40. Bharadwaj, P.; Kulkarni, A.; John, V. Impedance estimation of photovoltaic modules for inverter start-up analysis. *Sadhana—Acad. Proc. Eng. Sci.* **2017**, *42*, 1377–1387. [CrossRef]

41. Huang, J.H.; Lehman, B.; Qian, T. Submodule integrated boost DC-DC converters with no external input capacitor or input inductor for low power photovoltaic applications. In Proceedings of the IEEE Energy Conversion Congress and Exposition (ECCE), Milwaukee, WI, USA, 18–22 September 2016; pp. 1–7.

42. Duan, S.; Yan, G.; Jin, L.; Ren, J.; Wu, W. Design of photovoltaic power generation mppt controller based on SIC MOSFET. In Proceedings of the TENCON 2015–IEEE Region 10 Conference, Macao, China, 1–4 November 2015; pp. 1–5.

43. Kim, T.; Jang, M.; Agelidis, V.G. Practical implementation of a silicon carbide-based 300 kHz, 1.2 kW hard-switching boost-converter and comparative thermal performance evaluation. *IET Power Electron.* **2015**, *8*, 333–341. [CrossRef]

44. Mouli, G.R.C.; Schijffelen, J.H.; Bauer, P.; Zeman, M. Design and Comparison of a 10-kW Interleaved Boost Converter for PV Application Using Si and SiC Devices. *IEEE J. Emerg. Sel. Top. Power Electron.* **2017**, *5*, 610–623. [CrossRef]

45. Anthon, Y.Y.A.; Zhang, Z.; Andersen, M.A.E. A high power boost converter for PV Systems operating up to 300 kHz using SiC devices. In Proceedings of the International Power Electronics and Application Conference and Exposition, Shanghai, China, 5–8 November 2014; pp. 302–307.

46. Mostaghimi, O.; Wright, N.; Horsfall, A. Design and performance evaluation of SiC based DC-DC converters for PV applications. In Proceedings of the IEEE Energy Conversion Congress and Exposition (ECCE), Raleigh, NC, USA, 15–20 September 2012; pp. 3956–3963.

47. Fujii, K.; Noto, Y.; Oshima, M.; Okuma, Y. 1-MW solar power inverter with boost converter using all SiC power module. In Proceedings of the 17th European Conference on Power Electronics and Applications (EPE'15 ECCE-Europe), Geneva, Switzerland, 8–10 September 2015; pp. 1–10.

48. Agamy, M.S.; Chi, S.; Elasser, A.; Harfman-Todorovic, M.; Jiang, Y.; Mueller, F.; Tao, F. A High-Power-Density DC–DC Converter for Distributed PV Architectures. *IEEE J. Photovolt.* **2013**, *3*, 791–798. [CrossRef]

49. Akagi, H.; Yamagishi, T.; Tan, N.M.L.; Kinouchi, S.I.; Miyazaki, Y.; Koyama, M. Power-Loss Breakdown of a 750-V 100-kW 20-kHz Bidirectional Isolated DC–DC Converter Using SiC-MOSFET/SBD Dual Modules. *IEEE Trans. Ind. Appl.* **2015**, *51*, 420–428. [CrossRef]

50. Liserre, M.; Blaabjerg, F.; Dell'Aquila, A. Step-by-step design procedure for a grid-connected three-phase PWM voltage source converter. *Int. J. Electron.* **2004**, *91*, 445–460. [CrossRef]

51. Wu, W.; He, Y.; Tang, T.; Blaabjerg, F. A New Design Method for the Passive Damped LCL and LLCL Filter-Based Single-Phase Grid-Tied Inverter. *IEEE Trans. Ind. Electron.* **2013**, *60*, 4339–4350. [CrossRef]

52. Peña-Alzola, R.; Liserre, M.; Blaabjerg, F.; Sebastián, R.; Dannehl, J.; Fuchs, F.W. Analysis of the Passive Damping Losses in LCL-Filter-Based Grid Converters. *IEEE Trans. Power Electron.* **2013**, *28*, 2642–2646. [CrossRef]

53. Reznik, Y.Y.A.; Simões, M.G.; Al-Durra, A.; Muyeen, S.M. LCL Filter Design and Performance Analysis for Grid-Interconnected Systems. *IEEE Trans. Ind. Appl.* **2014**, *50*, 1225–1232. [CrossRef]

54. Liu, J.; Zhou, L.; Yu, X.; Li, B.; Zheng, C. Design and analysis of an LCL circuit-based three-phase grid-connected inverter. *IET Power Electron.* **2017**, *10*, 232–239. [CrossRef]

55. Parker, S.G.; McGrath, B.P.; Holmes, D.G. Regions of Active Damping Control for LCL Filters. *IEEE Trans. Ind. Appl.* **2014**, *50*, 424–432. [CrossRef]

56. Jayalath, S.; Hanif, M. Generalized LCL-Filter Design Algorithm for Grid-Connected Voltage-Source Inverter. *IEEE Trans. Ind. Electron.* **2017**, *64*, 1905–1915. [CrossRef]

57. Figueres, E.; Garcera, G.; Sandia, J.; Gonzalez-Espin, F.; Rubio, J.C. Sensitivity Study of the Dynamics of Three-Phase Photovoltaic Inverters With an LCL Grid Filter. *IEEE Trans. Ind. Electron.* **2009**, *56*, 706–717. [CrossRef]

58. Timbus, Y.Y.A.; Liserre, M.; Teodorescu, R.; Rodriguez, P.; Blaabjerg, F. Evaluation of Current Controllers for Distributed Power Generation Systems. *IEEE Trans. Power Electron.* **2009**, *24*, 654–664. [CrossRef]

59. Blaabjerg, F.; Teodorescu, R.; Liserre, M.; Timbus, A.V. Overview of Control and Grid Synchronization for Distributed Power Generation Systems. *IEEE Trans. Ind. Electron.* **2006**, *53*, 1398–1409. [CrossRef]

60. Vasquez, J.C.; Guerrero, J.M.; Savaghebi, M.; Eloy-Garcia, J.; Teodorescu, R. Modeling, Analysis, and Design of Stationary-Reference-Frame Droop-Controlled Parallel Three-Phase Voltage Source Inverters. *IEEE Trans. Ind. Electron.* **2013**, *60*, 1271–1280. [CrossRef]

61. Bosch, S.; Steinhart, H. Active power filter with model based predictive current control in natural and dq frame. In Proceedings of the 18th European Conference on Power Electronics and Applications (EPE'16 ECCE Europe), Karlsruhe, Germany, 5–9 September 2016; pp. 1–10.

62. Liserre, M.; Teodorescu, R.; Blaabjerg, F. Multiple harmonics control for three-phase grid converter systems with the use of PI-RES current controller in a rotating frame. *IEEE Trans. Power Electron.* **2006**, *21*, 836–841. [CrossRef]

63. Silicon carbide Power Module-BSM300D12P2E001. Available online: http://www.rohm.com/web/global/products/-/product/BSM300D12P2E001 (accessed on 21 February 2018).

64. CAS325M12HM2 1200V, 325A, Silicon Carbide High-Performance-Wolfspeed. Available online: https://www.wolfspeed.com/cas325m12hm2 (accessed on 21 February 2018).

65. SKM500MB120SC–SEMIKRON. Available online: https://www.semikron.com/products/product-classes/sic/full-sic/detail/skm500mb120sc-21919770.html (accessed on 21 February 2018).

66. SiC POWER MODULES. Available online: http://www.mitsubishielectric.com/semiconductors/catalog/pdf/sicpowermodule_e_201505.pdf (accessed on 21 February 2018).

67. *IEC Standard for Power Transformers*; IEC Standard IEC 60076-1:2011; IEC Webstore; IEC: Geneva, Switzerland, 2011.

68. *IEC Standard for Standard Voltages*; IEC Standard IEC 60038 2002-07; IEC Webstore; IEC: Geneva, Switzerland, 2002.

69. Kimball, J.W.; Zawodniok, M. Reducing Common-Mode Voltage in Three-Phase Sine-Triangle PWM with Interleaved Carriers. *IEEE Trans. Power Electron.* **2011**, *26*, 2229–2236. [CrossRef]

70. Feng, J.; Wang, H.; Xu, J.; Su, M.; Gui, W.; Li, X. A Three-Phase Grid-Connected Micro-Inverter for AC Photovoltaic Module Applications. *IEEE Trans. Power Electron.* **2017**. [CrossRef]

71. Shi, Y.; Wang, L.; Li, H. Stability Analysis and Grid Disturbance Rejection for a 60 kW SiC based Filter-less Grid-connected PV Inverter. *IEEE Trans. Ind. Appl.* **2017**. [CrossRef]

72. SpeedFit Design Simulator™, USA: Wolfspeed. Available online: https://www.wolfspeed.com/speedfit (accessed on 24 February 2018).

73. Cree, Inc. *CREE Application Note, Design Considerations for Designing with Cree SiC Modules*; Cree: Durham, NC, USA, 2013.

74. Metallized Polypropylene Film Capacitors (MKP). Available online: http://www.mouser.com/ds/2/136/MKP_B32774_778-19326.pdf (accessed on 24 February 2018).

75. Piegari, L.; Rizzo, R. Adaptive perturb and observe algorithm for photovoltaic maximum power point tracking. *IET Renew. Power Gener.* **2010**, *4*, 317–328. [CrossRef]

76. 1ED020I12_F2-DS-v02. Available online: https://www.infineon.com/dgdl/Infineon-1ED020I12_F2-DS-v02_00-en.pdf?fileId=db3a304330f68606013122ce5f3649cb (accessed on 24 February 2018).

77. *IEC Standard for Electromagnetic Compatibility (EMC)—Part 4-7: Testing and Measurement Techniques—General Guide on Harmonics and Interharmonics Measurements and Instrumentation, for Power Supply Systems and Equipment Connected Thereto*; IEC Standard IEC 61000-4-7:2002; IEC Webstore; IEC: Geneva, Switzerland, 2002.

78. *IEEE Standard for Recommended Practice and Requirements for Harmonic Control in Electric Power Systems*; IEEE Std. 519-2014; IEEE Standards Association: Piscataway, NJ, USA, 2014.

79. Magnetics® Inductor Design Tool, USA: Magnetics. Available online: https://www.mag-inc.com/Design/Design-Tools/Inductor-Design (accessed on 24 February 2018).

7

Open Circuit Fault Diagnosis and Fault Tolerance of Three-Phase Bridgeless Rectifier

Hong Cheng, Wenbo Chen *, Cong Wang and Jiaqing Deng

School of Mechanical Electronic & Information Engineering, China University of Mining and Technology Beijing, Ding No. 11 Xueyuan Road, Haidian District, Beijing 100083, China; chengh@cumtb.edu.cn (H.C.); wangc@cumtb.edu.cn (C.W.); dengjiaqingrae@163.com (J.D.)
* Correspondence: sglm.chen@gmail.com

Abstract: Bridgeless rectifiers are widely used in many applications due to a unity power factor, lower conduction loss and high efficiency, which does not need bidirectional energy transmission. In this case, the potential failures are threatening the reliability of these converters in critical applications such as power supply and electric motor driver. In this paper, open circuit fault is analyzed, taking a three-phase bridgeless as an example. Interference on both the input and output side are considered. Then, the fault diagnosis method including detection and location, and fault tolerance through additional switches are proposed. At last, simulation and experiments based on the hardware in loop technology are used to validate the feasibility of fault diagnosis and fault tolerance methodology.

Keywords: three-phase bridgeless rectifier; fault diagnosis; fault tolerant control; hardware in loop

1. Introduction

Multilevel converters have been widely used in middle- and high-voltage application fields in the past decades, such as renewable energy, adjustable speed drive, power transmission network, electric vehicle [1] etc. Topologies of these converters including H Bridge-based, neutral point clamping-based and bridgeless-based are most popular in literatures. Recently, bridgeless-based topologies have drawn increasing attention from industry and academia due to its high efficiency, low loss and simplification control strategy [2–6]. Compared with H Bridge-based converters, these bridgeless-based converters cannot work as inverters. However, considering that the applications are mostly pumps, fans and compressors which only need the power flowing unidirectionally [7], H-bridge-based converter has gradually been substituted by bridgeless-based converters as a pulse width modulation (PWM) rectifier in these fields. The bridgeless-based converter as a rectifier provides a sinusoidal input current at unity power factor and a controllable dc output voltage.

With the growing power switch numbers and power density, reliability of power electronic converters is increasingly important because the malfunctions are unacceptable and cause serious losses (e.g., nonscheduled downtime) in the critical applications. As a kind of electric energy conversion device, three-phase bridgeless converter (3-BLC) also endures high frequency voltage shock, over temperature impact, overload and improper driving signal. Semi-conductor devices, especially power switches, will fail more easily than other components. As discussed in Reference [8], the power switches contribute to 31% of failures, which are the most fragile components among capacitors, gate drivers, resistors and inductors. Power switches faults are usually caused by bond-wire lift-off or solder cracking, which will lead to an open circuit or short circuit of converters. The faults are named open circuit faults (OCF) and short circuit faults (SCF) respectively. An SCF will cause a large current and result in system shutdown, so hardware-based approaches such as fast fuses or breakers to transfer an SCF to an OCF are generally used. An OCF will not shutdown a system immediately but it degrades the performance inconspicuously. These may, in turn, cause secondary faults. Therefore, it is necessary to study a fault diagnosis and tolerance method for power switches OCF of the rectifier in this article.

In the past decade, numerous fault diagnosis and tolerance methods have been proposed for power electronic converters in the literature [9–16]. However, there are a few research works for ac-dc rectifiers, especially bridgeless rectifiers. For example, a diagnosis method based on a mixed logical dynamic model and residual generation was applied to a single-phase rectifier in a railway electrical traction drive system [17]. This method was fast, simple and stable, but it does not suit three-phase systems. For three-phase conditions, a current waveforms-based similarity analysis method for a three-phase PWM rectifier was proposed in Reference [18]; current waveforms were analyzed pairwise to diagnose an open circuit fault. There was a critical drawback of this method that it ignored the three-phase voltage imbalance, which is often seen for grid. In Reference [19], a fault tolerance control with additional devices for three-phase soft-switching mode rectifier is proposed. The circuit configuration had two extra center-tapped autotransformers and three more toggle switches compared with the traditional system. This method was more suitable for new design, but more retrofit cost was demanded for the existing systems. Considering the dc voltage decrease by OCF, Reference [20] proposed a fault tolerant method for the three-level rectifier in a wind turbine system, which was implemented by adding a compensation value to the reference voltages. The proposed method preserved the power factor under faulty conditions utilizing the redundancy of the switching devices, where the 3-BLC does not have such ability.

This study aims for an OCF diagnosis and tolerance method for bridgeless-based rectifiers [21–23], especially the 3-BLC. The contribution of this paper is to propose a fault feature extraction method for 3-BLC, and a fault tolerant method based on an extra two switches. The fault features were extracted from the three-phase currents. Load sudden change, source voltage imbalance or fluctuation and harmonic interference have been considered to prevent the impact on the proposed method. After that, the OCF is identified by a mixed logical model-based algorithm. When a fault was diagnosed, the drive signals of the faulty phase were redistributed artificially by the additional switches. Thus, it will maintain the current path in failure condition and make the 3-BLC still work as normal.

The rest of this paper is organized as follows. In Section 2, the mathematical model of single and three-phase bridgeless converters are analyzed. In Section 3, the fault diagnosis with an improvable feature extraction method is proposed. Section 4 details the fault tolerant implementation through additional devices. System validation using simulation and experiment data is provided in Section 5. Finally, conclusions are drawn in Section 6.

2. Basic Principles of Three-Phase Bridgeless Converters

2.1. Structure and Operation of 3-BLC

A three-phase bridgeless converter (3-BLC) is shown in Figure 1a. As can be seen, this three-phase converter is expanded from a single-phase dual-boost bridgeless structure, which has additional slow-recovery diodes D5 & D6 and two boost inductors L1 & L2 to reduce common mode noise [24,25]. Compared to the H bridge structure, bridgeless structure reduces 50% of fully controlled switches. Therefore, the control circuits, gate drivers, as well as protection units are greatly reduced, thus decreasing the system complexity and switching losses drastically [26]. The equivalent ac side circuit is depicted in Figure 1b and the mathematical model of 3-BLC can be expressed as

$$\begin{cases} L\frac{di_{sA}}{dt} = u_{sA} - u_{acA} - u_{NO} \\ L\frac{di_{sB}}{dt} = u_{sB} - u_{acB} - u_{NO} \\ L\frac{di_{sC}}{dt} = u_{sC} - u_{acC} - u_{NO} \end{cases} \tag{1}$$

where $u_{acA}, u_{acB}, u_{acC}$ are the ac voltages of phase A, B and C; u_{NO} is the neutral point voltage; L is the inductance of L1 and L2.

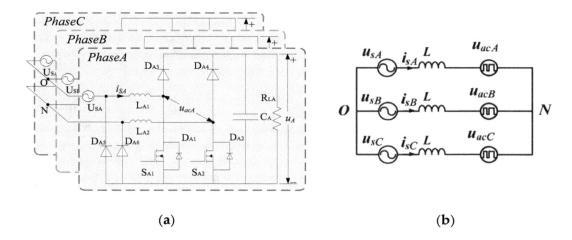

Figure 1. Three-phase bridgeless converter topology and equivalent ac side circuit. (**a**) Circuit topology; (**b**) the equivalent ac side circuit.

In order to clarify the analysis process, in this study, the slow-recovery diodes are temporarily substituted by the body diodes and the boost inductors are equivalent to an inductor L. Therefore, taking a single-phase bridgeless rectifier as an example, Figure 2 shows its principles on four operating modes. In this study, the power switches S1 and S2 are synchronously turned ON and OFF, which is named synchronous control scheme. Define S as a switch function of this converter. When $S = 1$, S1 and S2 are both turned ON, the power is transferred to power storage inductor L as shown in Figure 2a,c. When $S = 0$, S1 and S2 are both turned OFF, the power stored in inductor L is transferred to the load side as shown in Figure 2b,d. Then, a bulky electrolytic capacitor C is employed to buffer the power and, hence, smooth the output voltage. Thus, the steady-state mathematic model can be yielded by applying KVL and KCL as following.

$$\begin{cases} L\frac{di_L}{dt} = u_s - (1 - S)u_C \\ C\frac{du_C}{dt} = (1 - S)i_L - \frac{u_C}{R_L} \end{cases} \tag{2}$$

where L is the inductance of the power storage inductor, C is the capacitance of the output capacitor, R_L is the resistance of load, i_L is the input current, u_C is the capacitor voltage, u_s is the ac voltage, and $S^* = 1 - S$.

Applying the volt-second balance and ampere-second balance principles to Equation (2) derives

$$\begin{cases} i_L = \frac{u_s}{R_L(1-d)^2} \\ u_C = \frac{u_s}{1-d} \end{cases} \tag{3}$$

where d is the duty cycle.

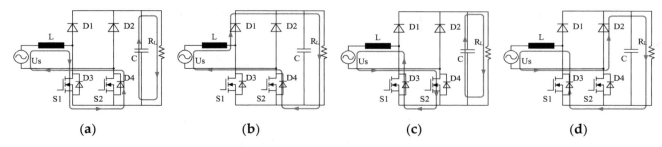

Figure 2. Basic bridgeless PFC rectifier topology and operating mode. (**a**) Mode 1 in positive ac cycle; (**b**) Mode 2 in positive ac cycle l; (**c**) Mode 3 in negative ac cycle; (**d**) Mode 4 in negative ac cycle.

2.2. Open Circuit Fault Analysis

As described above, there are four operating modes which generate through the ac voltage polarity and the switching states. When a switch fails due to physical damage or improperly driving signal, the corresponding operating mode no longer exists. This will cause changes on the related signals, which is called fault features, and is the theoretical foundation of fault detection and location. In the rest of this section, fault features of S1 OCF were analyzed as an example. Because power switches are the most fragile component and an SCF can be converted to an OCF by fast fuses immediately, the current and voltage waveforms shown in this section were acquired by simulations. The fault times were set in the positive and negative ac cycle of the input voltage respectively.

When an OCF occurs on switch S1, the power storage path as shown in Figure 2a is disconnected. Therefore, the converter works only in power discharging mode as shown in Figure 2b. However, in the negative ac cycle, the converter works properly because S1 is not in both a power charging and discharging path. The waveform of the input current and capacitor voltage during S1 open circuit fault at different half cycles are shown in Figure 3. Because it is a non-resonant circuit, the large impedance makes the input current fell into nearly zero after S1 failed in negative ac cycle, but it seems normal in positive ac cycles. The capacitor voltage resembles the input current which double frequency ripple disappears obviously after 0.5446 s, but it seems no change after 0.5346s until a positive ac cycle. The features of input current and capacitor voltage before and after switch S2 fails are just like S1.

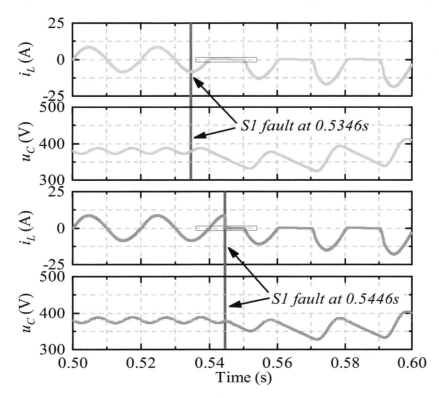

Figure 3. Input current and capacitor voltage waveforms at the time of S1 fault in positive and negative cycles.

The dynamic details of the input current inside the red frame of Figure 3 are also shown in Figure 4. According to Equation (3), the duty cycle of one switch will mutate to zero when it occurred an OCF. Since the switching period no longer existed, the assumption that the input voltage is constant during the switching cycle was no longer valid. Due to the impedance of the inductor to the low frequency signal, the input current became zero after the energy in the inductor was released. The load voltage was maintained by the DC capacitor until another ac cycle which was not affected by the fault.

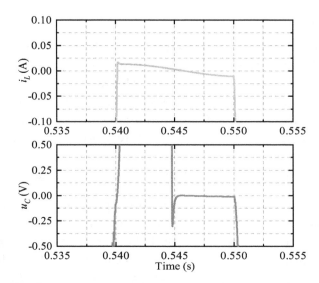

Figure 4. Dynamic details of input current and capacitor voltage waveforms at the time of S2 fault.

2.3. Input Side Interferences

Voltage fluctuations and harmonic pollution are unavoidable on the input side. They will affect the characteristics of current and voltage signals which have an impact on the diagnosis of the fault. Thus, this study discusses the features of input current and capacitor voltage under input voltage fluctuation and harmonic pollution, which are shown in Figure 5. About 7% harmonic was injected to input voltage at 0.3176 s, consequently, the THD of the input current increased from 4% to 8% but the impact on the capacitor voltage was limited. Still, from this figure, the input voltage fluctuated about 10% higher at 0.7273 s. The input current decreased a little and the capacitor voltage jittered rapidly at the same time. Different from the failure conditions, these varieties recovered in a short time. The features are significantly different from those failures discussed in the previous paragraphs.

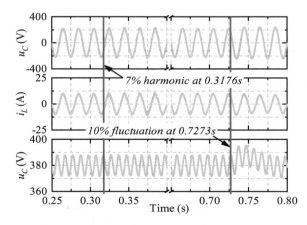

Figure 5. Input current and capacitor voltage waveforms on input voltage fluctuation and harmonic injection.

2.4. Load Side Interferences7

During the operation of the converters, the power on the load side was not always constant. The load sudden changed sometimes and the reference dc voltage also changed when transmitting different powers on the three-phase imbalance condition. This will lead to failure of fault diagnosis. In this study, output power variations due to reference voltage changed and load sudden change were also involved.

As shown in Figure 6, the reference output voltage increased from 300 V to 380 V at 0.5346 s, then the magnitude of the input current and capacitor voltage rose rapidly and became stable in two ac cycles. The output power saw a 60% growth from 400 W to 640 W. Still, in this figure, the load became heavier abruptly at the same time, which resulted in output power rising up to 1000 W with a 56% increment. The input current increased smoothly and rapidly, meanwhile the capacitor voltage decreased a little but restored fast with a higher ripple.

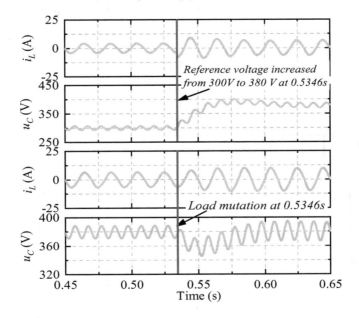

Figure 6. Input current and capacitor voltage during the time of load sudden change and reference voltage change.

In summary, an OCF will result in input current falling to about zero and increasing ripple of the capacitor voltage. The intensity of current oscillation depended on the values of L and C. The features of input or load side interferences were essentially similar, and did not affect the sinusoidal characteristic of the input current and capacitor voltage signals. It has to be noticed that in these analyses, when switch S1 failed in a negative ac cycle, the converter maintained regular operation until a positive ac cycle, and vice versa for switch S2. The duration of this condition lasted up to a half ac cycle, which is related to the time of failures.

3. Fault Diagnosis Technique

Fault diagnosis is the basis of fault tolerant control and consisted of fault detection and fault location [27]. It is important to detect and locate a malfunction switch rapidly, taking into account the fault features. In addition, the performance within fault tolerant behavior, such as redundancy control, is not only affected by fault diagnosing time directly, but also fault diagnosing accuracy including misdiagnosis and missed diagnosis [28]. Therefore, the fault diagnosis algorithm needs to be simple and effective.

3.1. Fault Features Extraction

As aforementioned, the sinusoidal characteristic of input current is damaged when an OCF occurs, but is reserved under other interferences. Therefore, it is feasible to select the input current as the characteristic signal of fault diagnosis. Generally, it is straightforward to utilize frequency domain characteristics as fault features [29]. However, most of the frequency domain methods require Fast Fourier Transform (FFT), which costs large amounts of computation. In order to improve the diagnostic efficiency and reduce the computational cost, this study used a direct time domain analysis method. Considering a three-phase converter, an $abc - \alpha\beta$ transformation, named Concordia transformation, is applied to analyze conveniently in a two-phase stationary coordinate system. This transformation can be expressed as

$$\begin{bmatrix} i_\alpha \\ i_\beta \end{bmatrix} = \begin{bmatrix} \sqrt{\frac{2}{3}} & -\frac{1}{\sqrt{6}} & -\frac{1}{\sqrt{6}} \\ 0 & \frac{1}{\sqrt{2}} & -\frac{1}{\sqrt{2}} \end{bmatrix} \begin{bmatrix} i_A \\ i_B \\ i_C \end{bmatrix} \tag{4}$$

Transformed input current signals are shown in Figure 7, which took i_α as the abscissa and i_β as the ordinate. The gradient of color represented the increase of time; therefore, the characteristics of i_α and i_β before and after failure were revealed in these figures. Single switch failures of each switch were shown in Figure 7, as well as normal and load side interference conditions. Input side interference was similar with the load side.

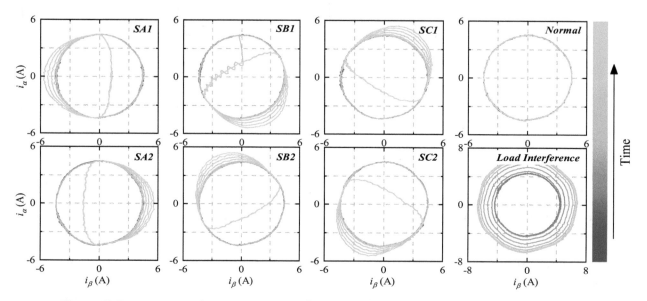

Figure 7. Input current features in a two-phase coordinate system under eight conditions.

3.2. Fault Detection and Location Method

According to the features shown in the above figure, i_α and i_β constitute a circular trajectory under normal and interference conditions. However, an OCF will change the trajectory and different fault location will result in different trajectories. In other words, vectors from $(0,0)$ to (i_α, i_β) contain distinguished features of OCFs. Define R as the length of this vector, an interval increment-based fault detection method was proposed and can be expressed as

$$S = \int_{t-t_s}^{t} |R_0 - R(t)| dt \qquad (5)$$
$$if\ S - S_{th} > res,\ then\ an\ OCF\ occurred$$

where S is the accumulative bias of R, S_{th} is a reference value for detecting an OCF, res is the threshold of residual for decision, and t_s is the length of the interval. $R_0 = \frac{1}{T} \int_{t-2T}^{t-T} R(t) dt$ is the reference of R which lags one ac cycle to increase the sensitivity, and T is the power frequency cycle.

Review the angle between this vector and the positive direction of the abscissa axis. It can be found that the abnormal R occurred at a specific angular interval corresponding to different fault locations. The central values of these intervals, defined as θ_{th}, are ideally taken as $\left(0, \pi, \frac{2\pi}{3}, \frac{5\pi}{3}, \frac{4\pi}{3}, \frac{\pi}{3}\right)$ corresponding to SA1, SA2, SB1, SB2, SC1, SC2 OCFs one by one. Then, a fault location method after fault detection was proposed based on this vector. It selected the minus central value between the angle when a fault was detected and θ_{th} as the location judgment, which can be expressed as

$$\min_i |\theta_{th}(i) - \theta_d| \quad \theta \in [0, 2\pi] \tag{6}$$

where $i = 1, 2, \cdots, 6$, which represents SA1, SA2, SB1, SB2, SC1, and SC2 respectively. θ_d is the angle when an OCF was detected.

The flow chart including fault detection and fault location algorithm are shown in Figure 8.

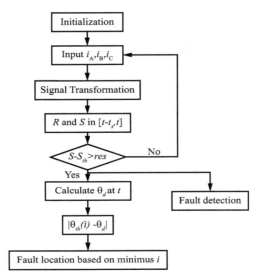

Figure 8. Fault detection and location algorithm flow chart based on current residual.

4. Fault Tolerance Method

An OCF will not cause the converter to shutdown immediately, but it slowly degrades the performance and reliability. Actually, a converter is desired to have the capability to operate in a quasi-normal condition in the post-failure period. Therefore, a fault tolerance method includes two aspects: fault tolerant topology with an extra two switches and a corresponding fault tolerant control method were proposed.

Also in the case of a single phase, the converter consists of two fast recovery diodes and two power switches with body diodes [25]. A current loop is broken due to an open circuit fault. Therefore, in order to obtain fault tolerant capability, additional devices must be added to restore the original current loop in the fault state. A fault tolerant topology with an additional two power switch which connected in parallel across the two fast recovery diodes is shown in Figure 9a. As aforementioned, more than two-thirds of ac-dc rectifiers in motor drives of industry only require a single direction for energy transmission. Therefore, considering the efficiency, loss, and control algorithms, this topology still operates as a bridgeless converter in the normal state, although it is similar to the H-bridge structure. The fault tolerant control diagram with drive signal distribution is shown in Figure 9b.

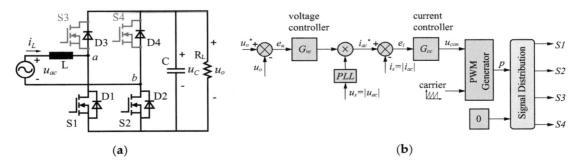

(a)   (b)

Figure 9. Fault tolerant control topology and control method. (a) Proposed fault tolerant topology with two additional switches; (b) block diagram of the fault tolerant control method with signal distribution.

The control method diagram includes two control loops: an inner current loop and an outer voltage loop. The role of the inner loop is to realize a unity power factor, and the role of the outer loop is to provide a controllable output DC voltage. The voltage controller G_{vc} and current controller G_{cc} are obtained as

$$G_{vc} = \frac{K_{Pv}s + K_{Iv}}{s}$$
$$G_{cc} = \frac{K_{Pi}s + K_{Ii}}{s} \tag{7}$$

The output of the double PI loop was compared with the carrier. Hence, in turn, the PWM signal was generated for signal distribution. At the same time, constant zero which means low level driving signal, was also generated.

The converter achieved fault tolerance through dynamic structure reconfiguration [30,31]. According to the location of the faulty device, there are three substructure types after an OCF. Each of them has a corresponding drive signal distribution rule to maintain the normal operation of the converter. The circuit reconfigurations are shown in Figure 10.

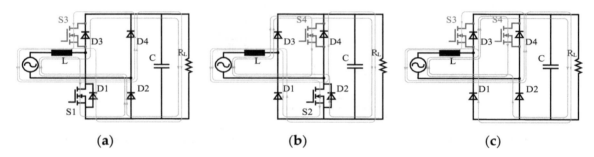

| (a) | (b) | (c) |

Figure 10. Circuit type reconfigurations. (**a**) Totem pole bridgeless; (**b**) Symmetry totem pole bridgeless; (**c**) Symmetry boost bridgeless.

In normal operation, both S3 and S4 are driven off by the low level, and S1 and S2 are driven by the PWM signal p in a synchronous drive mode. When it is detected that an open circuit fault occurs in S2, S3 will be driven by $-p$ which means S1 and S3 are driven by complementary signals. At this point, the circuit topology is converted into a totem pole bridgeless structure, as shown in Figure 10a. When an open circuit fault occurs in S1, the drive signal of S4 will also be replaced by $-p$, and S2 and S4 will continue to operate in a complementary signal drive mode. At this point, the circuit topology is converted to a symmetry totem pole bridgeless structure, as shown in Figure 10b. If an open circuit fault occurs in both S1 and S2, the circuit will be converted into a symmetry boost bridgeless structure consisting of S3 and S4, which will be driven by p synchronously. This situation is shown in Figure 10c. The current path for the different topologies is also shown in Figure 10, with orange representing the path for the positive ac cycle and blue for the negative. The drive signal distribution table corresponding to each fault state is shown in Table 1.

Table 1. Drive signal distribution of different fault switch.

Faulty Switch	Drive Signal			
	S1	**S2**	**S3**	**S4**
None	p	p	0	0
S1	/	p	0	$-p$
S2	p	/	$-p$	0
S1&S2	/	/	p	p

"0" respects low level, "p" respects the PWM signal, "/" respects an OCF.

5. Simulation and Experiment Results

It is essential to demonstrate the proposed fault diagnosis and tolerant control method function as expected in a real converter. However, failures of a real device will lead to uncontrollable consequences such as burning or explosion. Therefore, hardware in loop (HIL) simulation technology is suitable for device failure experiments. In this study, simulation and experiment-based MATLAB/Simulink and NI platform were realized, as can be seen in Figure 11a; and the details concerning the equipment used in this setup are shown in Figure 11b.

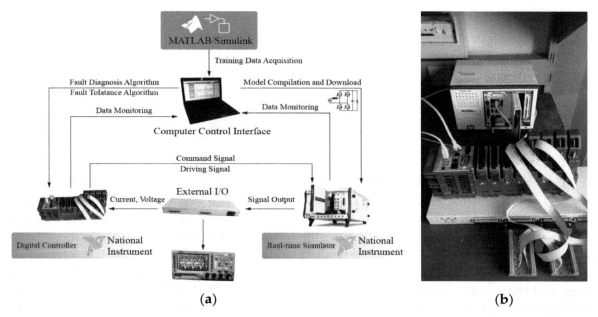

Figure 11. Diagram of simulation and experiment setup.

The simulation model of the three-phase bridgeless converter was built in MATLAB/Simulink and used to verify the proposed fault diagnosis and tolerant control method. Then HIL simulation based on NI-PXI platform was employed to emulate physical experiment, which is widely recognized and adopted in the field of power electronic device failure researches [32]. The power system model was set up as shown in Figure 1, and the parameters are presented in Table 2. The simulation and experiment results are revealed in two aspects: fault diagnosis results and fault tolerant control results. It is noticed that there was no real fault and all the OCFs were emulated by focusing the low level drive signal of specific switches.

Table 2. Specification of Simulation and Experiment.

Parameter	Value
Input ac voltage	220 V 50 Hz
Reference dc voltage	380 V
Boost inductor	5 mH
dc capacitor	330 μF
Normal power	2000 W
Switching frequency	10 kHz
Sampling frequency	10 kHz
K_{Pv}, K_{Iv}	0.021, 0.55
K_{Pi}, K_{Ii}	10.1, 200

For the diagnostic algorithm proposed in this paper, the selection of t_s has a direct impact on diagnostic resolution and diagnostic performance. Figure 12 shows the value of S when t_s is equal to 3 ms and 6 ms, plotted in red and green, respectively. Obviously, the diagnostic frequency (or resolution) of the proposed online real-time fault diagnosis is positively correlated with t_s. In the normal state, the value of S was about zero. When a fault occurred, S began to fluctuate greatly. The amplitude was also positively correlated with t_s. That is to say, a larger t_s meaning larger diagnostic interval or lower diagnostic frequency can make it easier to identify a fault trigger, reduce misdiagnosis or missed diagnosis.

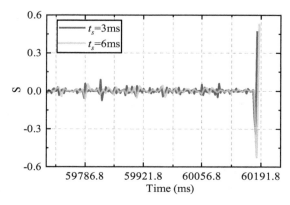

Figure 12. Plot of S with two different t_s until fault detection.

In order to make the method proposed in this study applicable with different system parameters, both i_α and i_β were normalized in the actual process. Furthermore, Figure 13 shows the results of 200 samples with different t_s from 2 to 6 ms, where the average detection time is connected in green and the average accuracy rates in red. In this figure, the influence of different S_{th} are also shown, in which the solid line represents $S_{th} = 0.2$ and the dashed dotted line represents $S_{th} = 0.4$. When $S_{th} = 0.2$, as can be seen, the fault detection time increased as t_s grew because it slowed down the diagnostic frequency, and the accuracy rates were maintained at around 98%. When $S_{th} = 0.4$, the detection time generally increased by about 20 ms, but there was greater decline in accuracy rates with t_s growing. The reason is that longer integration time makes the value of S closer to zero, and the larger threshold is gradually more unsuitable, resulting in a drop in the accuracy rate.

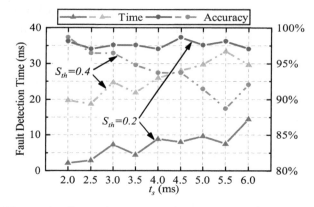

Figure 13. Detection time and accuracy with two S_{th}.

After selecting the appropriate parameters, the current signal i_A and corresponding trigger signals collected by the oscilloscope were as shown in Figure 14.

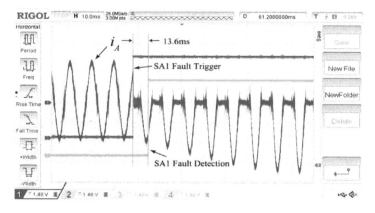

Figure 14. Input current and SA1 fault trigger and detection signals of phase A.

After a fault was detected, the fault location algorithm needed to be employed to locate the fault. Since the detected time t_d was not the actual time t_f at which a failure occurred. The θ_d calculated using $i_\alpha(t_d)$ and $i_\beta(t_d)$ was disorderly, and therefore, it was impossible to locate the failure in practice. In this study, this problem was solved by sacrificing a certain fault tolerant time that the value of θ_d was taking as the time minimizing R within a power cycle after t_d. It can be expressed as

$$\theta_d = \theta(t_{\min})$$
$$t_{\min} = \min_t R(t), t \in [t_d, t_d + T] \tag{8}$$

Figure 15 shows the scatter plots of $i_\alpha(t_d)$ and $i_\beta(t_d)$ for each of the 30 samples. The pink auxiliary line identifies the angles corresponding to each fault under ideal conditions. The data points generated by original θ_d were scattered and could not be used for fault location. The optimized data points were concentrated in the vicinity of the ideal values which could be used for accurate fault location.

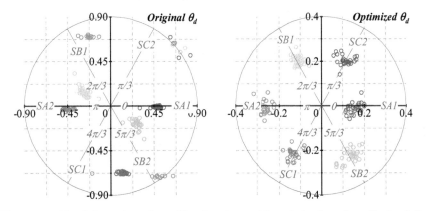

Figure 15. Fault location data distribution with original and optimized θ_d.

Under the condition of $t_s = 5$ ms, $S_{th} = 0.5$, 30 fault tolerance experiments were implemented where the faulty switch was randomly selected. The times from fault occurrence to fault tolerant control execution are shown in Figure 16. It can be found that the fault tolerance time increased by 20 ms relative to the detection time from Figure 13, where the delay of one ac cycle is caused by Equation (8).

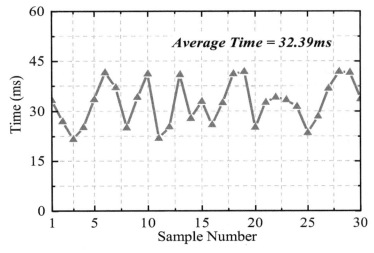

Figure 16. Fault tolerance times of thirty samples.

An example of fault diagnosis and tolerant control is shown in Figure 17. The signals were monitored by the host computer of the system built in this study.

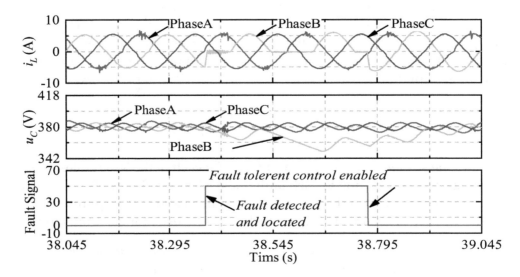

Figure 17. Signals collected by the host computer.

In summary, the proposed fault diagnosis and fault tolerant control method were verified in simulation and experimentation. An OCF will be detected within at least 20 ms in general conditions, and tolerated with a 20 ms delay. Various interferences have limited impact on this algorithm. The accuracy of fault diagnosis was over 98%, and the error rate of the fault location and fault tolerance was zero.

6. Conclusions

Reliability is one of the primary concerns for the three-phase bridgeless converter. This paper presented an open circuit fault diagnosis and tolerant control method to maintain the converter running. As the basis for fault tolerance, the open circuit fault is detected and located accurately within a few milliseconds, thus, the abnormal operation time is reduced. Only two additional switches are needed to maintain the normal operation by structure reconfiguration. A lookup table is built for switch drive signals reconfiguration. Finally, the feasibility and effect of the proposed method on reliability promotion was verified by simulations and experiments. Furthermore, the interference analysis in this paper is still insufficient that the proposed method is not robust enough in practice. This requires further work.

Author Contributions: W.C., H.C. and C.W. conceptualized the main idea of this project; W.C. proposed the methods and designed the work; W.C. conducted the experiments and analyzed the data; J.D. checked the results; W.C. wrote the whole paper; and H.C., C.W., and J.D. reviewed and edited the paper.

References

1. Blahnik, V.; Kosan, T.; Peroutka, Z.; Talla, J. Control of a Single-Phase Cascaded H-Bridge Active Rectifier under Unbalanced Load. *IEEE Trans. Power Electron.* **2018**, *33*, 5519–5527.
2. Wang, C.; Zhuang, Y.; Jiao, J.; Zhang, H.; Wang, C.; Cheng, H. Topologies and Control Strategies of Cascaded Bridgeless Multilevel Rectifiers. *IEEE J. Emerg. Sel. Top. Power Electron.* **2017**, *5*, 432–444. [CrossRef]
3. Kremes, W.D.J.; Font, C.H.I. Proposal of a three-phase bridgeless PFC SEPIC rectifier with MPPT for small wind energy systems. In Proceedings of the IEEE International Conference on Industry Applications, Curitiba, Brazil, 20–23 November 2016; pp. 1–8.
4. Silva, C.E.A.; Oliveira, D.S.; Barreto, L.H.S.C.; Bascopé, R.P.T. A novel three-phase rectifier with high power factor for wind energy conversion systems. In Proceedings of the Power Electronics Conference (COBEP '09), Bonito-Mato Grosso do Sul, Brazil, 27 September–1 October 2009; pp. 985–992.
5. Park, S.M.; Park, S. Versatile Control of Unidirectional AC–DC Boost Converters for Power Quality Mitigation. *IEEE Trans. Power Electron.* **2015**, *30*, 4738–4749. [CrossRef]

6. Whitaker, B.; Barkley, A.; Cole, Z.; Passmore, B.; Martin, D.; McNutt, T.R.; Lostetter, A.B.; Lee, J.S.; Shiozaki, K. A High-Density, High-Efficiency, Isolated On-Board Vehicle Battery Charger Utilizing Silicon Carbide Power Devices. *IEEE Trans. Power Electron.* **2014**, *29*, 2606–2617. [CrossRef]

7. Malinowski, M.; Gopakumar, K.; Rodriguez, J.; Pérez, M.A. A Survey on Cascaded Multilevel Inverters. *IEEE Trans. Ind. Electron.* **2010**, *57*, 2197–2206. [CrossRef]

8. Yang, S.; Bryant, A.; Mawby, P.; Xiang, D. An Industry-Based Survey of Reliability in Power Electronic Converters. *IEEE Trans. Ind. Appl.* **2011**, *47*, 1441–1451. [CrossRef]

9. Yu, Y.; Zhao, Y.; Wang, B.; Huang, X.; Xu, D. Current sensor fault diagnosis and tolerant control for VSI-based induction motor drives. *IEEE Trans. Power Electron.* **2018**, *33*, 4238–4248. [CrossRef]

10. Choi, U.; Lee, J.S.; Blaabjerg, F.; Lee, K.B. Open-Circuit Fault Diagnosis and Fault-Tolerant Control for a Grid-Connected NPC Inverter. *IEEE Trans. Power Electron.* **2016**, *31*, 7234–7247. [CrossRef]

11. Wang, T.; Xu, H.; Han, J.; Bouchikhi, E. Cascaded H-Bridge Multilevel Inverter System Fault Diagnosis Using a PCA and Multi-class Relevance Vector Machine Approach. *IEEE Trans. Power Electron.* **2015**, *30*, 7006–7018. [CrossRef]

12. Freire, N.; Estima, J.O.; Cardoso, A. A Voltage-Based Approach without Extra Hardware for Open-Circuit Fault Diagnosis in Closed-Loop PWM AC Regenerative Drives. *IEEE Trans. Ind. Electron.* **2014**, *61*, 4960–4970. [CrossRef]

13. Freire, N.; Estima, J.O.; Cardoso, A. Open-Circuit Fault Diagnosis in PMSG Drives for Wind Turbine Applications. *IEEE Trans. Ind. Electron.* **2013**, *60*, 3957–3967. [CrossRef]

14. Estima, J.O.; Cardoso, A.J.M. A New Approach for Real-Time Multiple Open-Circuit Fault Diagnosis in Voltage-Source Inverters. *IEEE Trans. Ind. Appl.* **2011**, *47*, 2487–2494. [CrossRef]

15. Zidani, F.; Diallo, D.; Benbouzid, M.E.H.; Nait-Said, R. A fuzzy-based approach for the diagnosis of fault modes in a voltage-fed PWM inverter induction motor drive. *IEEE Trans. Ind. Electron.* **2008**, *55*, 586–593. [CrossRef]

16. Aly, M.; Ahmed, E.; Shoyama, M. A New Single Phase Five-Level Inverter Topology for Single and Multiple Switches Fault Tolerance. *IEEE Trans. Power Electron.* **2018**, *33*, 9198–9208. [CrossRef]

17. Gou, B.; Ge, X.; Wang, S.; Feng, X. An Open-Switch Fault Diagnosis Method for Single-Phase PWM Rectifier using a Model-based Approach in High-Speed Railway Electrical Traction Drive System. *IEEE Trans. Power Electron.* **2016**, *31*, 3816–3826. [CrossRef]

18. Feng, W.; Jin, Z. Current Similarity Analysis Based Open-Circuit Fault Diagnosis for Two-Level Three-Phase PWM Rectifier. *IEEE Trans. Power Electron.* **2016**, *32*, 3935–3945.

19. Chao, K.; Hsieh, C. Mathematical Modeling and Fault Tolerance Control for a Three-Phase Soft-Switching Mode Rectifier. *Math. Probl. Eng.* **2013**, *2013*, 598130. [CrossRef]

20. Lee, J.S.; Lee, K.B. Open-Circuit Fault-Tolerant Control for Outer Switches of Three-Level Rectifiers in Wind Turbine Systems. *IEEE Trans. Power Electron.* **2016**, *31*, 3806–3815. [CrossRef]

21. Wang, C.; Zhuang, Y.; Kong, J.; Tian, C.; Cheng, H. Revised topology and control strategy of three-phase cascaded bridgeless rectifier. In Proceedings of the IEEE Industrial Electronics Society, Beijing, China, 29 October–1 November 2017; pp. 1499–1504.

22. Zhang, Y.; Kotecha, R.; Mantooth, H.A.; Balda, J.C.; Zhao, Y.; Farnell, C. Cascaded bridgeless totem pole multilevel converter with model predictive control for 400 V dc-powered data centers. In Proceedings of the Applied Power Electronics Conference and Exposition, Tampa, FL, USA, 26–30 March 2017; pp. 2745–2750.

23. Jiao, J.; Wang, C. The research of new cascade bridgeless multi-level rectifier. In Proceedings of the Power Electronics and Application Conference and Exposition, Shanghai, China, 5–8 November 2014; pp. 1513–1518.

24. Gopinath, M.; Prabakaran; Ramareddy, S. A brief analysis on bridgeless boost PFC converter. In Proceedings of the International Conference on Sustainable Energy and Intelligent Systems, Chennai, India, 20–22 July 2011; pp. 242–246.

25. Kolar, J.W.; Friedli, T. The Essence of Three-Phase PFC Rectifier Systems. In Proceedings of the IEEE Conference of Telecommunications Energy, Amsterdam, The Netherlands, 9–13 October 2011; pp. 179–198.

26. Wang, C.; Jiao, J.; Wang, C.; Jiang, X.; Cheng, H. Research on New Multi-level Cascaded Bridgeless Rectifier. *J. Power Supply* **2015**, *13*, 10–16.

27. Gao, Z.; Cecati, C.; Ding, S.X. A Survey of Fault Diagnosis and Fault-Tolerant Techniques-Part I: Fault Diagnosis with Model-Based and Signal-Based Approaches. *IEEE Trans. Ind. Electron.* **2015**, *62*, 3757–3767. [CrossRef]

28. Lezana, P.; Pou, J.; Meynard, T.A.; Rodriguez, J.; Ceballos, S.; Richardeau, F. Survey on Fault Operation on Multilevel Inverters. *IEEE Trans. Ind. Electron.* **2010**, *57*, 2207–2218. [CrossRef]

29. Qiao, W.; Lu, D. A Survey on Wind Turbine Condition Monitoring and Fault Diagnosis—Part II: Signals and Signal Processing Methods. *IEEE Trans. Ind. Electron.* **2015**, *62*, 6546–6557. [CrossRef]

30. Rahmani-Andebili, M.; Fotuhi-Firuzabad, M. An Adaptive Approach for PEVs Charging Management and Reconfiguration of Electrical Distribution System Penetrated by Renewables. *IEEE Trans. Ind. Inf.* **2018**, *14*, 2001–2010. [CrossRef]

31. Rahmani-Andebili, M. Dynamic and adaptive reconfiguration of electrical distribution system including renewables applying stochastic model predictive control. *IET Gener. Transm. Distrib.* **2017**, *11*, 3912–3921. [CrossRef]

32. Jamshidpour, E.; Poure, P.; Gholipour, E.; Saadate, S. Single-switch DC-DC converter with fault-tolerant capability under open-and short-circuit switch failures. *IEEE Trans. Power Electron.* **2015**, *30*, 2703–2712. [CrossRef]

8

Performance Evaluation of a Semi-Dual-Active-Bridge with PPWM Plus SPS Control

Ming Lu and Xiaodong Li *

Faculty of Information Technology, Macau University of Science and Technology, Taipa, Macau 999078, China; 1409853pii30001@student.must.edu.mo
* Correspondence: xdli@must.edu.mo

Abstract: In this paper, a semi-dual-active-bridge (S-DAB) DC/DC converter with primary pulse-width modulation plus secondary phase-shifted (PPWM + SPS) control for boost conversion is analyzed in detail. Under the new control scheme, all effective operation modes are identified at first. Then, the working principle, switching behaviour, and operation range in each mode are discussed. Compared with conventional secondary phase-shifted control, PPWM + SPS control with two controllable phase-shift angles can extend the zero-voltage switching (ZVS) range and enhance control flexibility. In addition, an effective control route is also given that can make the converter achieve at the global minimum root-mean-square (RMS) current across the whole power range and avoid the voltage ringing on the transformer secondary-side at a light load. Finally, a 200 W prototype circuit is built and tested to verify correctness and effectiveness of theoretical results.

Keywords: DC–DC conversion; zero-voltage switching (ZVS)

1. Introduction

With the development of modern technology, there is a constant rise in energy use. Therefore, the concerns regarding the availability of fossil energy and the associated pollution in the mining and consumption process is continuously growing too. In order to alleviate the energy crisis and environmental pollution, the use of renewable energy (solar energy, wind energy, etc.) has developed rapidly around the world. As an important component for the application of renewable energy, the DC/DC converter with higher performance has been one of the most popular research fields [1–5]. So far, a number of DC/DC converter topologies have been proposed according to the various application requirements. In these converters, the phase-shift full-bridge converter is more attractive due to high power density, electrical isolation, easy to realize soft-switching commutation, high efficiency and low electromagnetic interference (EMI) [6–13]. However, it still suffers from high voltage ringing, reverse recovery on the secondary-side rectifier diodes, limited zero-voltage switching (ZVS) range and duty cycle loss.

To extend the ZVS range, a series of the full-bridge converters with various resonant tanks are presented. Among them, the converters with LC or LLC resonant tank are more attractive [14–19]. Nevertheless, the parameters of resonant tank should be selected carefully to achieve higher performance. Meanwhile, the design of magnetic components becomes complicated. On the other hand, when the phase-shift full-bridge converter works at high-output voltage and high-power case, the reverse-recovery problem of the rectifier diodes becomes more serious. In order to solve this problem, two active switches are introduced into the secondary-side rectifier of the converter, which is named the semi-dual-active-bridge (S-DAB) converter [20–22]. On this basis, two modified S-DAB topologies in [23,24] are proposed to only reduce the voltage stress on primary-side and secondary-side semiconductor devices, respectively. Furthermore, the S-DAB converter with an LC resonant tank is also presented in [25,26]. However, when the S-DAB converters work in a discontinuous-current mode (DCM) for boost operation, the voltage ringing phenomenon is generated on the transformer secondary-side. In particular, at the high switching frequency, high power and voltage levels, the excessive ringing might result in strong EMI, distorted gating signals and abnormal high peak voltages across the switches. In addition, the amount of power loss from snubber/parasitic capacitor will also increase during the ringing process [27,28]. Although a customized RC snubber is helpful to alleviate this problem, extra loss will be introduced in continuous-current mode (CCM) operation and the overall efficiency would be lower.

To the authors' best knowledge, the voltage ringing problem in S-DAB has yet to be resolved. In this paper, PPWM + SPS control is applied on an S-DAB converter for boost operation to further improve performance, which also avoids the voltage ringing problem. The rest of this paper is organized as follows: in Section 2, each steady-sate mode of an S-DAB converter with PPWM + SPS control is analyzed comprehensively, including working principle, switching behaviour, and operation range. In Section 3, an effective control route across the whole power range is presented in order to achieve minimum root-mean-square (RMS) current and no voltage ringing. Experimental results are provided in Section 4. Conclusions are drawn in Section 5.

2. Operation Principle of an S-DAB Converter with PPWM + SPS Control

2.1. Basic Operation Principle

The schematic of an S-DAB converter is shown in Figure 1. The primary H-bridge consists of four switches ($M_1 - M_4$), while the secondary H-bridge is realized by a diode leg (D_{S1} and D_{S2}) and a switch leg (M_5 and M_6). The high frequency transformer T with a turns ratio of $n_t : 1$ not only provides galvanic isolation, but also matches voltage level. The voltage gain is defined as $M = n_t V_o / V_{in}$, and $M > 1$ refers to boost operation. The inductor L_s includes the leakage inductance of the transformer and an external inductance. The filter capacitor C_o is connected in parallel with the load R_{load} to depress the output voltage ripple. In this paper, PPWM + SPS control with two phase shift angles is employed on an S-DAB converter. All switches operate at the same frequency f_s with 50% duty cycle, and switches in each switch leg are turned on/off complementarily. α is defined as the inner phase-shift by which the gating signal of M_4 lags that of M_1. Similarly, ϕ is defined as the outer-phase-shift by which the gating signal of M_6 lags that of M_1. Two pulse-width-modulated voltages v_{AB} and v_{CD} are generated by the two bridges, respectively. The pulse-width of v_{AB} is determined by α solely. However, the waveform of v_{CD} is associated with not only phase-shift angles but also the load level.

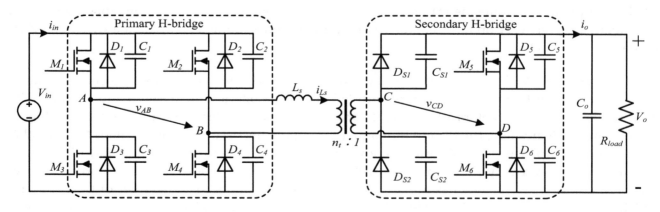

Figure 1. The circuit configuration of a S-DAB converter.

Depending on the relationship between two phase-shifts, an S-DAB converter with PPWM + SPS control can operate in three steady-state working modes, including one CCM (Mode A) and two DCMs, (Mode B and C). In the following part, each mode will be analyzed in detail one by one. In order to simplify the analysis process, four assumptions are made as follows:

1. All components, such as switches, diodes are ideal and lossless.
2. The magnetizing inductance of the transformer is infinity.
3. The snubber/parasitic capacitors and dead-times influence are neglected.
4. The filter capacitor is large enough to maintain constant voltage on the load.

2.2. Steady-State Analysis of Continuous-Current Mode

The ideal steady-state waveforms in Mode A are shown in Figure 2, where β denotes the first zero-crossing points referred to the turn-on moment of M_1. It can be seen that Mode A is featured with $\alpha < \beta < \phi < \pi$; and there are eight different intervals in one switching period. The corresponding equivalent circuits of the first four intervals are presented in Figure 3, respectively. The other four intervals are almost the same except for the directions of voltage/current and involved conducting devices.

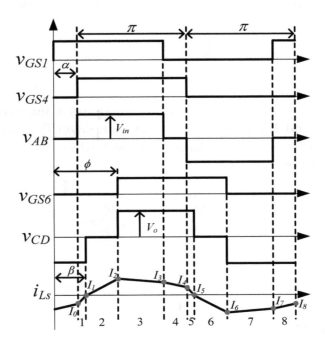

Figure 2. Steady-state waveforms in Mode A.

Interval 1 [Figure 3a]: At the beginning, M_2 is turned off and M_4 is turned on with ZVS. In this interval, the conducting devices are M_1 and M_4, M_5 and D_{s2}. Thus, the voltage across the inductor is clamped at $(V_{in} + n_t V_o)$, the value of i_{Ls} decreases linearly from the negative value I_0 to I_1. The power stored in the inductor is delivered to input DC power and load during this interval:

$$I_1 = I_0 + \frac{V_{in} + n_t V_o}{2\pi f_s Ls}(\beta - \alpha) = 0. \tag{1}$$

Interval 2 [Figure 3b]: At $\beta - \alpha$, the polarity of i_{Ls} is changed, the current flowing diode-leg is shifted naturally from D_{s2} to D_{s1}, i.e., D_{s2} is turned off with zero current. The secondary-side of transformer is shorted now by M_5 and D_{s1}. Meanwhile, the primary current flows from V_{in} to L_s through primary switches M_1 and M_4. Thus, the voltage across the inductor is clamped at V_{in}, the value of i_{Ls} increases linearly from I_1 to the positive maximum I_2. The power is being stored in the inductor L_s during this interval:

$$I_2 = I_1 + \frac{V_{in}}{2\pi f_s L_s}(\phi - \beta). \tag{2}$$

Interval 3 [Figure 3c]: At $(\phi - \alpha)$, M_5 is turned off and M_6 is turned on with ZVS. In this interval, the situation on the primary side does not change; the secondary current is shifted to D_{s2} and M_6, flowing to the load. Thus, during this interval, the voltage across inductor is clamped at $(V_{in} - n_t V_o)$, and the power is transmitted to the load. Due to $(V_{in} < n_t V_o)$, the value of i_{Ls} starts to decrease linearly from I_2 to the positive value I_3:

$$I_3 = I_2 + \frac{V_{in} - n_t V_o}{2\pi f_s L_s}(\pi - \phi). \tag{3}$$

Interval 4 [Figure 3d]: At $(\pi - \alpha)$, M_1 is turned off and M_3 is turned on with ZVS. In this interval, the primary side is shorted by M_3 and M_4; and no change happens on the secondary side. Thus, the voltage across inductor is clamped at $-n_t V_o$. During this interval, the value of i_{Ls} starts to decrease linearly until it reaches $-I_0$. The power stored in the inductor is delivered to load:

$$I_4 = I_3 - \frac{n_t V_o}{2\pi f_s L_s}\alpha = -I_0. \tag{4}$$

Based on Equations (1)–(4), the instantaneous current values at the moments of transition can be calculated as functions of α and ϕ. Furthermore, the output power P_o and the inductor RMS current $I_{Ls,rms}$ can be obtained too. These results are listed in Table 1, where the current and power values are normalized by the following base values:

$$I_b = \frac{V_{in}}{2\pi f_s L_s}, P_b = \frac{V_{in}^2}{2\pi f_s L_s}. \tag{5}$$

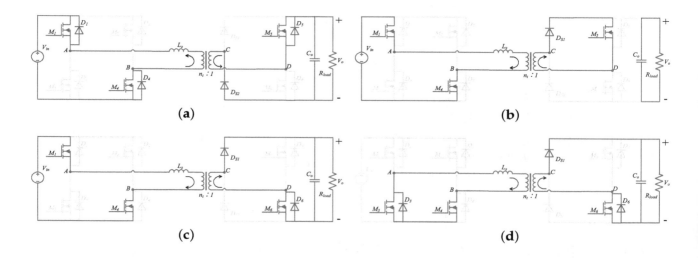

(a)   (b)

(c)   (d)

Figure 3. Equivalent circuits corresponding to the first four intervals in Mode A: (a) Interval 1 $[0, \beta - \alpha]$; (b) Interval 2 $[\beta - \alpha, \phi - \alpha]$; (c) Interval 3 $[\phi - \alpha, \pi - \alpha]$; (d) Interval 4 $[\pi - \alpha, \pi]$.

Table 1. Theoretical values of inductor current and output power in Mode A.

Value	Expression
$I_{0,pu}$	$\dfrac{(1+M)(\alpha-\pi+\alpha M-\phi M+\pi M)}{2+M}$
$I_{1,pu}$	0
$I_{2,pu}$	$\pi - \dfrac{\alpha-2\phi+3\pi}{2+M}$
$I_{3,pu}$	$\pi + (\pi-\phi)(1-M) - \dfrac{\alpha-2\phi+3\pi}{2+M}$
β	$\dfrac{\pi+\alpha+\phi M-\pi M}{2+M}$
$P_{o,pu}$	$\dfrac{M}{2\pi(2+M)^2}\left(\begin{array}{l}2\alpha\phi M^2+4\pi\phi M^2+4\alpha\phi M-\alpha^2 M^2-2\pi\alpha M^2-2\phi^2 M^2-2\pi^2 M^2\\+4\pi\phi M+\pi^2 M-4\phi^2 M-3\alpha^2 M-2\pi\alpha M\\+\pi^2+4\alpha\phi+4\pi\phi-3\alpha^2-2\pi\alpha-4\phi^2\end{array}\right)$
$I_{Ls,rms,pu}$	$\dfrac{1}{\sqrt{3\pi(M+2)^2}}\sqrt{\begin{array}{l}M^3\alpha^3-3M^3\alpha^2\phi+3M^3\alpha^2\pi+3M^3\alpha\phi^2-6M^3\alpha\phi\pi\\+3M^3\alpha\pi^2-2M^3\phi^3+6M^3\phi^2\pi-6M^3\phi\pi^2+2M^3\pi^3\\+3M^2\alpha^3-6M^2\alpha^2\phi+3M^2\alpha^2\pi+3M^2\alpha\phi^2-3M^2\alpha\pi^2\\-2M^2\phi^3+6M^2\phi\pi^2-3M^2\pi^3+4M\alpha^3-6M\alpha^2\phi+\\6M\alpha\phi^2-6M\alpha\phi\pi-4M\phi^3+6M\phi^2\pi+2\alpha^3-3\alpha^2\pi+\pi^3\end{array}}$

2.3. Steady-State Analysis of Discontinuous-Current Mode

Different from CCM, the inductor current i_{Ls} in DCM remains at zero for a small duration in each switching period. Steady-state waveforms of two DCMs are shown in Figure 4, where γ denotes the second zero-crossing points referring to the turn-on moment of M_1. It can be found that the difference between those two DCMs can be concluded as: $\alpha < \phi < \pi < \gamma$ for Mode B and $\alpha < \phi < \gamma < \pi$ for Mode C.

Mode B [Figure 4a]

According to the steady-state waveforms in Mode B, the equivalent circuits in the first four intervals are shown in Figure 5. It can be seen that the first three intervals in Mode B are almost the same as Intervals 2–4 in Mode A, except that the inductor current at the end of Interval 3 (Mode B) can arrive again at zero.

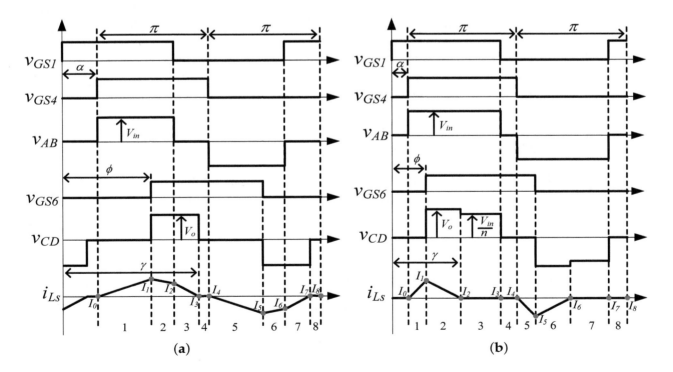

Figure 4. Steady-state waveforms in: (**a**) Mode B; (**b**) Mode C.

Interval 4 [Figure 5d]: After $\gamma - \alpha$, all secondary diodes are reversed biased, which will result in the secondary-side of transformer being open-circuited. Meanwhile, the primary switches M_3 and M_4 are still conducting. Thus, the transformer secondary voltage is clamped at 0, and i_{Ls} is kept at zero. This interval ends up with M_2 turned on at zero current. In this zero-current interval, there is no power transferring in the converter.

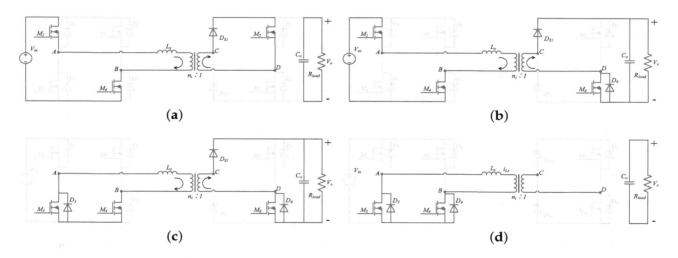

(a)   **(b)**

(c)   **(d)**

Figure 5. Equivalent circuits corresponding to the first four intervals in Mode B: **(a)** Interval 1 $[0, \phi - \alpha]$; **(b)** Interval 2 $[\phi - \alpha, \pi - \alpha]$; **(c)** Interval 3 $[\pi - \alpha, \gamma - \alpha]$; **(d)** Interval 4 $[\gamma - \alpha, \pi]$.

Mode C [Figure 4b]

Similarly, the equivalent circuits corresponding to first four intervals in Mode C are shown in Figure 6, respectively. It can be seen that the first second intervals in Mode C are almost the same as Intervals 2 and 3 in Mode A, except that the inductor current at end of Interval 2 (Mode C) can arrive again at zero.

Interval 3 [Figure 6c] in Mode C is different from those aforementioned intervals. Although the input DC source V_{in} is applied on the primary-side of transformer by the switches M_1 and M_4, there is no flowing current in the converter. The secondary-side of transformer is open-circuited since all secondary diodes are reversed biased. Thus, the secondary-side transformer voltage is clamped at $\frac{V_{in}}{n}$. Interval 3 ends up with M_3 turned on at zero current. This interval also belongs to the zero-current interval, and there is no power transferring. Interval 4 is the same as Interval 4 in Mode B.

Based on the steady-state analysis in each DCM, the instantaneous current values at the moment of transition can be calculated. Similarly, the output power and RMS current across inductor can be also obtained. These theoretical results are listed in Table 2.

Table 2. Theoretical values of inductor current and output power in DCM.

Value	Mode B	Mode C
$I_{0,pu}$	0	0
$I_{1,pu}$	$\phi - \alpha$	$\phi - \alpha$
$I_{2,pu}$	$\pi - \alpha + \phi M - \pi M$	0
$I_{3,pu}$	0	0
γ	$\frac{\pi - \alpha + \phi M}{M}$	$\frac{M\phi - \alpha}{M - 1}$
$P_{o,pu}$	$\frac{\alpha^2 + \pi^2 + 2\pi\phi M - 2\pi\alpha - \phi^2 M - \pi^2 M}{2\pi}$	$\frac{M(\phi - \alpha)^2}{2\pi(M-1)}$
$I_{Ls,rms,pu}$	$\frac{1}{\sqrt{3\pi M}}\sqrt{\begin{array}{c}3M^2\phi^2\pi - M^2\phi^3 - M^2\phi\pi^2 + M^2\pi^3 - M\alpha^3 \\ +3M\alpha^2\phi - 6M\alpha\phi\pi + 3M\alpha\pi^2 + 3M\phi\pi^2 - \\ 2M\pi^3 - \alpha^3 + 3\alpha^2\pi - 3\alpha\pi^2 + \pi^3\end{array}}$	$\sqrt{\frac{M(\phi - \alpha)^3}{3\pi(M-1)}}$

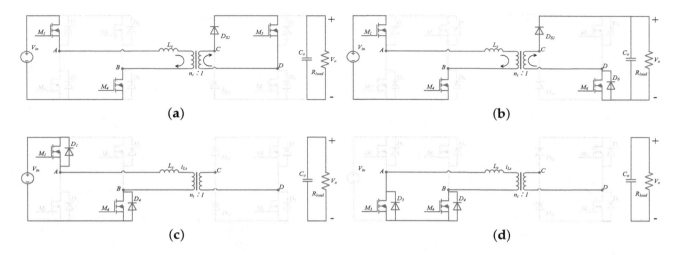

Figure 6. Equivalent circuits corresponding to the first four intervals in Mode C: (**a**) Interval 1 $[0, \phi - \alpha]$; (**b**) Interval 2 $[\phi - \alpha, \gamma - \alpha]$; (**c**) Interval 3 $[\gamma - \alpha, \pi - \alpha]$; (**d**) Interval 4 $[\pi - \alpha, \pi]$.

2.4. Switching Behaviour

Since the converter may work in three different steady-state modes, the switching behaviour of all switches and diodes also vary with the operation modes. According to the current polarity at switching moment of all switches and diodes, the switching behaviour in each mode are concluded in Table 3. First of all, the diodes in the secondary H-bridge can be turned on/off at zero current in any mode. In Mode A, all switches operate with ZVS and each diode is turned on/off with zero-current. Compared with Mode A, the switching loss in two DCMs is slightly increased due to the partial switch losing ZVS, and the switching loss in Mode C is higher than those in Mode B. Thus, Mode A should be selected as the main operation mode.

Table 3. Switching behavior in different modes.

Mode	M_1, M_3	M_2, M_4	M_5, M_6	D_{S1}, D_{S2}
A	ZVS	ZVS	ZVS	zero-current-on/off
B	ZVS	zero-current-on/off	ZVS	zero-current-on/off
C	zero-current-on/off	zero-current-on/off	ZVS	zero-current-on/off

2.5. Operation Range of Each Mode

Through comparing three steady-state modes, it can be found that Mode B is an in-between mode, and there are two boundary conditions existing between Mode B and the other two modes. Thus, the operating range of two controllable phase-shifts α and ϕ will be different in each mode. Knowing these conditions and range is helpful for the design of the converter.

When $\gamma = \pi$ in Mode B, the converter works at the boundary condition (6) between Mode A and B. At this boundary, the secondary H-bridge works in synchronous rectification mode:

$$\phi = \frac{\alpha + \alpha M + \pi M - \pi}{M}. \tag{6}$$

When $\gamma = \pi - \alpha$ in Mode B, it works at another boundary condition (7) between those two DCMs, in which the second zero-crossing happens at the moment of the switch M_3 being turned on:

$$\phi = \frac{\alpha + \pi M - \pi}{M}. \tag{7}$$

According to both boundary conditions, the operating range of each mode is shown in Figure 7 for $M = 1.5$ and $M = 2$, respectively.

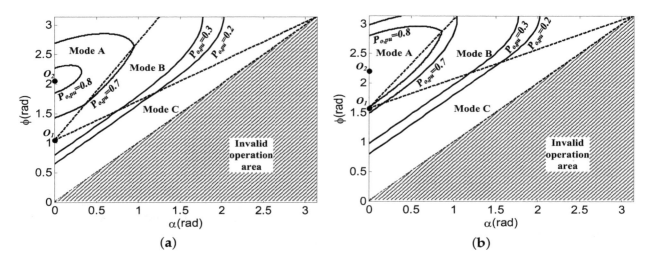

Figure 7. Operating range of each boost mode with suspended power contours for $M = 1.5$ and $M = 2$: (a) $M = 1.5$; (b) $M = 2$.

In Figure 7, the shaded area $\alpha \geq \phi$ represents an invalid operation area. The boundary of the neighboring modes is plotted by two dashed lines using (6) and (7), respectively, and these two boundary lines intersect at the point O_1. Meanwhile, the power contours with the same normalized power $P_{o,pu}$ value are shown by solid curves, in which the maximum load capacity of the converter is located at the point O_2, $P_{o,pu,max} = \frac{\pi M(M+1)}{2(M^2+2M+2)}$. The operating range with the conventional secondary phase-shifted control is only along the ϕ-axis. Compared with conventional control, PPWM + SPS can expand regulating range of output power and enhance flexibility of phase-shift control.

3. Proposed Control Route of an S-DAB Converter with PPWM + SPS Control

It is obvious that countless control routes exist from full power at O_2 to zero power at $\phi = \alpha$ for an S-DAB converter with PPWM + SPS control in Figure 7. Therefore, in order to select a reasonable control route, theoretical analysis of the inductor current is carried out to achieve lower conducting loss.

Based on Tables 1 and 2, the relationship between normalized inductor RMS current $I_{Ls,rms,pu}$ and phase-shift α at different power contours are shown in Figure 8, with $M = 1.5$ as an example. It can be found that, at the high power levels, the converter may work in Mode A. $I_{Ls,rms,pu}$ values can arrive at the minimum values when the converter is operated at conventional secondary phase-shifted control from O_2 to O_1 along the ϕ-axis. At the low power levels from O_1 to zero power O_0, the converter may operate in Mode C, in which $I_{Ls,rms,pu}$ is minimum and constant for the same power level.

In practical application, each switch and diode has its own snubber/parasitic capacitor. It is possible to get the voltage ringing on the transformer secondary-side when the converter is working in a zero-current interval. In Mode C, there are two zero-current Intervals (3 and 4) in the half period. Taking a capacitor into account, Interval 3 can be equivalent to the new circuit as Figure 9. In this interval, a resonance circuit is formed by a power inductor and the snubber/parasitic capacitor of the diode leg. Thus, the voltage ringing will be introduced into the transformer secondary-side. However, there is no voltage ringing in Interval 4. The main reason is that the DC source V_{in} in Figure 9 is short-circuited in interval 4. Compared with Mode C, there is only one zero-current interval in Mode B and it is free of voltage ringing, which is same as Interval 4 in Mode C. Considering that voltage ringing will potentially bring up system instability and damage the semiconductor devices, the control route for low power is put on the boundary line between Mode B and C. Thus, the selected route from full power to zero power with PPWM + SPS control is given as Equations (8) and (9), which is a piecewise function. Under the proposed route control, the converter can achieve at the minimum RMS current for the full power range and is free of voltage ringing across the transformer.

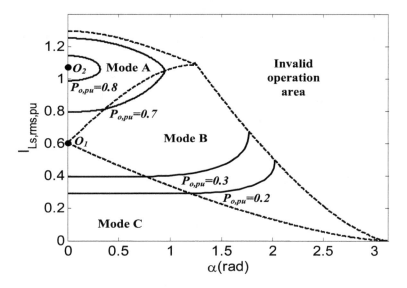

Figure 8. Normalized inductor RMS current vs. phase-shift α at different power contours for $M = 1.5$.

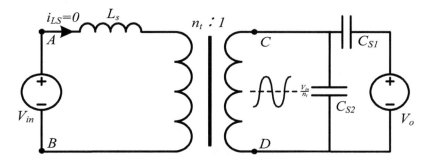

Figure 9. New equivalent circuit corresponding to interval 3 in Mode C.

When $P_{o,pu} \in \left[\frac{\pi(M-1)}{2M}, P_{o,pu,max} \right]$,

$$
\begin{cases}
\phi = \pi - \dfrac{(2+M)\sqrt{2\pi M \left(\pi M^2 + \pi M - 2M^2 P_{o,pu} - 4M P_{o,pu} - 4P_{o,pu} \right)} + X_1}{2M^3 + 4M^2 + 4M} \\
\alpha = 0.
\end{cases}
\tag{8}
$$

When $P_{o,pu} \in \left[0, \frac{\pi(M-1)}{2M} \right]$,

$$
\begin{cases}
\phi = \pi - \dfrac{X_2 \sqrt{P_{o,pu}}}{M} \\
\alpha = \pi - X_2 \sqrt{P_{o,pu}}
\end{cases},
\tag{9}
$$

where $X_1 = 2\pi M^2 + 2\pi M$, $X_2 = \frac{\sqrt{2\pi M(M-1)}}{M-1}$.

4. Experimental Verifications

To verify the theoretical analysis above, a 200 W S-DAB prototype is built, as shown in Figure 10. The specifications of the lab-scale converter are listed in Table 4. The gating signals of S-DAB are implemented using a TMS320F28335 DSP from TI (Texas Instruments, Dallas, TX, USA) and the switching frequency is set at 100 kHz.

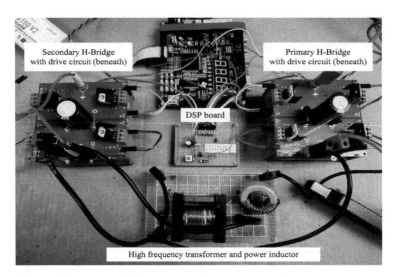

Figure 10. The layout of a 200 W S-DAB laboratory prototype.

Table 4. Specifications of a 200 W S-DAB converter.

Component	Parameter
Input DC voltage V_{in}	80 V
Load DC voltage V_o	120 V
Power inductance L_s	38 μH, CM400125/MPPcore
Transformer turns ratio n_t	15:15, ETD49/N97
Filter capacitor C_o	470 μF, 1 electrolytic cap
Switch $M_1 \sim M_6$	STP40NF20, $R_{ds} = 38$ mΩ
Diode D_{s1}, D_{s2}	MBR40250TG, $V_F = 0.86$ V

Based on the analysis of each operation mode in Section 2, a set of experimental waveforms corresponding to three effective modes are obtained and shown in Figure 11, respectively. These experimental results match the theoretical prediction closely. In Figure 11c, the voltage ringing shows up on the transformer secondary-side v_{CD} after i_{Ls} decreases to zero and remains a small duration until $v_{AB} = 0$. As expected, the voltage ringing is not identified in Figure 11b.

A series of experimental tests are then performed along the proposed minimized rms current route. The boundary power between the two stages of the control route is calculated to be 52% load, i.e., 104 W. Thus, the converter works in CCM (Mode A) at the high power of 200 W and 150 W. Two phase-shift angles are calculated as $\alpha = 0°$, $\phi = 90.25°$ (200 W) and $\alpha = 0°$, $\phi = 63.76°$ (150 W) according to (8). Experimental results of 200 W and 150 W are shown in Figure 12a,b, respectively. It can be seen from those two figures that the inductor current i_{Ls} is continuous. The related waveforms satisfy the operation condition of $\phi > \beta > \alpha = 0$ and all switches can operate at ZVS.

When the output power is lower than 52% load, the converter is operated at the boundary between Mode B and C. Using (9), two phase-shift angles are obtained as follows: (1) 100 W, $\alpha = 28.06°$, $\phi = 78.71°$; (2) 50 W, $\alpha = 72.46°$, $\phi = 108.3°$. Experimental results of 100 W and 50 W are shown in Figure 12c,d, respectively. It can be seen that the transition moment of v_{AB} from V_{in} to zero happens at the zero-crossing point of the inductor current and there is a small zero-current duration. The operation conditions of two experimental results match the boundary feature between both DCMs. In addition, the voltage ringing on the waveform v_{CD} is prevented in comparison to DCM under conventional secondary phase-shifted control (Figure 13). Based on these experimental results, the values of the RMS current, peak current, and efficiency are listed in Table 5, where the highest efficiency can arrive at 95.53% for 150 W. For 50 W operation with conventional control in Figure 13, the measured efficiency is 88.46%. It is seen that the efficiency using the proposed control is improved slightly since the current values and the switching behaviour under the two control methods are almost the same and the loss due to the ringing accounts for a small portion of total loss. However, the removal of the ringing phenomenon depresses EMI so that the risk of distortion in gating signals is reduced and the operation stability is improved consequently.

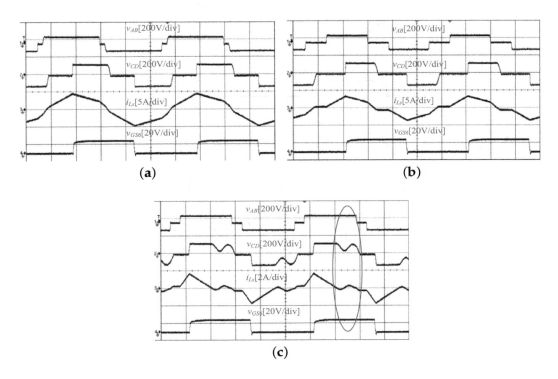

Figure 11. Experimental waveforms of v_{AB}, v_{CD}, i_{Ls} and V_{GS6} (from **top–bottom**), time scale: (2 μs/div), (**a**) Mode A; (**b**) Mode B; (**c**) Mode C.

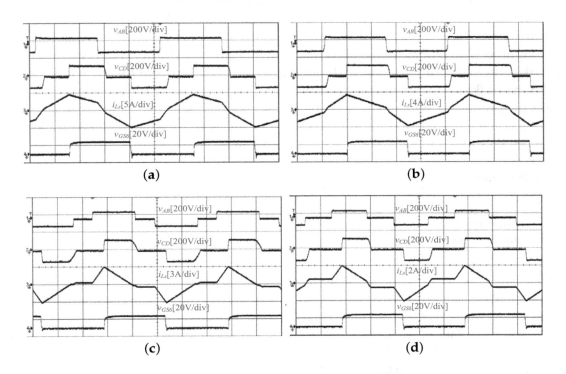

Figure 12. Experimental waveforms at four different power levels under the proposed route for $V_{in} = 80$ V and $V_o = 120$ V, time scale: (2 μs/div), (**a**) 200 W; (**b**) 150 W; (**c**) 100 W; (**d**) 50 W.

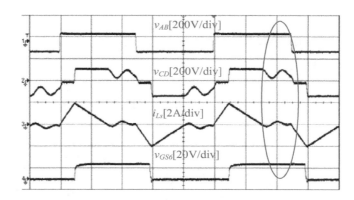

Figure 13. Experimental waveforms at 50 W with conventional secondary phase-shifted control.

Table 5. Measured results at $V_{in} = 80$ V and $V_o = 120$ V.

Power		$I_{Ls,rms}$ (A)	$I_{Ls,peak}$ (A)	η (%)
200 W	theor.	2.9	4.52	-
	exp.	3.08	4.7	94.97%
150 W	theor.	2.14	3.63	-
	exp.	2.19	3.63	95.53%
100 W	theor.	1.57	2.96	-
	exp.	1.58	2.92	92.13%
50 W	theor.	0.94	2.1	-
	exp.	0.95	2.07	88.95%

5. Conclusions

In this work, PPWM + SPS control with two controllable phase-shifts is applied on an S-DAB converter for boost operation and all effective steady-state modes are identified. Based on the characteristics of each mode, a reasonable control route is developed and implemented on a lab-scale S-DAB prototype. The experimental results show the consistency with the theoretical analysis results. Compared with conventional secondary phase-shifted control, the proposed control route not only makes the converter operate with the minimized RMS current for the whole power range, but also eliminates the voltage ringing on the secondary-side of the HF transformer completely. More importantly, the proposed hybrid control can be also applied on the other S-DAB converters in [22–24] to prevent the voltage ringing and improve stability. In addition, other optimization objectives could be developed according to various application requirements.

Author Contributions: M.L. did theoretical analysis, derivation, circuit implementation, experimental test and paper writing. X.L. was responsible for planning, coordination and proofreading.

Acknowledgments: The authors would like to acknowledge Song Hu, Guo Chen for their support in the preparation of circuit implementation.

References

1. Blaabjerg, F.; Chen, Z.; Kjaer, S.B. Power electronics as efficient interface in dispersed power generation systems. *IEEE Trans. Power Electron.* **2004**, *19*, 1184–1194. [CrossRef]
2. Tahir, S.; Wang, J.; Baloch, M.; Kaloi, G. Digital control techniques based on voltage source inverters in renewable energy applications: A review. *Electronics* **2018**, *7*, 18. [CrossRef]
3. Chen, Z.; Guerrero, J.M.; Blaabjerg, F. A review of the state of the art of power electronics for wind turbines. *IEEE Trans. Power Electron.* **2009**, *24*, 1859–1875. [CrossRef]
4. Abdelsalam, A.K.; Massoud, A.M.; Ahmed, S.; Enjeti, P.N. High-performance adaptive perturb and observe MPPT technique for photovoltaic-based microgrids. *IEEE Trans. Power Electron.* **2011**, *26*, 1010–1021. [CrossRef]

5.   Almalaq, Y.; Matin, M. Three topologies of a non-isolated high gain switched-inductor switched-capacitor step-up cuk converter for renewable energy applications. *Electronics* **2018**, *7*, 94. [CrossRef]

6.   Lee, W.J.; Kim, C.E.; Moon, G.W.; Han, S.K. A new phase-shifted full-bridge converter with voltage-doubler-type rectifier for high-efficiency PDP sustaining power module. *IEEE Trans. Ind. Electron.* **2008**, *55*, 2450–2458.

7.   Cha, H.; Chen, L.; Ding, R.; Tang, Q.; Fang, Z.P. An alternative energy recovery clamp circuit for full-bridge PWM converters with wide ranges of input voltage. *IEEE Trans. Power Electron.* **2008**, *23*, 2828–2837.

8.   Chen, W.; Ruan, X.; Zhang, R. A novel zero-voltage-switching PWM full bridge converter. *IEEE Trans. Power Electron.* **2008**, *23*, 793–801. [CrossRef]

9.   Jang, Y.; Jovanovic, M.M.; Chang, Y.-M. A new ZVS-PWM full-bridge converter. *IEEE Trans. Power Electron.* **2003**, *18*, 1122–1129. [CrossRef]

10. Yoon, H.K.; Han, S.K.; Choi, E.-S.; Moon, G.-W.; Youn, M.-J. Zero-voltage switching and soft-commutating two-transformer full-bridge PWM converter using the voltage-ripple. *IEEE Trans. Ind. Electron.* **2008**, *55*, 1478–1488. [CrossRef]

11. Lee, I.O.; Moon, G.W. Phase-shifted PWM converter with a wide ZVS range and reduced circulating current. *IEEE Trans. Power Electron.* **2013**, *28*, 908–919. [CrossRef]

12. Jang, Y.; Jovanovic, M.M. A new family of full-bridge ZVS converters. *IEEE Trans. Power Electron.* **2004**, *19*, 701–708. [CrossRef]

13. Ruan, X.; Yan, Y. A novel zero-voltage and zero-current-switching PWM full-bridge converter using two diodes in series with the lagging leg. *IEEE Trans. Ind. Electron.* **2001**, *48*, 777–785. [CrossRef]

14. Lo, Y.K.; Lin, C.Y.; Hsieh, M.T.; Lin, C.Y. Phase-shifted full-bridge series-resonant DC-DC converters for wide load variations. *IEEE Trans. Ind. Electron.* **2011**, *58*, 2572–2575. [CrossRef]

15. Gautam, D.S.; Bhat, A.K.S. A comparison of soft-switched DC-to-DC converters for electrolyzer application. *IEEE Trans. Power Electron.* **2013**, *28*, 54–63. [CrossRef]

16. Ali, K.; Das, P.; Panda, S.K. Analysis and design of APWM half-bridge series resonant converter with magnetizing current assisted ZVS. *IEEE Trans. Ind. Electron.* **2017**, *64*, 1993–2003. [CrossRef]

17. Mumtahina, U.; Wolfs, P.J. Multimode optimization of the phase shifted LLC eries resonant converter. *IEEE Trans. Power Electron.* **2018**, *1*. [CrossRef]

18. Lee, I.O.; Moon, G.W. The *k*-Q Analysis for an LLC series resonant converter. *IEEE Trans. Power Electron.* **2014**, *29*, 13–16. [CrossRef]

19. Fang, X.; Hu, H.; Chen, F.; Somani, U.; Auadisian, E.; Shen, J.; Batarseh, I. Efficiency-oriented optimal design of the LLC resonant converter based on peak gain placement. *IEEE Trans. Power Electron.* **2013**, *28*, 2285–2296. [CrossRef]

20. Zhang, J.; Zhang, F.; Xie, X.; Jiao, D.; Qian, Z. A novel ZVS DC/DC converter for high power applications. *IEEE Trans. Power Electron.* **2004**, *19*, 420–429. [CrossRef]

21. Mishima, T.; Nakaoka, M. Practical evaluations of a ZVS-PWM DC-DC converter with secondary-side phase-shifting active rectifier. *IEEE Trans. Power Electron.* **2011**, *26*, 3896–3907. [CrossRef]

22. Kulasekaran, S.; Ayyanar, R. Analysis, design, and experimental results of the semidual-active-bridge converter. *IEEE Trans. Power Electron.* **2014**, *29*, 5136–5147. [CrossRef]

23. Li, W.; Zong, S.; Liu, F.; Yang, H.; He, X.; Wu, B. Secondary-side phase shift controlled ZVS DC/DC converter with wide voltage gain for high input voltage applications. *IEEE Trans. Power Electron.* **2013**, *28*, 5128–5139. [CrossRef]

24. Wu, H.; Chen, L.; Xing, Y. Secondary-side phase-shift-controlled dual-transformer-based asymmetrical dual-bridge converter with wide voltage gain. *IEEE Trans. Power Electron.* **2015**, *30*, 5381–5392. [CrossRef]

25. Hu, S.; Li, X. Performance evaluation of a semi-dual-bridge resonant DC/DC converter with secondary phase-shifted control. *IEEE Trans. Power Electron.* **2017**, *32*, 7727–7738. [CrossRef]

26. Hu, S.; Li, X.; Lu, M.; Luan, B.-Y. Operation modes of a secondary-side phase-shifted resonant converter. *Energies* **2015**, *8*, 12314–12330. [CrossRef]

27. Park, K.B.; Kim, C.E.; Moon, G.W.; Youn, M.J. Voltage oscillation reduction technique for phase-shift full-bridge converter. *IEEE Trans. Ind. Electron.* **2007**, *54*, 2779–2790. [CrossRef]

28. Garabandic, D.; Dunford, W.G.; Edmunds, M. Zero-voltage-zero-current switching in high-output-voltage full-bridge PWM converters using the interwinding capacitance. *IEEE Trans. Power Electron.* **1999**, *14*, 343–349. [CrossRef]

9

Performance Improvement for PMSM DTC System through Composite Active Vectors Modulation

Tianqing Yuan and Dazhi Wang *

School of Information Science and Engineering, Northeastern University, Shenyang 110819, China; tqyuan@stumail.neu.edu.cn
* Correspondence: noblefuture@163.com

Abstract: In this paper, a novel direct torque control (DTC) scheme based on composite active vectors modulation (CVM) is proposed for permanent magnet synchronous motor (PMSM). The precondition of the accurate compensations of torque error and flux linkage error is that the errors can be compensated fully during the entire control period. Therefore, the compensational effects of torque error and flux linkage error in different operating conditions of the PMSM are analyzed firstly, and then, the operating conditions of the PMSM are divided into three cases according to the error compensational effects. To bring the novel composite active vectors modulation strategy smoothly, the effect factors are used to represent the error compensational effects provided by the applied active vectors. The error compensational effects supplied by single active vector or synthetic voltage vector are analyzed while the PMSM is operated in three different operating conditions. The effectiveness of the proposed CVM-DTC is verified through the experimental results on a 100-W PMSM drive system.

Keywords: direct torque control (DTC); composite active vectors modulation (CVM); permanent magnet synchronous motor (PMSM); effect factors

1. Introduction

Permanent magnet synchronous motors (PMSM) have a lot of merits such as high reliability, high efficiency, simple construction, and good control performance, and thus, it has been applied in various control systems including electrical drives, industrial applications, and medical devices in recent years [1–7]. Direct torque control (DTC) and field-oriented control (FOC) are two widely applied high-performance control strategies for the PMSM. Different from the decoupled-analyzing method in FOC, torque and flux linkage are controlled directly in DTC, and therefore, the quickest dynamic response can be obtained in the PMSM driven by DTC. However, as only six active vectors can be selected to compensate the errors of flux linkage and torque in conventional DTC (CDTC), the PMSM suffers from some drawbacks, such as large torque and flux linkage ripples. To improve the steady-state performance of the PMSM, many researchers have attempted to reduce these ripples by adding the amount of the active vectors through different methods.

With more appropriate active vectors selected in each control period, a novel DTC-fed PMSM system is proposed on the basis of a three-level inverter [8,9], and thus, the ripples of torque and flux linkage in PMSM can be suppressed effectively. In References [10,11], a novel DTC strategy using a matrix converter is proposed. Four enhanced switching tables are designed for the selection of switching states, and therefore, the ripples of the PMSM can be reduced effectively. Despite the fact that multiple active vectors can be supplied by three-level inverter or matrix converter, the cost of the DTC system is inevitably increased.

In fact, torque error and flux linkage error are tiny in most cases, and therefore, these errors will be over-compensated if the selected vector is applied over the whole control period. To solve these problems, duty ratio modulation strategy is introduced into the DTC-fed PMSM. Different duty ratio modulation methods for DTC (DDTC) are studied in References [12–16], and the ripples of torque and flux linkage can be reduced effectively without degrading the fast dynamic response in CDTC. In Reference [17], a new suboptimal control algorithm applying dynamic programming and a ramp trajectory method is proposed to DB-DTFC. In DB-DTFC, the maximum torque changes in one inverter switching period are used to determine the number of quantized stages for the minimum-time ramp trajectory method. It can be found that, the error compensational effects are considered in DB-DTFC, and therefore, torque and flux linkage command trajectories can be developed in different shapes according to the desired objectives, and fast dynamic responses can be achieved easily.

The stator flux linkage currents are decoupled in d-q axes and controlled independently in FOC, and thus, the outstanding operating performance of the PMSM can be obtained easily. The decoupled-analysis method is adopted in the novel DTC based on a space vector modulation (SVM) strategy with simple proportional-integral (PI) regulator or sliding mode observer [18–25]. With the independent control of torque and flux linkage in SVM-DTC, the amplitude and the phase of the wanted active vector can be determined accurately, and the errors of torque and flux linkage can be compensated precisely. However, the introduced PI regulator or sliding mode observer will degrade the dynamic response of the system.

To improve the operation performance of PMSM effectively, a novel DTC scheme utilizing composite active vectors modulation (CVM) strategy is presented in this paper. The compensational effects of torque error and flux linkage error in the PMSM driven by different control strategies are analyzed. Subsequently, the precondition of the accurate error compensation is obtained, and then, the precondition is adopted to determine the applied control strategy for the PMSM in different operation conditions, which is ignored in SVM-DTC and CDTC.

It should be noted that the most complicated control process is the transient-state. The large error component should be compensated fully, and the low error component should not be over-compensated; therefore, the duty ratio direct torque control strategy which, considering the active angles and the impact angles in Reference [1], can be used. The effectiveness of the proposed CVM-DTC scheme is validated through the experimental results. It should be noted that the steady-state performance and the dynamic response of the PMSM driven by CDTC, DDTC, and SVM-DTC are also studied in this paper.

The rest of this paper comprises the following sections. The principles of the conventional DTC are analyzed in Section 2. The compensational effects of torque error and flux linkage error in the SVM-DTC system are also illustrated in Section 2. The dividing process of the PMSM operation conditions and the error compensational effect supplied by different vectors in different operation conditions are described in Section 3 and the precondition of the accurate error compensations are also analyzed in Section 3. The description of experimental setup and discussions on experimental results are given in Section 4. The conclusion is analyzed in Section 5.

2. Principle of the Conventional DTC and SVM-DTC

2.1. Principle of the Conventional DTC

In the PMSM DTC system driven by a two-level voltage source inverter, eight voltage vectors can be applied to compensate the errors of torque and flux linkage, including six active vectors V_n (n = 1, 2, 3, 4, 5, 6) and two null vectors (V_0 and V_7). The spatial placements of the six active vectors in $\alpha\beta$-reference frames are shown in Figure 1. The whole rotation space of stator flux linkage φ_s can be divided into six sectors through the section boundary lines l_i (i = 1, 2, 3, 4, 5, 6), as shown in Figure 1. The six sectors are represented with number "N", and the sector vector vs. represents the active vector in every sector which the stator flux linkage φ_s is located in.

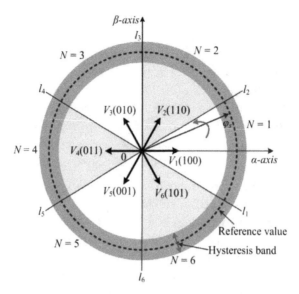

Figure 1. Active vectors in DTC system.

The torque error e_T is obtained by the comparison between the reference value T_{ref} and the real-time value T. The hysteresis comparator is used to determine the property ε_T of torque error e_T. The property ε_T value is 1 or -1, which indicates torque T needs to be increased if the value of property ε_T is 1, while the torque needs to be decreased if the value of property ε_T is -1. The determination methods of another parameter flux linkage φ is in the same way. The active vector selection rules in the SV-CDTC system are described in Table 1.

Table 1. Conventional switching table.

Sector number N		Torque (ε_T)	
		1	**−1**
Stator flux linkage (ε_F)	1	V_{N+1}	V_{N-1}
	−1	V_{N+2}	V_{N-2}

2.2. Compensations of Torque Error and Flux Linkage Error in SVM-DTC System

The number and the direction of the active vectors are fixed in CDTC, and therefore, the errors of torque and flux linkage are difficult to be compensated effectively, leading to large ripples. To improve the steady-state performance of the PMSM, the decoupling control strategy adopted in FOC is introduced into DTC. The PI controllers are used to obtain the amplitude of the torque vector and flux linkage vector on the basis of torque error and flux linkage error; then, the space vector modulation (SVM) is used to determine the precise vectors. The schematic diagram of SVM-DTC is shown in Figure 2.

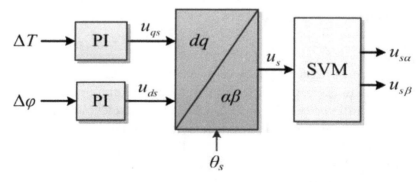

Figure 2. Schematic diagram of SVM-DTC.

Figure 2 shows the error compensations provided by a synthesis voltage vector. It can be found that torque error ΔT and flux linkage error $\Delta \varphi$ can be compensated through vectors u_{qs} and u_{ds}, respectively. Consequently, the synthesis voltage vector u_s can be obtained on the basis of the rotor position θ_s. As shown in Figure 3, the voltage vectors $u_{s\alpha}$ and $u_{s\beta}$ will be obtained through coordinate transformation based on u_s.

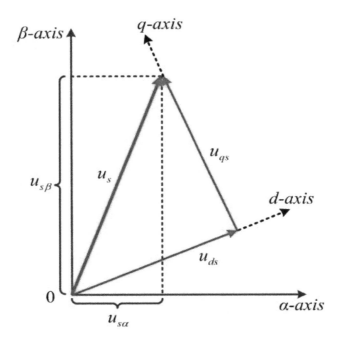

Figure 3. Synthesis voltage vectors in d-q reference frame.

From the aforementioned analyses, it can be observed that the switching table in CDTC is replaced by the PI controllers in SVM-DTC. Therefore, the active vectors used to compensate the errors of torque and flux linkage are not limited to the six basic active vectors. Additionally, the ideal steady-state performance of the PMSM can be obtained easily. Despite the fact that the precision of the wanted synthetic vector can be ensured with the using of SVM strategy in SVM-DTC, the dynamic response of the PMSM is affected inevitably.

3. Analysis of Error Compensations

The ripples of torque and flux linkage in the PMSM driven by SVM-DTC are relatively minor while the PMSM is operated in the steady-state condition. On the other hand, the dynamic performance will be affected by the complicated calculations of the synthesis voltage vector and the over-modulation process while the PMSM is operated in the dynamic response condition. This is the main reason that the fast dynamic performance of permanent magnet synchronous motor driven by SVM-DTC is degraded.

To improve the steady-state performance of PMSM, and maintain the fast dynamic response at the same time, appropriate control strategy should be selected and applied to the system according to the operation conditions, including CDTC, DDTC, and SVM-DTC. Therefore, the differences of the error compensational effects provided by the synthesis voltage vector and single active vector under different operation conditions should be analyzed firstly.

3.1. Operation Conditions

The stator flux linkage φ_s changes from φ_{s1} to φ_{s2} during one control period and the variation of the stator flux linkage φ_s is $\Delta \varphi_s$, which can be decoupled into $\Delta \varphi_{sd}$ and $\Delta \varphi_{sq}$ in the d-q axis. In this control period, the errors of torque and flux linkage are ΔT and $\Delta \varphi$, respectively. Hence, the torque component variation of the stator flux linkage is $\Delta\varphi_{sq}$, and the amplitude component variation of the stator flux linkage is $\Delta\varphi_{sd}$, as shown in Figure 4.

The torque component variation and the amplitude component variation of the stator flux linkage can be expressed as

$$\Delta\varphi_{sq} = \frac{2L_s}{3p} \cdot \frac{1}{\varphi_f} \cdot \Delta T \tag{1}$$

$$\Delta\varphi_{sd} = \Delta\varphi \tag{2}$$

where L_s is the stator inductance, p is the number of pole pairs, and φ_f is the permanent magnet flux linkage.

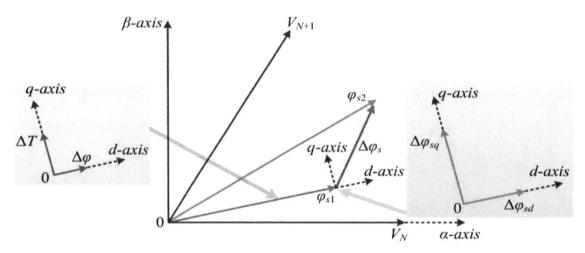

Figure 4. Analysis of error compensational effects.

The compensational effect of the stator flux linkage supplied by single active vector V_N and synthesis voltage vector u_s are $\Delta\varphi'_{s1}$ and $\Delta\varphi'_s$, respectively, as shown in Figure 5.

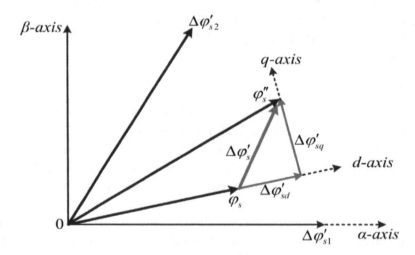

Figure 5. Analysis of error compensational effects.

The torque component compensation of the stator flux linkage supplied by synthesis voltage vector u_s is $\Delta\varphi'_{sq}$, and the amplitude component compensation of the stator flux linkage provided by synthesis voltage vector u_s is $\Delta\varphi'_{sd}$. It is obvious that the parameters of $\Delta\varphi'_s$, $\Delta\varphi'_{sq}$, and $\Delta\varphi'_{sd}$ are fixed during each control period in the system. While the torque error ΔT and the flux linkage error $\Delta\varphi$ will vary with the variation of the stator flux linkage location in different control period. Therefore, the real values of $\Delta\varphi_{sq}$ and $\Delta\varphi_{sd}$ are also different.

The relationships between the real error compensations of the stator flux linkage and the errors can be described in the following way.

First item, the actual compensations are greater than the errors:

$$\left(\begin{array}{l} \Delta\varphi'_{sd} > \Delta\varphi_{sd} \\ \Delta\varphi'_{sq} > \Delta\varphi_{sq} \end{array}\right. \tag{3}$$

Second item, the actual compensations are less than the errors:

$$\left(\begin{array}{l} \Delta\varphi'_{sd} < \Delta\varphi_{sd} \\ \Delta\varphi'_{sq} < \Delta\varphi_{sq} \end{array}\right. \tag{4}$$

Third item, the actual compensation of the amplitude component is less than the error while the actual compensation of torque component is greater than the error:

$$\left(\begin{array}{l} \Delta\varphi'_{sd} < \Delta\varphi_{sd} \\ \Delta\varphi'_{sq} > \Delta\varphi_{sq} \end{array}\right. \tag{5}$$

Fourth item, the actual compensation of the amplitude component is greater than the error while the actual compensation of torque component is less than the error:

$$\left(\begin{array}{l} \Delta\varphi'_{sd} > \Delta\varphi_{sd} \\ \Delta\varphi'_{sq} < \Delta\varphi_{sq} \end{array}\right. \tag{6}$$

The operation conditions of the PMSM can be divided into three items in accordance with the errors and the actual compensations, as shown in Table 2.

Table 2. Operation conditions

Operation Conditions	Reference Equation
Steady-state	(3)
Transient-state	(5) and (6)
Dynamic-state	(4)

3.2. Error Compensation Analysis in Steady-State Case

The values of torque error ΔT and flux linkage error $\Delta\varphi_s$ are relatively low in steady-state case [6]. The angle between the stator flux linkage φ_s and the active vector V_N is θ_1, as shown in Figure 6. It is also shown in Figure 6 that both the compensations of ΔT and $\Delta\varphi_s$ are bigger than the errors. Consequently, the errors will be over-compensated if the active vector or the synthesized voltage vector is applied during the entire control period.

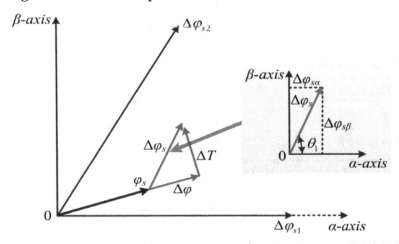

Figure 6. Analysis of error compensational effects in steady-state.

In DDTC-fed PMSM, the applied time of the active vector is modulated by duty ratio modulation strategy. As a result, the over-compensation of the errors can be avoided; nevertheless, the fixed active vectors limit the compensational effects.

In the PMSM driven by SVM-DTC, the adjacent active vectors V_N and V_{N+1} are selected as the benchmark vectors to obtain the synthesized voltage vector u_s. Furthermore, the applied time of V_N and V_{N+1} are T_1 and T_2, respectively. The error compensations can be evaluated by

$$\Delta \varphi_{s1} = V_N \cdot T_1 \tag{7}$$

$$\Delta \varphi_{s2} = V_{N+1} \cdot T_2 \tag{8}$$

The modulation process of the active vectors can be expressed as

$$\Delta \varphi_{s1} + \Delta \varphi_{s2} \cdot \cos \frac{\pi}{3} = \Delta \varphi_{s\alpha} = \Delta \varphi_s \cdot \cos \theta_1 \tag{9}$$

$$\Delta \varphi_{s2} \cdot \sin \frac{\pi}{3} = \Delta \varphi_{s\beta} = \Delta \varphi_s \cdot \sin \theta_1 \tag{10}$$

$$T_1 + T_2 + T_0 = T_s \tag{11}$$

where T_0 is the zero voltage vector applied time.

From the aforementioned analyses, it can be found that the errors of torque and flux linkage can be compensated accurately through SVM strategy while the PMSM is operated in steady-state.

3.3. Error Compensation Analysis in Dynamic-State Case

The errors of torque or flux linkage may become greater in the dynamic-state while the speed or the torque changes. As shown in Figure 7, the stator flux linkage error is $\Delta \varphi_s$; the angle between the stator flux linkage error $\Delta \varphi_s$ and the active vector V_N is θ_2. It can be found that the torque error and the flux linkage error are greater than the error compensations.

It can be found that the torque error ΔT and the flux linkage error $\Delta \varphi$ cannot be compensated fully by any single active vector or the synthesized voltage vector in the next control period. The differences of the error compensation effect supplied by the single active vector or the synthesized voltage vector are described in following parts.

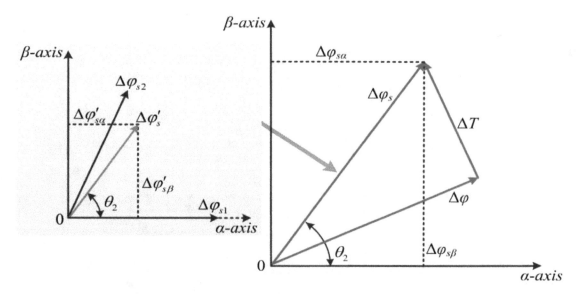

Figure 7. Analysis of error compensational effects in dynamic-state.

3.3.1. Synthetic Voltage Vector

The adjacent active vectors V_N and V_{N+1} are selected as the benchmark vectors. The applied time of V_N and V_{N+1} are T_1 and T_2, respectively. Therefore, the error compensations can be calculated as

$$\Delta\varphi'_{s1} = V_N \cdot T_1 \tag{12}$$

$$\Delta\varphi'_{s2} = V_{N+1} \cdot T_2 \tag{13}$$

$$\Delta\varphi'_{s1} + \Delta\varphi'_{s2} \cdot \cos\frac{\pi}{3} = \Delta\varphi_{s\alpha} = \Delta\varphi_s \cdot \cos\theta_2 \tag{14}$$

$$\Delta\varphi'_{s2} \cdot \sin\frac{\pi}{3} = \Delta\varphi_{s\beta} = \Delta\varphi_s \cdot \sin\theta_2 \tag{15}$$

Since the actual compensations are smaller than the errors, therefore

$$T_1 + T_2 > T_s \tag{16}$$

The applied time of the applied active vectors can be over-modulated as

$$T'_1 = \frac{T_1}{T_1 + T_2} \cdot T_s \tag{17}$$

$$T'_2 = \frac{T_2}{T_1 + T_2} \cdot T_s \tag{18}$$

The applied time of the applied active vectors can be rewritten as

$$T_1 = k_1 \cdot T_s \tag{19}$$

$$T_2 = k_2 \cdot T_s \tag{20}$$

where k_1 and k_2 are the duty ratio values of applied time of V_N and V_{N+1}, respectively.

Therefore, the actual compensation of the stator flux linkage in one control period is

$$\Delta\varphi'_{s\alpha} = V_N \cdot \frac{k_1}{k_1 + k_2} \cdot T_s + V_{N+1} \cdot \cos\frac{\pi}{3} \cdot \frac{k_2}{k_1 + k_2} \cdot T_s \tag{21}$$

$$\Delta\varphi'_{s\beta} = V_{N+1} \cdot \sin\frac{\pi}{3} \cdot \frac{k_2}{k_1 + k_2} \cdot T_s \tag{22}$$

which can be simplified as

$$\left(\begin{array}{l} \Delta\varphi'_{s\alpha} = V_N \cdot T_s \cdot \left(\frac{k_1}{k_1 + k_2} + \frac{1}{2} \cdot \frac{k_2}{k_1 + k_2} \right) \\ \Delta\varphi'_{s\beta} = V_N \cdot T_s \cdot \frac{\sqrt{3}}{2} \cdot \frac{k_2}{k_1 + k_2} \end{array} \right. \tag{23}$$

Therefore, the compensation of the stator flux linkage supplied by synthesized voltage vector u_s can be given as

$$|\Delta\varphi'_s| = \sqrt{(\Delta\varphi'_{s\alpha})^2 + (\Delta\varphi'_{s\beta})^2} = V_N \cdot T_s \cdot \sqrt{\frac{k_1^2 + k_1 k_2 + k_2^2}{(k_1 + k_2)^2}} < V_N \cdot T_s \tag{24}$$

3.3.2. Single Active Vector

The compensations of the stator flux linkage supplied by adjacent vectors V_N and V_{N+1} during the whole control period are

$$\Delta\varphi''_{s1} = V_N \cdot T_s \tag{25}$$

$$\Delta\varphi''_{s2} = V_{N+1} \cdot T_s \tag{26}$$

Therefore, the compensations of the stator flux linkage provided by the single active vector can be expressed as

$$\Delta\varphi''_s = \Delta\varphi''_{s1} = \Delta\varphi''_{s2} = V_N \cdot T_s \tag{27}$$

The comparison result of the stator flux linkage compensations supplied by the different vectors can be described as

$$\Delta\varphi'_s < \Delta\varphi''_s \tag{28}$$

From the aforementioned analyses, it can be observed that the compensational effects of the stator flux linkage supplied by the synthesized voltage vector is weaker than the single active vector. Therefore, the SVM strategy is not required to compensate the errors of torque and flux linkage while the PMSM is operated in the dynamic state. To simplify the calculations of the system, the appropriate active vector can be selected from a conventional switching table and be used in the system over the entire control period.

In short, CDTC strategy should be used to reduce the ripples of torque and flux linkage in the PMSM when the PMSM is operated in a dynamic state. Hence, the delayed dynamic response caused by the PI controller can be eliminated, and the ripples' depressing effects of the PMSM driven by CDTC are the same as that driven by SVM-DTC.

3.4. Error Compensation Analysis in Transient-State Case

The operation condition of the PMSM may deviate the steady-state due to external disturbance. Therefore, the PMSM may operate in a transient-state if one parameter of torque error and flux linkage error is large while another parameter is relatively low.

The torque error ΔT is high and the flux linkage error $\Delta\varphi$ is relatively low as shown in Figure 8. The variation of the stator flux linkage is $\Delta\varphi_s$.

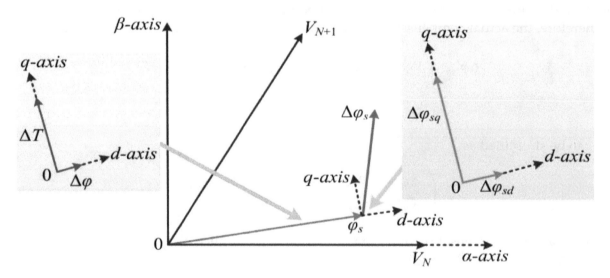

Figure 8. Analysis of error compensational effects in transient-state.

The differences of the error compensation effect supplied by single active vector or synthesized voltage vector are described in the following section.

3.4.1. Synthetic Voltage Vector

Figure 9 shows the error compensational effects provided by different active vectors.

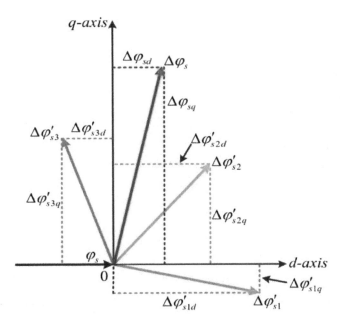

Figure 9. Analysis of error compensational effects provided by different active vectors.

As shown in Figure 9, the adjacent active vectors V_{N+1} and V_{N+2} are selected as the benchmark vectors. The applied time of V_{N+1} and V_{N+2} are T_1 and T_2, respectively. Therefore, the error compensations can be evaluated by

$$\Delta\varphi'_{s2q} \cdot T_1 + \Delta\varphi'_{s3q} \cdot T_2 = \Delta\varphi_{sq} \cdot T_s \tag{29}$$

$$\Delta\varphi'_{s2d} \cdot T_1 - \Delta\varphi'_{s3d} \cdot T_2 = \Delta\varphi_{sq} \cdot T_s \approx 0 \tag{30}$$

3.4.2. Single Active Vector V_n

The stator flux linkage error $\Delta\varphi_s$ is located in the middle of error compensations $\Delta\varphi'_{s2}$ and $\Delta\varphi'_{s3}$, as shown in Figure 9. To compensate the error $\Delta\varphi_{sq}$ effectively and avoid the over-compensation of the error $\Delta\varphi_{sd}$ at the same time, the adjacent vectors V_{N+1} and V_{N+2} can be selected and applied to half of the control period.

From the above analysis, it can be found that the torque error can be compensated fully supplied by a single active vector while the PMSM is operated in the transient-state; however, the flux linkage error cannot be compensated fully. Despite the fact that the torque error and the flux linkage error can be compensated fully by synthetic voltage vector, the calculations of the system are inevitably increased. It should be noted that the novel DDTC strategy based on the active angle in Reference [1] has solved the problem while one parameter is large and another parameter is relatively small. Therefore, the DDTC strategy can be used to improve the performance of the system while the PMSM is operated in a transient-state.

3.5. Novel Composite Active Vectors Modulation Strategy

To improve the operation performance of the PMSM effectively, a novel composite active vectors modulation DTC (CVM-DTC) strategy considering the precondition of the accurate errors compensations is presented in this section. The schematic diagram of the presented CVM-DTC system is shown in Figure 10. The parameters in CVM-DTC are defined by:

u_{abc}: Stator voltage;
i_{abc}: Stator currents;
U_{DC}: DC bus voltage;
n: Actual rotor speed;

n_{ref}: Reference rotor speed;
σ: Rotor position;
T_{ref}: Reference torque;
φ_{ref}: Reference flux linkage;
ΔT: *Reference torque compensation;
$\Delta \varphi$: *Reference flux linkage compensation;
V_n: Single active vector;
d: Duty ratio value of applied time;
u_s: Synthetic voltage vector.

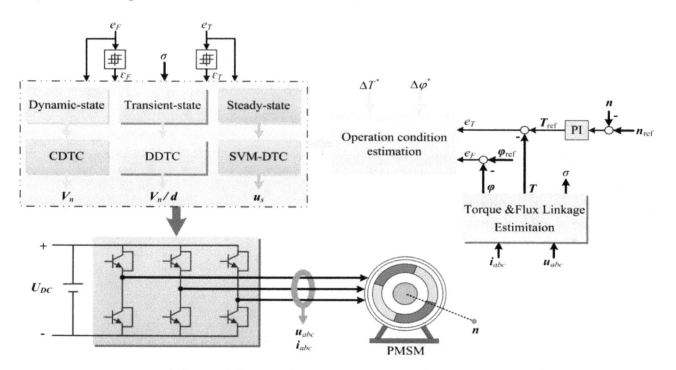

Figure 10. Schematic diagram of the CVM-DTC for PMSM.

In order to maintain the fast dynamic response in CDTC and obtain the minimum ripples of the system, the applied control strategy should adjust according to the operation conditions of the PMSM.

The precondition of the accurate compensations of torque error and flux linkage error is that the torque error and the flux linkage error can be compensated and fully supplied by the applied active vector in the whole control period. However, this precondition is ignored in the SVM-DTC system. Therefore, the torque error and the flux linkage error will be analyzed through decoupled calculations through PI controllers, while the compensational effects of the stator flux linkage in SVM-DTC and CDTC when the PMSM is operated in non-steady-state are nearly the same. As a result, the error compensational effects are not satisfied and the dynamic response will be affected without considering the operation conditions of the PMSM.

3.6. Determining of the Operation Condition through Effect Factors

The relationship between the active vector V_n and the stator flux linkage variation $\Delta \varphi_s$ in each control period is

$$\Delta \varphi_s = V_n \cdot T_s \qquad (31)$$

During the whole control period, the max compensations of $\Delta \varphi_{sq}$ and $\Delta \varphi_{sd}$ can be expressed as

$$\Delta \varphi_{sq-max} = \Delta \varphi_s = V_n \cdot T_s \qquad (32)$$

$$\Delta \varphi_{sd-\max} = \Delta \varphi_s = V_n \cdot T_s \qquad (33)$$

The max compensation of the torque is

$$\Delta T_{\max} = \frac{3p}{2L_s} \cdot \varphi_f \cdot V_n \cdot T_s \qquad (34)$$

And the max compensation of the flux linkage is

$$\Delta \varphi_{d\max} = V_n \cdot T_s \qquad (35)$$

Defining the reference values of torque variation and flux linkage variation are ΔT^* and $\Delta \varphi^*$, respectively, which can be expressed as

$$\Delta T^* = \Delta T_{\max} = \frac{3p}{2L_s} \cdot \varphi_f \cdot V_n \cdot T_s \qquad (36)$$

$$\Delta \varphi^* = \Delta \varphi_{d\max} = V_n \cdot T_s \qquad (37)$$

The effect factors of torque and flux linkage are k_T and k_φ, respectively, which can be given as

$$k_T = \frac{\Delta T}{\Delta T^*} \qquad (38)$$

$$k_\varphi = \frac{\Delta \varphi}{\Delta \varphi^*} \qquad (39)$$

The introduced effect factors can be obtained through the errors and the reference values of the variation in any control period. The operation conditions of the PMSM can be classified into three cases: steady-state, transient-state, and dynamic-state. The relationships between the effect factors and the operation conditions are shown in Table 3.

Table 3. Effect factors for different operation conditions.

Effect Factors		Operation Conditions
k_T	k_φ	
$(-\infty, -1)$	$(-\infty, -1)$	Dynamic-state
	$(-1, 1)$	Transient-state
	$(1, +\infty)$	Dynamic-state
$(-1, 1)$	$(-\infty, -1)$	Transient-state
	$(-1, 1)$	Steady-state
	$(1, +\infty)$	Transient-state
$(1, +\infty)$	$(-\infty, -1)$	Dynamic-state
	$(-1, 1)$	Transient-state
	$(1, +\infty)$	Dynamic-state

4. Experimental Analysis

4.1. Experimental System Setup

Experimental studies are carried out on a 100-W PMSM drive system to validate the feasibility and effectiveness of the proposed CVM-DTC strategy. The experimental hardware setup is illustrated in Figure 11. The parameters of the PMSM are given as follows: $R_s = 0.76\ \Omega$; $L_s = 0.00182$ H; the number of pole pairs $p = 4$. The DC voltage is 36 V. This study compares the steady-state and the dynamic response performance of CDTC, DDTC, SVM-DTC, and CVM-DTC. The experiments are implemented in a TMS320F28335 DSP control system with a sampling period of 100 µs.

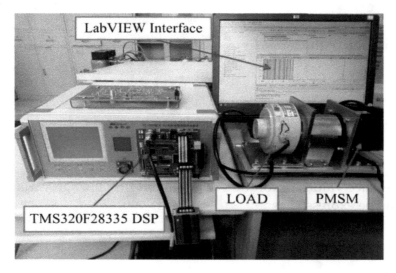

Figure 11. Experimental setup of control system.

4.2. Steady-State Performance

The steady-state performances of CDTC, DDTC, SVM-DTC, and CVM-DTC are compared under the same operating conditions. The PMSM is operated at 500 rpm and the reference values of torque and flux linkage are 0.8 N·m and 0.3 Wb, respectively. The torque and flux linkage waveforms of the PMSM are driven by different control strategies as shown in Figure 12.

From these experimental results, it can be found that the torque ripples of CDTC, SV-DDTC, SVM-DTC, and CVM-DTC are 0.56, 0.4, 0.32, and 0.34 N·m, respectively, and the flux linkage ripples of the four control system are 0.08, 0.06, 0.04, and 0.038 Wb, respectively. Therefore, compared with CDTC, DDTC and SVM-DTC can reduce the torque ripple by at least 28% and 42%, respectively, and reduce the flux linkage ripple at least 25% and 50%, respectively. While the steady-state performances of the PMSM driven by CVM-DTC in the setting operation conditions are nearly the same as SVM-DTC. The experimental results show that the errors of torque and flux should be compensated through SVM-DTC strategy, which indicates that the applied control strategy in CVM-DTC in the steady-state condition is appropriate.

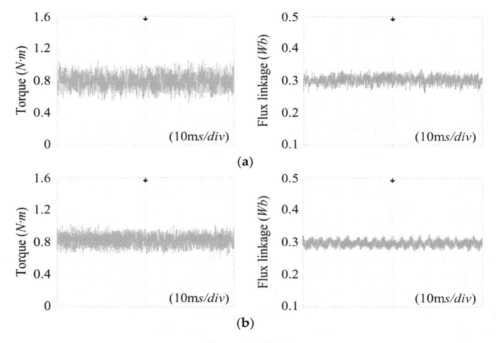

Figure 12. *Cont.*

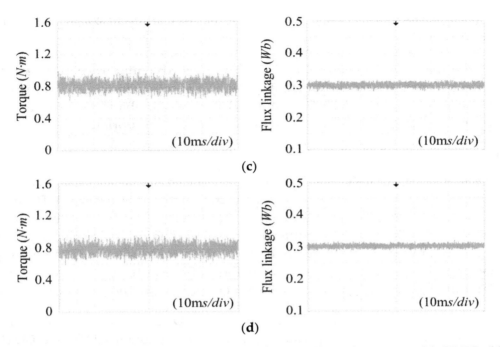

Figure 12. Experimental torque and flux linkage of the PMSM when using: (**a**) CDTC; (**b**) DDTC; (**c**) SVM-DTC; (**d**) CVM-DTC.

4.3. Dynamic Performance

To validate the fast dynamic response of the proposed novel CVM-DTC, the speed responses of the PMSM driven by the four control strategies are tested when the torque is set as 0.5 N·m. In these tests, a step change from 200 to 400 rpm is applied on the speed reference, as shown in Figure 13.

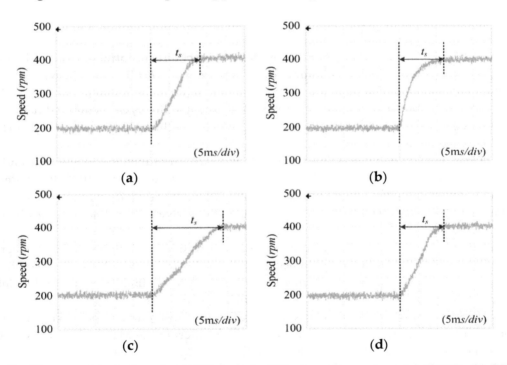

Figure 13. The speed trajectory from 200 rpm to 400 rpm when using: (**a**) CDTC; (**b**) DDTC; (**c**) SVM-DTC; (**d**) CVM-DTC.

It can be seen that the ripple of the rotor speed is 35 rpm when using CDTC, while the speed ripples of the PMSM can be reduced to 30, 25, and 24 rpm with the use of DDTC, SVM-DTC, and CVM-DTC.

Moreover, the settling times of the rotor speed using the four different control strategies are 0.013, 0.012, 0.019, and 0.012 s.

Therefore, the main advantage of CDTC, i.e., the fast dynamic response, is maintained in CVM-DTC. The experimental results show that dynamic response has a higher priority than ripples in dynamic-state condition, hence, DDTC or SVM-DTC should be abandoned. In short, the applied control strategy in CVM-DTC in the dynamic-state condition is appropriate.

5. Conclusions

The precondition of the accurate compensations of torque error and flux linkage error is considered in the proposed novel CVM-DTC scheme in this paper, which is ignored in CDTC and SVM-DTC. Therefore, the compensational effects of torque error and flux linkage error provided by the single active vector or synthetic voltage vector in different operation conditions are analyzed firstly, and then, the operating conditions of the PMSM are divided into three cases according to the compensational effects (effect factors). To improve the performance of the PMSM effectively, the applied control strategy for the PMSM in different sampling periods will vary on the basis of the introduced effect factors.

Experimental results clearly indicate that the novel CVM-DTC scheme exhibits excellent control of torque and flux linkage with lower steady-state ripples when compared to CDTC and DDTC, and faster transient response performances when compared to SVM-DTC.

Author Contributions: This paper was a collaborative effort between the authors. T.Y. and D.W. proposed the original idea; T.Y. wrote the full manuscript and carried out the experiments.

Acknowledgments: This work was supported in part by National Natural Science Foundation of China under Grant 61,433,004 and 51467017, and in part by National Key Research and Development Program of China under Grant 2017YFB1300900.

References

1. Cheema, M.A.M.; Fletcher, J.E.; Xiao, D.; Rahman, M.F. A direct thrust control scheme for linear permanent magnet synchronous motor based on online duty ratio control. *IEEE Trans. Power Electron.* **2016**, *31*, 4416–4428. [CrossRef]

2. Abosh, A.; Zhu, Z.Q.; Ren, Y. Reduction of torque and flux ripples in space vector modulation-based direct torque control of asymmetric permanent magnet synchronous machine. *IEEE Trans. Power Electron.* **2017**, *32*, 2976–2986. [CrossRef]

3. Zhou, Y.Z.; Chen, G.T. Predictive DTC Strategy with fault-tolerant function for six-phase and three-phase PMSM series-connected drive system. *IEEE Trans. Ind. Electron.* **2018**, *65*, 9101–9112. [CrossRef]

4. Shinohara, A.; Inoue, Y.; Morimoto, S.; Sanada, M. Direct calculation method of reference flux linkage for maximum torque per ampere control in DTC-based IPMSM drives. *IEEE Trans. Power Electron.* **2016**, *32*, 2114–2122. [CrossRef]

5. Alsofyani, I.M.; Idris, N.R.N.; Lee, K.B. Dynamic hysteresis torque band for improving the performance of lookup-table-based DTC of induction machines. *IEEE Trans. Power Electron.* **2018**, *33*, 7959–7970. [CrossRef]

6. Yuan, T.Q.; Wang, D.Z.; Li, Y.L. Duty ratio modulation strategy to minimize torque and flux linkage ripples in IPMSM DTC system. *IEEE Access.* **2017**, *5*, 14323–14332. [CrossRef]

7. Putri, A.K.; Rick, S.; Franck, D.; Hameyer, K. Application of sinusoidal field pole in a permanent-magnet synchronous machine to improve the NVH behavior considering the MTPA and MTPV operation area. *IEEE Trans. Ind. Appl.* **2016**, *52*, 2280–2288. [CrossRef]

8. Tatte, Y.N.; Aware, M.V.; Pandit, J.K.; Nemade, R. Performance improvement of three-level five-phase inverter-fed DTC-controlled five-phase induction motor during low-speed operation. *IEEE Trans. Ind. Appl.* **2018**, *54*, 2349–2357. [CrossRef]

9. Payami, S.; Behera, R.K.; Iqbal, A. DTC of three-level NPC inverter fed five-phase induction motor drive with novel neutral point voltage balancing scheme. *IEEE Trans. Power Electron.* **2018**, *33*, 1487–1500. [CrossRef]

10. Xia, C.; Zhao, J.; Yan, Y.; Shi, T. A novel direct torque and flux control method of matrix converter-fed PMSM drives. *IEEE Trans. Power Electron.* **2014**, *29*, 5417–5430. [CrossRef]

11. Yan, Y.; Zhao, J.; Xia, C.; Shi, T. Direct torque control of matrix converter-fed permanent magnet synchronous motor drives based on master and slave vectors. *IET Power Electron.* **2015**, *8*, 288–296. [CrossRef]

12. Mohan, D.; Zhang, X.; Foo, G.H.B. A simple duty cycle control strategy to reduce torque ripples and improve low-speed performance of a three-level inverter fed DTC IPMSM drive. *IEEE Trans. Ind. Electron.* **2017**, *64*, 2709–2721. [CrossRef]

13. Niu, F.; Wang, B.; Babel, A.S.; Li, K.; Strangas, E.G. Comparative evaluation of direct torque control strategies for permanent magnet synchronous machines. *IEEE Trans. Power Electron.* **2016**, *31*, 1408–1424. [CrossRef]

14. Zhang, Y.; Zhu, J.; Xu, W.; Guo, Y. A simple method to reduce torque ripple in direct torque-controlled permanent-magnet synchronous motor by using vectors with variable amplitude and angle. *IEEE Trans. Ind. Electron.* **2011**, *58*, 2848–2859. [CrossRef]

15. Ren, Y.; Zhu, Z.Q.; Liu, J. Direct torque control of permanent-magnet synchronous machine drives with a simple duty ratio regulator. *IEEE Trans. Ind. Electron.* **2014**, *61*, 5249–5258. [CrossRef]

16. Niu, F.; Li, K.; Wang, Y. Direct torque control for permanent magnet synchronous machines based on duty ratio modulation. *IEEE Trans. Ind. Electron.* **2015**, *62*, 6160–6170. [CrossRef]

17. Lee, J.S.; Lorenz, R.D. Deadbeat direct torque and flux control of IPMSM drives using a minimum time ramp trajectory method at voltage and current limits. *IEEE Trans. Ind. Appl.* **2014**, *50*, 3795–3804. [CrossRef]

18. Mohan, D.; Zhang, X.; Foo, G. Generalized DTC strategy for multilevel inverter fed IPMSMs with constant inverter switching frequency and reduced torque ripples. *IEEE Trans. Energy Convers.* **2017**, *32*, 1031–1041. [CrossRef]

19. Zhang, Z.; Zhao, Y.; Qiao, W.; Qu, L. A space-vector-modulated sensorless direct-torque control for direct-drive PMSG wind turbines. *IEEE Trans. Ind. Appl.* **2014**, *50*, 2331–2341. [CrossRef]

20. Zhang, X.; Foo, G.H. Over-modulation of constant switching frequency based DTC for reluctance synchronous motors incorporating field-weakening operation. *IEEE Trans. Ind. Electron.* **2019**, *66*, 37–47. [CrossRef]

21. Berzoy, A.; Rengifo, J.; Mohammed, O. Fuzzy predictive DTC of induction machines with reduced torque ripple and high performance operation. *IEEE Trans. Power Electron.* **2018**, *33*, 2580–2587. [CrossRef]

22. Do, T.D.; Choi, H.H.; Jung, J.W. Nonlinear Optimal DTC Design and Stability Analysis for Interior Permanent Magnet Synchronous Motor Drives. *IEEE/ASME Trans. Mechatron* **2716**, *20*.

23. Choi, Y.-S.; Choi, H.H.; Jung, J.W. Feedback linearization direct torque control with reduced torque and flux ripples for IPMSM drives. *IEEE Trans. Power Electron.* **2016**, *31*, 3728–3737. [CrossRef]

24. Zhang, Z.; Wei, C.; Qiao, W.; Qu, L. Adaptive saturation controller-based direct torque control for permanent-magnet synchronous machines. *IEEE Trans. Power Electron.* **2016**, *31*, 7112–7122. [CrossRef]

25. Liang, D.L.; Li, J.; Qu, R.H.; Kong, W.B. Adaptive second-order sliding-mode observer for PMSM sensorless control considering VSI nonlinearity. *IEEE Trans. Power Electron.* **2018**, *33*, 8994–9004. [CrossRef]

10

Series Active Filter Design Based on Asymmetric Hybrid Modular Multilevel Converter for Traction System

Muhammad Ali [1],*, Muhammad Mansoor Khan [1], Jianming Xu [2], Muhammad Talib Faiz [1], Yaqoob Ali [1], Khurram Hashmi [1] and Houjun Tang [1]

[1] School of Electronics, Information & Electrical Engineering (SEIEE), Smart Grid Research & Development Centre, Shanghai Jiao Tong University, Shanghai 200240, China; mansoor@sjtu.edu.cn (M.M.K.); talib_faiz@sjtu.edu.cn (M.T.F.); yaqoob.ali@uetpeshawar.edu.pk (Y.A.); khurram_hashmi@sjtu.edu.cn (K.H.); hjtang@sjtu.edu.cn (H.T.)

[2] Changzhou Power Supply Company, Changzhou 213176, Jiangsu, China; x_jianming@outlook.com

* Correspondence: engrmak.ee@sjtu.edu.cn

Abstract: This paper presents a comparative analysis of a new topology based on an asymmetric hybrid modular multilevel converter (AHMMC) with recently proposed multilevel converter topologies. The analysis is based on various parameters for medium voltage-high power electric traction system. Among recently proposed topologies, few converters have been analysed through simulation results. In addition, the study investigates AHMMC converter which is a cascade arrangement of H-bridge with five-level cascaded converter module (FCCM) in more detail. The key features of the proposed AHMMC includes: reduced switch losses by minimizing the switching frequency as well as the components count, and improved power factor with minimum harmonic distortion. Extensive simulation results and low voltage laboratory prototype validates the working principle of the proposed converter topology. Furthermore, the paper concludes with the comparison factors evaluation of the discussed converter topologies for medium voltage traction applications.

Keywords: hybrid converter; multi-level converter (MLC); series active filter; power factor correction (PFC)

---

1. Introduction

An electrified ac railway system, being an energy-efficient and governmentally-friendly medium of mass transportation, achieved great demand in many countries. However, to maximize the traffic, it still requires developing highly efficient, reliable, and compact traction systems with reduced cost, minimum time delay and less vibrations to assure passengers comfort [1]. The efficiency of railway traction systems can be improved with a regenerative braking system which enables transforming the kinetic energy of the rail vehicle in slowing down the speed into electrical power. Using a bidirectional converter, the electric power generated due to the regenerative braking system can be harvested for reuse [2].

A high efficiency and low production cost is required mostly by industrial processes, which can be achieved by increasing the power rating of electrical components/equipment with the reduced installation size. The power can be increased either by developing semiconductor devices with a capability to withstand high voltage or by introducing multilevel converters which will allow to connect the converter system directly to medium-voltage line. Recently, continuous development in power semiconductors devices i.e., high-voltage insulated-gate bipolar transistors (IGBTs) and integrated-gate commutated thyristors (IGCTs) and their use in self-commutated converters increases the nominal voltage and power ratings of the converter system.

Lower medium voltage high power railway traction system multi-level converters (MLC) have significant advantages over classical two-level and three-level converters due to their lower current harmonics distortion, less electromagnetic interference, increased output voltage level, and high power density. Furthermore, due to the simple layout of classical two-level converters, connecting single power semiconductor switch to medium grid voltages directly is not appropriate [3,4]. The key conventional MLC topologies i.e., neutral point clamped (NPC), flying capacitors (FC), and the cascaded H-bridge (CHB) have been reported in [4–8]. The MLC contains numerous power semiconductor switches and a dc-link capacitor which are arranged according to the required output voltage level. However, increasing the number of components accounts for attaining a high voltage level to increase the control complexity, which will affect the efficiency and reliability of the converter [6]. The requirement for MLC includes the dc-link voltage value at reference voltage level, unity power factor on the ac side and low harmonics distortion in the injected current to ac grid [9]. Therefore, an appropriate control scheme is required to track the current signal and maintain the dc-link capacitor voltages of MLC at their respective desired reference. A pulse width modulation (PWM) technique which is used by various MLC drive systems has been investigated in [10,11].

Recently, arrangement of different conventional MLC topologies known as hybrid multilevel converters has been introduced in [12,13]. In medium and high voltage range application, numerous multilevel topologies share the market for industrial applications. Hybrid MLC topologies have been a focus of interest due to substantial advantages which include, a redundant converter design, wide operating range, minimum line harmonics, and improved power factor. Some of these arrangements exist in the literature, such as NPC-H module [14], FC-H module [15], NPC-FC module [16] and H-cascaded module [17]. Furthermore, unequal dc link voltages are used which minimizes the redundant switching states and increases the output voltage level of the converter. Such converters are known as asymmetric converters in the literature [17,18].

The non-linearity of the converter injects harmonic currents in the ac grid, which adversely affects stability and the power quality of the ac grid [19]. Various techniques are available to mitigate high harmonics; a modulation scheme based on selected harmonic elimination pulse width modulation is studied in [12]. Various passive filters are introduced in [18,20,21] to overcome the harmonics issues in converters, but due to its system parameters, the performance is limited and may cause a resonance problem [22]. A compensator based on an electronics converter known as series active power filter (APF) converter is studied in the literature [23–25] to tackle the limit imposed by the high harmonics. Moreover, a converter based compensator in the existing research needs improvement in a supply system for the electrified traction system.

This paper presents a comparative study of the recently proposed MLC converter for the medium voltage-high power traction system which is examined through simulation results. Among all converters, a new topology based on an asymmetric hybrid modular multilevel converter (AHMMC), which is a series arrangement of a classical H-bridge and FCCM voltage source PWM converters, is investigated in more detail. The design of AHMMC topology, its corrective current and its voltage control methods are demonstrated. The overall system's controllability, total harmonic distortion (voltage and current), power factor and voltage stability on the dc-link capacitors is analysed through extensive simulation results. The proposed topology is practically validated through low-power laboratory prototype.

The paper is organized as follows. Section 2 includes a comparative analysis of five-level converter topologies and their output results are examined. Section 3 first investigates the proposed converter model, afterwards, its modulation strategy and control scheme are discussed. The simulation and experimental results of the proposed converter are presented and discussed in Section 4. An evaluation of comparison factors of the discussed converter topologies are examined in Section 5. Finally, the study is concluded in Section 6.

2. Comparative Analysis of 5-Level Converter Topologies

A comparative analysis of a few MLC converter arrangements i.e., NPC, hybrid NPC with H-bridge module, CHB based NPC, and HNPC with a cascaded module for lower-medium voltage application are discussed in this section.

2.1. NPC Converter

NPC converters are most widely used in industrial application from the last two decades among the other high power converters [26,27] in the range of 2.3 kV, 4.16 kV and 6 KV applications [28]. Figure 1 shows a NPC voltage source converter (VSC), is the most commonly used topology for self-commutated medium voltage converters (MVCs). The distinct features of NPC VSC ensures a better output voltage quality compared to the conventional two-level converter.

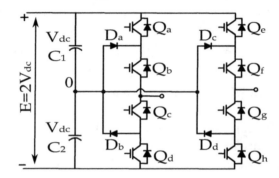

Figure 1. Five-level NPC converter topology.

The NPC converter circuit shown in Figure 1, contains eight power semiconductor switches, four clamping diodes $D_{(a-d)}$, a dc-bus voltage of E = 2Vdc and two dc-link capacitors, each of the dc-link capacitor is charged to half of the dc-bus voltage i.e., Vca = Vcb = $\frac{1}{2}$E. The output voltage of the NPC converter is set to five level $\pm2V_{dc}$, $\pm V_{dc}$, and 0. The modulation of converter switches can be achieved by three methods: (1) Carrier Based Pulse Width Modulation (CBPWM); (2) Selective Harmonic Elimination (SHE); and (3) Space Vector Modulation (SVM). In this study a CBPWM based modulation is used for a 3L-NPC, for which the maximum current is achieved at the maximum modulation index (m = 1). Based on topological design, consider that the NPC converter half bridge is under voltage stress equal to half of the dc-bus voltage, whereas the half bridge of the H-bridge converter is stressed equal to dc-bus voltages. To correlate the system voltage and required semiconductor voltage in the three-level NPC converter, a 3.3 KV ratings devices are equipped in 2.3 kV converter without series connection of devices. Using 3L-NPC VSC, an output voltage of up to 4.16 kV can be achieved without any series connection of devices [29–31]. Currently, 3L-NPC converters are available in market with a variety of devices. Siemens^{TM}, a European manufacturer, utilizes 3.3 and 6.5 KV IGBT modules devices where higher voltage levels are attained through series connection of semiconductor devices, to cover a range of 2.3 to 7.2 kV converters [32–34]. ABB^{TM} use 4.5 kV IGCTs [35] to offer 3.3 KV converters and Converteam^{TM} recently launched a 3.3 KV NPC product using the press-pack 4.5 kV IGBT technology [36].

2.2. Hybrid NPC with H-Bridge

A hybrid modular converter based on the cascaded arrangement of NPC five level cell (having two three-level NPC leg) with a three-level H-bridge cell and H-NPC module with a H-bridge as shown in Figure 2, are investigated in [37–39] for different voltage ratios of the dc-link among the two power cells. In the study, a converter cell which gives a fundamental component support operates at low frequency, whereas a converter cell as a series active filter operates at high frequency for high harmonics minimization and coupling inductor size reduction.

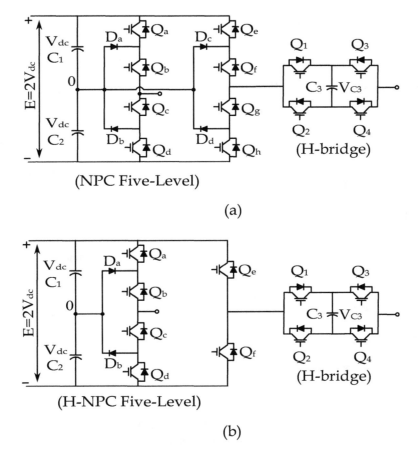

Figure 2. Five level NPC converter with H-bridge (**a**) NPC with H-bridge module, and (**b**) H-NPC with H-bridge module.

The power losses across the semiconductor switches are calculated and grouped by the power cell of each converter. The total power loss(P_t) includes, switching loss (P_{sw}) and conduction loss (P_c). The conduction losses of the H-NPC cell are less due to the minimum number of power semiconductor devices. The switching losses of the harmonic compensation converter will be higher because of the high voltage stress and high average switching frequency.

2.3. CHB Based NPC Converter

As discussed above, the hybrid NPC with H-bridge configuration will cause a high power loss in the voltage stressed power semiconductor switches because of the high average switching frequency. To overcome the said issues, the dc-link voltage is equally distributed by replacing an H-bridge cell with a five-level cascaded H-bridge (CHB) as shown in Figure 3. However, the topology with a five level CHB arrangement requires an increased number of power semiconductor switches which will cause high conduction losses and require a greater number of drive circuits.

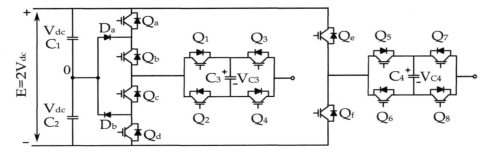

Figure 3. CHB based converter topology.

2.4. H-NPC with Cascaded Converter Module

Figure 4 depicts an arrangement for H-NPC with a cascaded module that is constructed in this study for lower medium voltage application. The topology has the benefit of a lower number of power semiconductor devices in comparison with the CHB based converter topology discussed in Section 2.3. The HNPC with cascaded module will minimise the total switch loss and number of drive circuits.

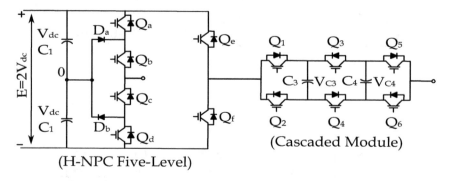

Figure 4. Five-level H-NPC with cascaded converter module.

2.5. Topological Analysis Based on Simulation Results

To verify the converter system performance, a few of these arrangements, including the NPC module, hybrid H-NPC with H-bridge module and H-NPC with cascaded converter module are simulated in Matlab/Simulink environment. A five-level and seven-level converter output voltage and the inductor current is obtained. The output voltage and inductor current of the converters are further analysed by fast Fourier transform (FFT). The simulation parameters of the NPC Converter are listed in Table 1.

Table 1. Simulation Parameters of NPC Converter.

Parameters	Value
Grid Voltage	1750 V (rms)
DC bus Voltage	3000 V
DC link Capacitor C_1, C_2	1500 V
Converter Current	70 Amp (rms)
L	0.6 mH
Modulation Depth	1
$Freq_{pwm}$	2 kHz

The simulation results of an NPC converter are shown in Figure 5. In Figure 5a,b the output voltage with its harmonics spectrum is shown using conventional PWM control technique with a modulation index m = 1. Figure 5c,d shows that the converter current is in phase with the grid voltage, ensuring the rectifier operation mode of the converter.

The simulation parameters of seven-level H-NPC with H-bridge module and H-NPC with cascaded converter module are presented in Table 2. By operating H-NPC with H-bridge module converter at maximum modulation index m = 1.06, a maximum seven-level of output voltage is achieved as shown in Figure 6a. The figure depicts the output voltage of the H-NPC, H-bridge and overall converter, whereas Figure 6b shows its respective harmonics spectrum by using conventional carrier based PWM control technique. Figure 7 shows H-NPC converter with H-bridge module operating in rectifier mode. It can be seen from Figure 7a that the converter current has the same phase with the grid voltage shown in Figure 7b. Figure 8 shows the simulation results of a 7-L HNPC with a cascaded module. Figure 8a depicts the output voltage of the H-NPC converter, cascaded module and overall converter, whereas Figure 8b shows its respective harmonics spectrum by using conventional carrier based PWM control technique with a modulation index m = 1.06. Figure 9 shows the converter operating in rectifier mode. It can be seen from Figure 9a that the converter current has the same phase with the grid voltage shown in Figure 9b.

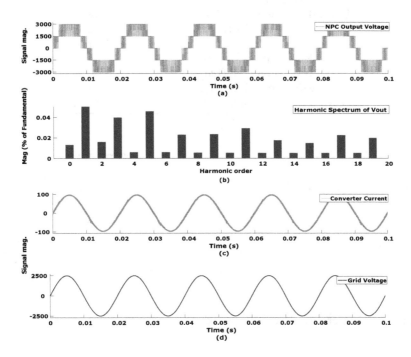

Figure 5. 5-L NPC converter topology (**a**) output voltage, (**b**) harmonic spectrum of output voltage, (**c**) converter current, and (**d**) grid voltage.

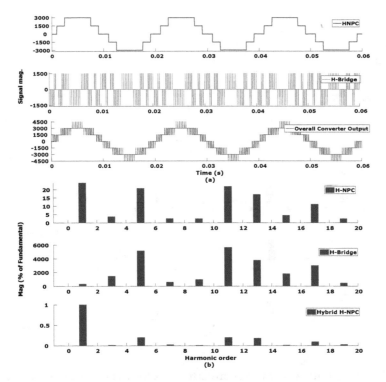

Figure 6. 7-L hybrid H-NPC with H-bridge converter topology (**a**) output voltage, and (**b**) harmonic spectrum of output voltage.

Table 2. Simulation parameters of Hybrid H-NPC converter.

Parameters	Value
Grid Voltage	1750 V (rms)
DC bus Voltage	3000 V
DC link Capacitor C_1, C_2, C_3, C_4	1500 V
Converter Current	70 Amp (rms)
L	0.6 mH
Modulation Depth	1.06
H-NPC converter $freq_{pwm}$	350 Hz
H-Bridge cell $freq_{pwm}$	2 kHz
Cascaded module converter $freq_{pwm}$	2 kHz

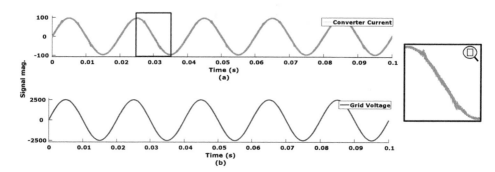

Figure 7. Converter operating in rectifier mode (**a**) converter current, and (**b**) grid voltage.

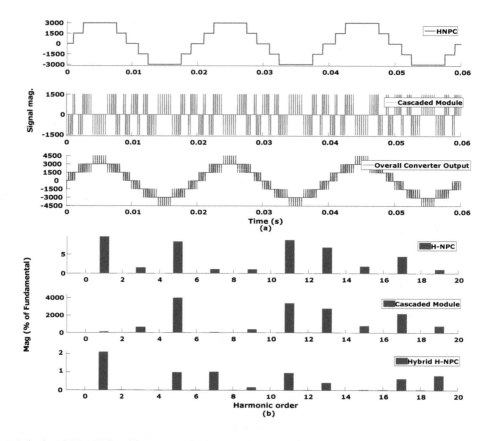

Figure 8. 7-L hybrid H-NPC with a cascaded converter topology (**a**) output voltage, and (**b**) harmonic spectrum of output voltage.

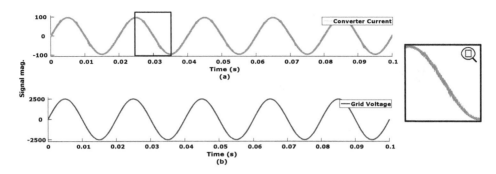

Figure 9. Converter operating in rectifier mode (**a**) converter current, and (**b**) grid voltage.

3. Proposed AHMMC Converter

3.1. Structure and Working Principle

Figure 10 shows the proposed AHMMC topology. The basic circuit configuration contains an FCCM in series with an H-bridge cell. An H-bridge converter contains four power semiconductor switches Q_{a-d}, a dc-bus voltage of $E = 2V_{dc}$, and a single dc-link capacitors C_a charged to the dc-bus voltage. The output voltage of the H-bridge converter $(V_{(AB)})$ is set to three level $\pm V_{dc}$, and 0. The FCCM which is the origin of the AHMMC topology contains six power semiconductor switches Q_{1-6} and two dc-link capacitors V_{c1} and V_{c2}, each of the dc-link is charged to $1/2\,E$ of the dc-bus voltage. The output voltage of FCCM $(V_{(BC)})$ is synthesized which generates 5-level that are $\pm 2V_{dc}, \pm V_{dc}$, and 0. The output voltage of the proposed AHHMC converter can be achieved by Equation (1):

$$V_{(AHMMC)} = V_{(AB)} + V_{(BC)} \tag{1}$$

The mathematical model of the proposed AHMMC grid tied converter can be expressed by the following set of Equations (2)–(4):

$$\frac{di_s}{dt} = \frac{1}{L}(S_x V_{c1} + S_y V_{c2} \mp S_z 2V_{dc} \pm V_g) \tag{2}$$

$$\frac{dV_{c1}}{dt} = \frac{S_x i}{C_1} \tag{3}$$

$$\frac{dV_{c2}}{dt} = \frac{S_y i}{C_2} \tag{4}$$

where $V_g = E\sin(wt)$ is a grid voltage, S_x, S_y and S_z are the switching functions of the AHMMC which can be expressed by the following Equations (5)–(7):

$$S_x = Q_1 Q_4 - Q_2 Q_3 \tag{5}$$

$$S_y = Q_4 Q_5 - Q_3 Q_6 \tag{6}$$

$$S_z = Q_a Q_d - Q_b Q_c \tag{7}$$

The eight distinct operating modes of FCCM are achieved by three complementary pairs of the power switches (Q_2, Q_4, Q_6 are the complement pair of Q_1, Q_3, Q_5, respectively) are listed in Table 3. Figure 11 depicts an output of seven-level $\pm 3V_{dc}, \pm 2V_{dc}, \pm V_{dc}$, and 0 which is achieved by operating the converter modules with the combination of switching states. Table 4 summarises the seven-level output voltage of the AHMMC converter.

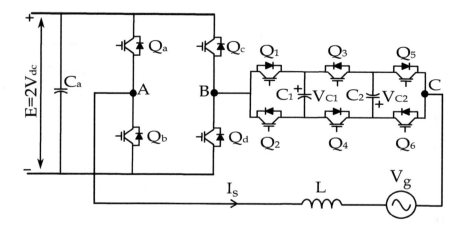

Figure 10. Proposed AHMMC converter topology.

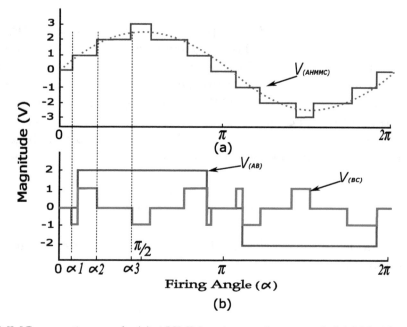

Figure 11. AHMMC operation mode (**a**) AHMM output voltage, and (**b**) H-bridge output voltage ($V_{(AB)}$) and FCCM output voltage ($V_{(BC)}$).

Table 3. Switching modes of FCCM module.

Modes	Switching States			Capacitor	Capacitor	FCCM Output Voltage
-	Q_1	Q_4	Q_5	C_1	C_2	$V_{(BC)}$
				Table (a) when $i_s > 0$		
1	1	1	1	discharge	discharge	$2V_{dc}$
2	1	1	0	discharge	by pass	V_{dc}
3	0	1	0	by pass	by pass	0
4	1	0	0	by pass	charge	$-V_{dc}$
5	0	0	0	charge	charge	$-2V_{dc}$
				Table (b) when $i_s > 0$		
1	1	1	1	discharge	discharge	$2V_{dc}$
2	0	1	1	by pass	discharge	V_{dc}
3	1	0	1	by pass	by pass	0
4	0	0	1	charge	by pass	$-V_{dc}$
5	0	0	0	charge	charge	$-2V_{dc}$

Table 4. AHMMC Converter Output Voltage.

Angle	$V_{(AB)}$	$V_{(BC)}$	$V_{(AHMMC)} = (V_{(AB)}) + (-V_{(BC)})$
$0 \leq \theta \leq \alpha_1$	0	0	0
$\alpha_1 \leq \theta \leq \alpha_2$	0	$-V_{dc}$	V_{dc}
$\alpha_1 \leq \theta \leq \alpha_2$	$2V_{dc}$	V_{dc}	V_{dc}
$\alpha_2 \leq \theta \leq \alpha_3$	$2V_{dc}$	0	$2V_{dc}$
$\alpha_3 \leq \theta \leq \pi$	$2V_{dc}$	$-V_{dc}$	$3V_{dc}$

3.2. Modulation Technique

A phase opposition disposition (POD) multi-carriers modulation strategy is implemented as a PWM technique for the proposed converter. In order to minimize the switching loss, the converter switches are categorised in two sets (i) high frequency switching of low voltage switches; (ii) low frequency switching of high voltage switches. Design of the heat dissipation of the system and switch type are the parameter for optimal selection of high frequency switches. In the proposed AHMMC converter, the switches Q_a and Q_c are under high voltage stress and therefore, they are operated with the fundamental switching frequency to minimise the switching loss. The voltage stress on the switches of the FCCM converter is not symmetric. Within the FCCM, the middle leg switches Q_3 and Q_4 are under high voltage stress and therefore, for the design consideration the PWM frequency of these switches is restricted below 1 kHz which will reduce the switching loss, whereas the PWM frequency for the outer leg switches Q_2 and Q_6 are set to 2 kHz (1 kHz effective switching frequency) due to the permissible operating range of high voltage devices up-to 2 kV. Moreover, the modulation scheme studied in this work has the benefit of a wide modulation index range. Figure 12 shows the graph of modulation index versus dc-link voltage value for series active filter (FCCM) at various operating angle.

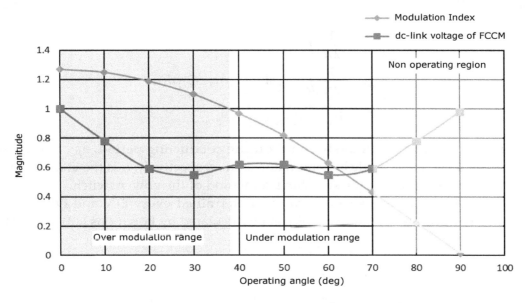

Figure 12. Modulation index versus dc-link voltage of FCCM.

3.2.1. Over Modulation Range (1~1.27)

In the proposed AHMMC converter, the H-bridge cell will provide a fundamental support whereas the FCCM module will cancel harmonics produced by the H-bridge cell. The overall output voltage level of the proposed AHMMC converter depends on the modulation index (m) which is expressed by Equation (8). Where θ is the initial phase angle of the h-bridge cell which is calculated using Equation (9).

$$m = \frac{4}{\pi}\cos(\theta) \tag{8}$$

$$\theta = arccos\frac{\pi}{4}\frac{V_g}{V_{dc}} \tag{9}$$

At $\theta = 0$, a 27% maximum modulation index is achieved which results in the higher output voltage of the proposed converter. Figure 12 shows the over modulation range along with the dc-link voltage required for harmonic compensation. When the theta lies between ($0 \leq \theta < 38$) degree, maximum seven-level output voltage is achieved which will lower the current value for the same power rating of the converter.

3.2.2. Switch Stress and Modulation Range (0.4~1)

Operating the proposed converter in the modulation range of m = 0.4~1, a five level converter output voltage is achieved with approximately half of the dc-bus voltage required for series active filter (FCCM). This will lower the voltage stress on the semiconductor switches. The voltage stress on the middle leg of the FCCM will be 0.6 E, whereas, voltage stresses on the upper legs will be 0.3E. Therefore, the switching frequency can be varied according to the system design.

3.3. Switch Losses

The switch losses were divided into the conduction and switching power losses. The total power switch losses P_T are grouped by power converters. The proposed AHMMC topology studied in this paper consists of 10 power switches with anti-parallel diodes. The conduction loss of the switch depends on the power conduction through the switch due to direction of the current. The calculation of total conduction losses P_{cd} at any instant across the power switch (p_{sw}) and diode (d) can be given as Equations (10)–(12):

$$P_{c\_m(p_{sw})} = \frac{1}{2\pi}\int_0^{2\pi} [(m_{p_{sw}} \times v_{p_{sw}})i_s \cdot Q_l]\, d(\omega t) \tag{10}$$

$$P_{c\_m(d)} = \frac{1}{2\pi}\int_0^{2\pi} [(m_d \times v_d)i_s \cdot Q_l]\, d(\omega t) \tag{11}$$

$$P_{cd} = P_{c\_m(p_{sw})} + P_{c\_m(d)} \tag{12}$$

where $m_{p_{sw}}$ and m_d are switch and diode count during conducting period, $v_{p_{sw}}$ and v_d are voltage drops of the power switch and diode in conduction state, respectively, i_s is the conduction current of the power switch device, and Q_l is the switching command of the power switch.

The switching losses of active switch can be determined every ON and OFF state during a reference period. The overall switching losses can be calculated by Equations (13) and (14):

$$P_{sw} = P_{sw\_on} + P_{sw\_off} \tag{13}$$

$$= \sum_{j=1}^{2n+2} \left[\frac{1}{6}v_{b(j)} \cdot i \cdot (t_{on} + t_{off})f_j\right] \tag{14}$$

where, v_b, i and f_j is the blocking voltage, a conduction current, and the switching frequency of the active switch, respectively.

The total power switch losses across the converter can be computed as:

$$P_T = P_{cd} + P_{sw} \tag{15}$$

3.4. Control Scheme

For a stable operation of the proposed converter, corrective current and voltage control techniques shown in Figure 13 are implemented to track the reference current and reference voltage to ensure the unity power factor and balanced dc-link voltage, respectively. An inner current control loop is implemented at the FCCM module; since the operating frequency of this converter is high it will respond quickly to any change.

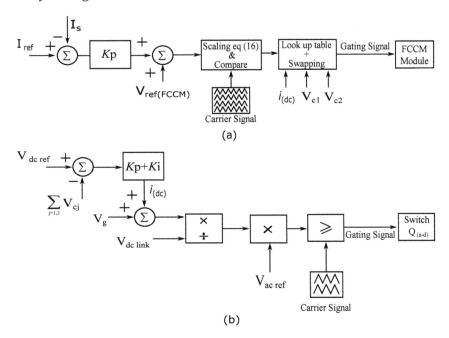

Figure 13. Corrective control scheme (**a**) current control loop, and (**b**) voltage control loop.

Figure 13a, shows a current control loop, in which an error signal is achieved by comparison of measured current I_s and I_{ref} which is further compensated by a proportional gain K_p and then added in the modulating signal for the FCCM converter. The dc link voltage variation impact in the current is minimized through dc voltage feed forward loop control by adjusting the PWM accordingly as described in Equation (16):

$$d = \left(p_1 \frac{|V_{out}|}{V_{c_1}} + p_2 \frac{|V_{out}|}{V_{c_2}} + p_3 \frac{|V_{out}|-V_{c_1}}{V_{c_2}} + p_4 \frac{|V_{out}|-V_{c_2}}{V_{c_1}}\right) \times Sign(V_{out}) \qquad (16)$$

where d and $p_{(1-4)}$ is the duty cycle and PWM output conditions for FCCM, respectively. Table 5 presents PWM output conditions, whereas the direction of PWM output is determined by $Sign(V_{out})$ which is expressed by Equation (17):

$$Sign(V_{out}) = \begin{cases} 1, if \ V_{out} \geq 0 \\ -1, if \ V_{out} < 0 \end{cases} \qquad (17)$$

Figure 13b shows the voltage control loop implemented for the proposed converter. In order to attain the desired output voltage level of the converter, The PWM allows to conduct through one node of the capacitor, which leads to the unbalancing of the capacitor V_{c1} and V_{c2}. Moreover, some of the power dissipates across the switches of the FCCM module which is fed by dc-link capacitors; therefore, a voltage control is needed to maintain balanced voltage on the dc-link. The voltage control loop is categorised into two parts (i) common mode voltage control (ii) voltage balancing among the capacitors. The common mode voltage is controlled by modifying the operating angle of the fundamental component of the H-bridge converter. For a stable operation of the converter, the difference among the dc-link capacitor voltages needs to be balanced, which is achieved by the swapping technique, using the redundant switching states given in Table 3. It is presented in Table 3 that the voltage level $\pm V$ and 0 can be achieved in multiple ways with their respective states of charging and discharging of capacitor. The charge swapping using redundant switching states is based on the following set of rules.

- When $(i_{(s)} \times V_{ref}) < 0$, if $V_{C1} < V_{C2}$, then Table 3b will be selected.
- When $(i_{(s)} \times V_{ref}) > 0$, if $V_{C1} > V_{C2}$, then Table 3a will be selected.

where $i_{(dc)}$ is a direction of the dc reference current. Using the above relation of selecting tables, the balancing of capacitor voltage is achieved.

Table 5. PWM operating conditions.

Operating Condition	p_1	p_2	p_3	p_4
V_{c1} is a lower capacitor				
$\|V_{out}\| \leq V_{c1}$	1	0	0	0
$V_{c1} < \|V_{out}\| < (V_{c1} + V_{c2})$	0	0	1	0
V_{c2} is a lower capacitor				
$\|V_{out}\| \leq V_{c2}$	0	1	0	0
$V_{c2} < \|V_{out}\| < (V_{c1} + V_{c2})$	0	0	0	1

4. Simulation and Experimental Results

The AHMMC converter is simulated in MATLAB/Simulink environment to demonstrate the system's effectiveness, and performance of the modulation and control scheme. The AHMMC converter is designed for lower medium voltage application, hence efficiency greater then 97% is expected. However, an experimental validation is performed at low voltage laboratory prototype in which the proposed converter efficiency is determined by conduction losses of the semiconductor switches, which is not a very good indicator of real power losses appearing on the system, operating at a voltage range greater than 1 KV. To verify the system results, simulation results are scaled accordingly to a low voltage laboratory prototype using the system's parameters presented in Table 6, which are the same for both simulation and experimental models. The selection of all component values has been done by extensive simulations. The criteria for capacitors selection are based on the voltage ripple and stability consideration of a converter i.e., the capacitor capacity should be high enough to withstand imbalance to a couple of cycles. Through simulation analysis, approximately 2% capacitor ripple is designed, whereas the inductor L value is selected for a current ripple of 0.5% (peak to peak) under full rated current.

Table 6. System parameters of AHMMC.

Parameter	Value
Inductance	0.6 mH
DC-link Capacitor	20 μF
FCCM Switching Frequency	2 kHz
Capacitor Voltage	50 V_{dc}
Grid Voltage	110 (rms)
Current I_s	14 A (rms)
DC-bus Voltage	100 V

Figure 14 depicts the simulation and practical results of the AHMMC converter output voltage at m = 1.26. The three angles per quarter cycle operation gives an output of seven-level. Figure 14a shows simulation results, whereas Figure 14b shows experimental results of the stable converter operation because of balanced capacitor voltages. An active power is transferred by compensating the harmonics produced by H-bridge through series active filter (FCCM).

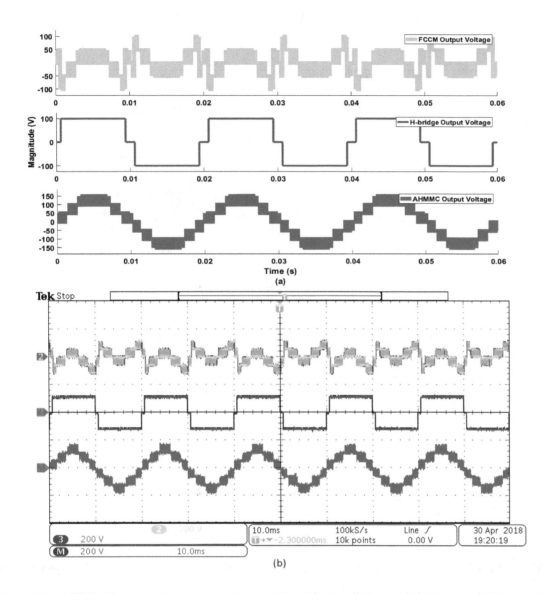

Figure 14. AHMMC converter output at m = 1.26, (**a**) simulation validation, and (**b**) experimental validation, traces channels 2 and 3 (200 V/div).

Figure 15 depicts the simulation and practical results of the AHMMC converter output voltage at m = 0.89. The two angles per quarter cycle operation gives an output of five-level. Figure 15a shows simulation results, whereas Figure 15b shows experimental results of the stable converter operation due to the balanced dc-link capacitor voltages. The situation is similar when an active power is transferred by compensating the harmonics produced by the H-bridge through series active filter (FCCM).

Figure 16 shows simulation and practical validation of the balance operation of the converter with balanced capacitors voltage due to voltage swapping technique by redundant switching state selection presented in Table 3 together with converter current. Figure 17 depicts the simulation and practical results of the grid voltage and converter current. The signals are in-phase which tells that the converter works as a rectifier. Figure 18 depicts the low voltage laboratory prototype of a proposed AHMMC converter. Table 7 presents the simulation and experimental current THD value of the converter operating at different power ratings.

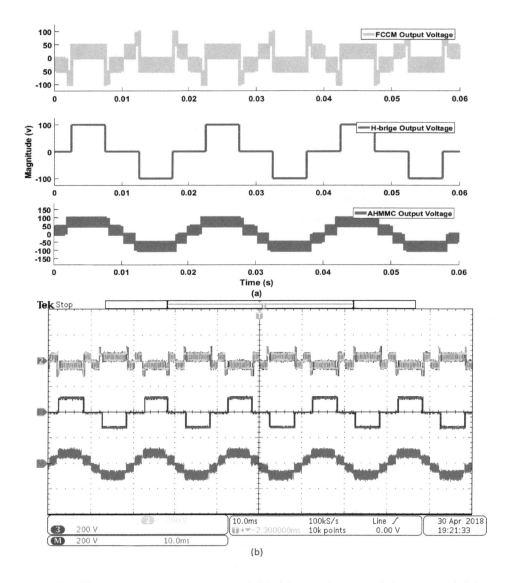

Figure 15. AHMMC converter output at m = 0.89, (**a**) simulation validation, and (**b**) experimental validation, traces chanels 2 and 3 (200 V/div).

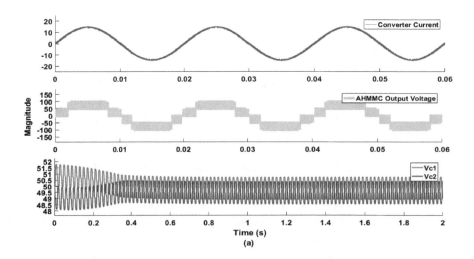

Figure 16. *Cont.*

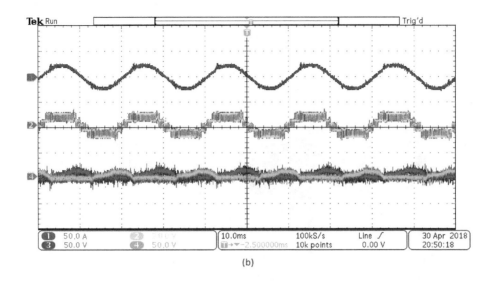

(b)

Figure 16. AHMMC converter output, (**a**) simulation validation of capacitor voltage balancing togather with a unity power factor, and (**b**) experimental validation, traces channels 1, 3, 4 (50 V/div), and 2 (200 V/div).

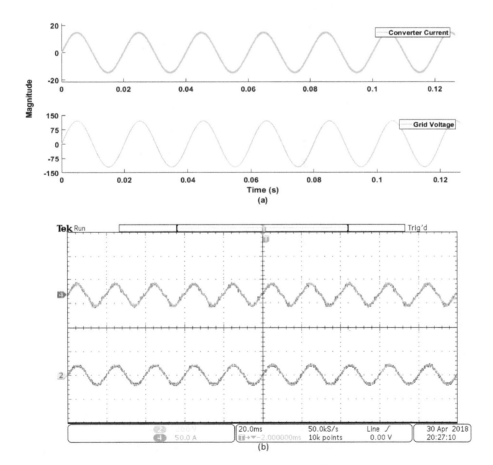

Figure 17. Low distortion in input current, (**a**) simulation validation of minimum distorion in current injected to ac main, and (**b**) experimental validation of minimum distortion in current injected to ac main. traces channels 2 (200 V/div), and 4 (50 A/div).

Table 7. Current THD of AHMMC operating at differnt power rating.

Current THD	Power Rating		
	25%	**50%**	**100%**
Simulation THD	6.55	3.31	1.66
Experiment THD	7.95	4.87	2.98

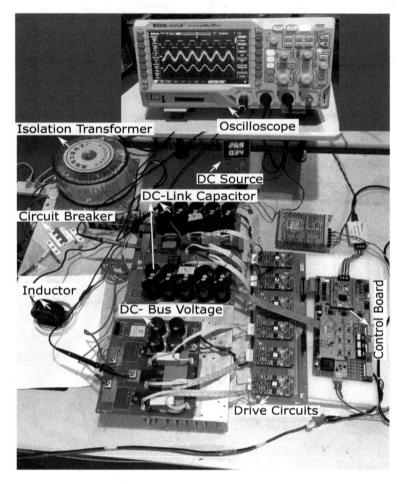

Figure 18. Experimental setup.

5. Comparision and Discussion

In this paper, the important characteristics of five different converter configurations were examined. A comparison is carried out among these topologies under power semiconductor devices (IGBT) available in different voltage levels (2.3, 3.3, and 4.16 kV) [30].

The NPC VSC has superior features compared to other hybrid converters presented here for upper voltage range up-to 3 kV for present devices operating at 1 kHz average switching frequency [40]. The key features include low conduction losses, better voltage and current THD, low components count and equal voltage stress on the active switches. However, due to the unequal loss distribution, the result is unsymmetrical temperature among the semiconductor devices.

Considering the dc bus voltage up-to 5 kV, the structural design of hybrid H-NPC with H-bridge has the benefit of seven-level output voltage by two more active switches which is inevitable. The topology can be used for higher voltage range as the dc link required for harmonic compensation is 0.35 E at maximum modulation index. Moreover, the switching losses is reduced by modulating the high voltage stressed switch at fundamental switching frequency. The structural drawback of the H-NPC converter is the high voltage stresses on half bridge leg of H-NPC converter. The voltage stresses can be higher across the H-bridge for the topology used in higher range voltage application.

A CHB based converter is introduced to tackle the limit imposed by high voltage stresses which is reduced by splitting the dc link voltage of the H-bridge converter using five-level CHB for higher voltage application. However, the high component count will lead to a high system cost. A topology based on a H-NPC and cascaded module achieves better output voltage with reduced active switches and minimum conduction loss in comparison with CHB based topology.

The proposed converter topology is designed for a lower medium voltage range up-to 5 kV with the maximum output voltage of seven-level. Considering the same system parameters, the proposed topology has certain advantages over other hybrid converter topologies, for example, the wide operating rang, low current THD presented in Table 11, and the lower component count in comparison with the hybrid converter topologies listed in Table 8. The switching losses are addressed by operating the high stressed switch with low switching frequency; the switching loss of the proposed converter is lower which results in higher efficiency in comparison with NPC VSC as shown in Figure 19 based on the converter parameters given in Table 9. However, the current THD of the NPC converter is less than the AHMMC converter.

Table 8. Component count of converter.

Converter	Components Count		
	Active Switches	Diode	Dc-Link Capacitor
NPC	8	4	2
HNPC with H-bridge	10	2	3
HNPC with cascaded module	12	2	4
Proposed AHMMC converter	10	-	3

Table 9. Converter parameter for efficiency comparison.

Parameter	Value			
DC bus voltage	3 kV			
Grid voltage	1.75 kV (rms)			
Modulation index	Maximum modulation index			
Reactor inductance	0.6 mH			
	V_{CES}	V_{CESat}	Rise time	Fall time
IGBT	1.7 kV	2.45 V	0.29 μs	0.29 μs
IGBT	4.5 kV	2.84 V	0.35 μs	0.35 μs

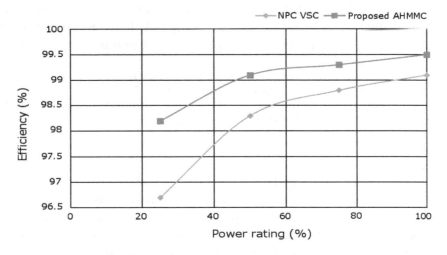

Figure 19. Converter's efficency.

Table 10 presents voltage stress on the semiconductor switches. The proposed topology has four active switches (in the outer leg of FCCM) which are under low voltage stress; thus, they are operated at high frequency to minimize the current ripple together with inductor size. The current total harmonic distortion (THD) of the converter operating at different power ratings is given in Table 11, which shows that the proposed converter current THD is lower than the other hybrid converter topologies discussed here.

Table 10. Voltage stress on semiconductor devices.

Semiconductor Devices	Voltage Stress	
	Active Switches	Diode
Half bridge NPC	E/2	E/2
Half bridge	E/2	–
Cascaded module centre cell	E	–
Cascaded module outer cell	E/2	–
Half Bridge (AHMMC)	E	–
Cascaded module centre cell	E	–
Cascaded module outer cell	E/2	–

Table 11. Converters current THD operating at different power ratings.

Converter	Power Rating		
	25%	50%	100%
NPC	5.31	2.83	1.38
HNPC with H-bridge	6.71	3.43	1.73
HNPC with cascaded module	6.59	3.39	1.71
Proposed AHMMC	6.55	3.31	1.66

6. Conclusions

In this paper, a comparative analysis of MLC topologies for medium voltage traction application has been demonstrated. The study will be helpful to select a suitable converter among the converters discussed for specific voltage ranges. Among all the converters, the proposed AHMMC configuration is studied in detail for medium voltage high power application due to its various features including a better output voltage, reduced switch losses by minimizing the switching frequency as well as the component count, and improved power factor with low THD. The proposed system performance is validated through simulation results using MATLAB/Simulink. Furthermore, a low voltage laboratory prototype is developed to test the performance of the proposed converter in practice.

Author Contributions: M.A. and M.M.K. proposed the idea for writing the manuscript. M.M.K. and H.T. suggested the literature and supervised in writing the manuscript. M.T.F. and K.H. helped M.A. in writing and formatting. Y.A. helped in modifying the figures and shared the summary of various credible articles to be included in this manuscript. X.J. helped in practical system design.

References

1. Feng, J.; Chu, W.Q.; Zhang, Z.; Zhu, Z.Q. Power Electronic Transformer-Based Railway Traction Systems: Challenges and Opportunities. *IEEE J. Emerg. Sel. Top. Power Electron.* **2017**, *5*, 1237–1253. [CrossRef]
2. Flinders, F.; Oghanna, W. Energy efficiency improvements to electric locomotives using PWM rectifier technology. In Proceedings of the 1995 International Conference on Electric Railways in a United Europe, Amsterdam, The Netherlands, 27–30 March 1995; pp. 106–110. [CrossRef]

3. Rodriguez, J.; Lai, J.S.; Peng, F.Z. Multilevel inverters: A survey of topologies, controls, and applications. *IEEE Trans. Ind. Electron.* **2002**, *49*, 724–738. [CrossRef]
4. Nordvall, A. *Multilevel Inverter Topology Survey*; Chalmers University of Technology: Goteborg, Swenden, 2011.
5. Chauhan, N.; Jana, K.C. Cascaded multilevel inverter for underground traction drives. In Proceedings of the 2012 IEEE International Conference on Power Electronics, Drives and Energy Systems (PEDES), Bengaluru, India, 16–19 December 2012; pp. 1–5. [CrossRef]
6. Pulikanti, S.R.; Konstantinou, G.S.; Agelidis, V.G. Generalisation of flying capacitor-based active-neutralpoint-clamped multilevel converter using voltage-level modulation. *IET Power Electron.* **2012**, *5*, 456–466. [CrossRef]
7. Malinowski, M.; Gopakumar, K.; Rodriguez, J.; Perez, M.A. A Survey on Cascaded Multilevel Inverters. *IEEE Trans. Ind. Electron.* **2010**, *57*, 2197–2206. [CrossRef]
8. Salinas, F.; Gonzalez, M.A.; Escalante, M.F. Voltage balancing scheme for flying capacitor multilevel converters. *IET Power Electron.* **2013**, *6*, 835–842. [CrossRef]
9. Cecati, C.; Dell'Aquila, A.; Liserre, M.; Monopoli, V.G. A passivity-based multilevel active rectifier with adaptive compensation for traction applications. *IEEE Trans. Ind. Appl.* **2003**, *39*, 1404–1413. [CrossRef]
10. Tarisciotti, L.; Zanchetta, P.; Watson, A.; Bifaretti, S.; Clare, J.C.; Wheeler, P.W. Active DC Voltage Balancing PWM Technique for High-Power Cascaded Multilevel Converters. *IEEE Trans. Ind. Electron.* **2014**, *61*, 6157–6167. [CrossRef]
11. Konstantinou, G.; Capella, G.J.; Pou, J.; Ceballos, S. Single-Carrier Phase-Disposition PWM Techniques for Multiple Interleaved Voltage-Source Converter Legs. *IEEE Trans. Ind. Electron.* **2018**, *65*, 4466–4474. [CrossRef]
12. Pulikanti, S.R.; Konstantinou, G.; Agelidis, V.G. Hybrid Seven-Level Cascaded Active Neutral-Point-Clamped-Based Multilevel Converter Under SHE-PWM. *IEEE Trans. Ind. Electron.* **2013**, *60*, 4794–4804. [CrossRef]
13. Ren, L.; Gong, C.; He, K.; Zhao, Y. Modified hybrid modulation scheme with even switch thermal distribution for H-bridge hybrid cascaded inverters. *IET Power Electron.* **2017**, *10*, 261–268. [CrossRef]
14. Chattopadhyay, S.K.; Chakraborty, C.; Pal, B.C. Cascaded H-Bridge amp; neutral point clamped hybrid asymmetric multilevel inverter topology for grid interactive transformerless photovoltaic power plant. In Proceedings of the IECON 2012—38th Annual Conference on IEEE Industrial Electronics Society, Montreal, QC, Canada, 25–28 October 2012; pp. 5074–5079. [CrossRef]
15. Sneineh, A.A.; Wang, M.Y. A Novel Hybrid Flying-Capacitor-Half-Bridge Cascade 13-Level Inverter for High Power Applications. In Proceedings of the 2007 2nd IEEE Conference on Industrial Electronics and Applications, Harbin, China, 23–25 May 2007; pp. 2421–2426. [CrossRef]
16. Teymour, H.R.; Sutanto, D.; Muttaqi, K.M.; Ciufo, P. A Novel Modulation Technique and a New Balancing Control Strategy for a Single-Phase Five-Level ANPC Converter. *IEEE Trans. Ind. Appl.* **2015**, *51*, 1215–1227. [CrossRef]
17. Ali, M.; Mansoor, M.; Tang, H.; Rana, A. Analysis of a seven-level asymmetrical hybrid multilevel converter for traction systems. *IET Power Electron.* **2017**, *10*, 1878–1888. [CrossRef]
18. Carnielutti, F.; Pinheiro, H. Hybrid Modulation Strategy for Asymmetrical Cascaded Multilevel Converters Under Normal and Fault Conditions. *IEEE Trans. Ind. Electron.* **2016**, *63*, 92–101. [CrossRef]
19. Wang, S.; Song, W.; Zhao, J.; Feng, X. Hybrid single-carrier-based pulse width modulation scheme for single-phase three-level neutral-point-clamped grid-side converters in electric railway traction. *IET Power Electron.* **2016**, *9*, 2500–2509. [CrossRef]
20. Hu, H.; He, Z.; Gao, S. Passive Filter Design for China High-Speed Railway With Considering Harmonic Resonance and Characteristic Harmonics. *IEEE Trans. Power Deliv.* **2015**, *30*, 505–514. [CrossRef]
21. Faiz, M.T.; Khan, M.M.; Jianming, X.; Habib, S.; Tang, H. Parallel feedforward compensation based active damping of LCL-type grid connected inverter. In Proceedings of the 2018 IEEE International Conference on Industrial Technology (ICIT), Lyon, France, 20–22 February 2018; pp. 788–793. [CrossRef]
22. He, Z.; Hu, H.; Zhang, Y.; Gao, S. Harmonic Resonance Assessment to Traction Power-Supply System Considering Train Model in China High-Speed Railway. *IEEE Trans. Power Deliv.* **2014**, *29*, 1735–1743. [CrossRef]
23. Jianben, L.; Shaojun, D.; Qiaofu, C.; Kun, T. Modelling and industrial application of series hybrid active power filter. *IET Power Electron.* **2013**, *6*, 1707–1714. [CrossRef]

24. Pereda, J.; Dixon, J. 23-Level Inverter for Electric Vehicles Using a Single Battery Pack and Series Active Filters. *IEEE Trans. Veh. Technol.* **2012**, *61*, 1043–1051. [CrossRef]

25. Ortuzar, M.E.; Carmi, R.E.; Dixon, J.W.; Moran, L. Voltage-source active power filter based on multilevel converter and ultracapacitor DC link. *IEEE Trans. Ind. Electron.* **2006**, *53*, 477–485. [CrossRef]

26. Nabae, A.; Takahashi, I.; Akagi, H. A New Neutral-Point-Clamped PWM Inverter. *IEEE Trans. Ind. Appl.* **1981**, *IA-17*, 518–523. [CrossRef]

27. Baker, R.H. Bridge Converter Circuit. U.S. Patent 4270163, 26 May 1981.

28. Rodriguez, J.; Bernet, S.; Steimer, P.K.; Lizama, I.E. A Survey on Neutral-Point-Clamped Inverters. *IEEE Trans. Ind. Electron.* **2010**, *57*, 2219–2230. [CrossRef]

29. Sayago, J.A.; Bruckner, T.; Bernet, S. How to Select the System Voltage of MV Drives; A Comparison of Semiconductor Expenses. *IEEE Trans. Ind. Electron.* **2008**, *55*, 3381–3390. [CrossRef]

30. Bernet, S. State of the art and developments of medium voltage converters-an overview. *Przeglkad Elektrotechn.* **2006**, *82*, 1–10.

31. Rodriguez, J.; Bernet, S.; Wu, B.; Pontt, J.O.; Kouro, S. Multilevel Voltage-Source-Converter Topologies for Industrial Medium-Voltage Drives. *IEEE Trans. Ind. Electron.* **2007**, *54*, 2930–2945. [CrossRef]

32. D12, Siemens. Sinamics Catalogue. In *Automation and Drives*; Siemens AG: Nuremberg, Germany, 2006.

33. D12-Supplement 04/2007, Siemens. Sinamics Catalogue. In *Automation and Drives*; Siemens AG: Nuremberg, Germany, 2007.

34. Dietrich, C.; Gediga, S.; Hiller, M.; Sommer, R.; Tischmacher, H. A new 7.2kV medium voltage 3-Level-NPC inverter using 6.5kV-IGBTs. In Proceedings of the 2007 European Conference on Power Electronics and Applications, Aalborg, Denmark, 2–5 September 2007; pp. 1–9. [CrossRef]

35. ABB Industry. *Drive ACS6000 Brochure and Datasheet Rev. B*; ABB Switzerland Ltd.: Turgi, Switzerland, 2005.

36. Jakob, R.; Keller, C.; Mohlenkamp, G.; Gollentz, B. 3-Level high power converter with press pack IGBT. In Proceedings of the 2007 European Conference on Power Electronics and Applications, Aalborg, Denmark, 2–5 September 2007; pp. 1–7. [CrossRef]

37. Sousa, R.P.R.; Jacobina, C.B.; Bahia, F.A.C.; Barros, L.M. Comparative analysis of cascaded inverters based on 5-level and 3-level H-bridges. In Proceedings of the 2017 IEEE Energy Conversion Congress and Exposition (ECCE), Cincinnati, OH, USA, 1–5 October 2017; pp. 2161–2167. [CrossRef]

38. Xu, Y.; Zou, Y. Research on a novel hybrid cascade multilevel converter. In Proceedings of the 2007 International Power Engineering Conference (IPEC 2007), Singapore, 3–6 December 2007; pp. 1081–1085.

39. Kai, D.; Yunping, Z.; Lei, L.; Zhichao, W.; Hongyuan, J.; Xudong, Z. Novel Hybrid Cascade Asymmetric Inverter Based on 5-level Asymmetric inverter. In Proceedings of the 2005 IEEE 36th Power Electronics Specialists Conference, Recife, Brazil, 16 June 2005; pp. 2302–2306. [CrossRef]

40. Sanchez-Ruiz, A.; Mazuela, M.; Alvarez, S.; Abad, G.; Baraia, I. Medium Voltage–High Power Converter Topologies Comparison Procedure, for a 6.6 kV Drive Application Using 4.5 kV IGBT Modules. *IEEE Trans. Ind. Electron.* **2012**, *59*, 1462–1476. [CrossRef]

New Fault-Tolerant Control Strategy of Five-Phase Induction Motor with Four-Phase and Three-Phase Modes of Operation

Sonali Chetan Rangari [1,*], Hiralal Murlidhar Suryawanshi [2] and Mohan Renge [1]

[1] Department of Electrical Engineering, Shri Ramdeobaba College of Engineering and Management, Nagpur 440013, India; rangaris@rknec.edu

[2] Department of Electrical Engineering, Visvesvaraya National Institute of Technology, Nagpur 44013, India; hms_1963@rediffmail.com

* Correspondence: renkey10@yahoo.co.in

Abstract: The developed torque with minimum oscillations is one of the difficulties faced when designing drive systems. High ripple torque contents result in fluctuations and acoustic noise that impact the life of a drive system. A multiphase machine can offer a better alternative to a conventional three-phase machine in faulty situations by reducing the number of interruptions in industrial operation. This paper proposes a unique fault-tolerant control strategy for a five-phase induction motor. The paper considers a variable-voltage, variable-frequency control five-phase induction motor in one- and two-phase open circuit faults. The four-phase and three-phase operation modes for these faults are utilized with a modified voltage reference signal. The suggested remedial strategy is the method for compensating a faulty open phase of the machine through a modified reference signal. A modified voltage reference signal can be efficiently executed by a carrier-based pulse width modulation (PWM) system. A test bench for the execution of the fault-tolerant control strategy of the motor drive system is presented in detail along with the experimental results.

Keywords: five-phase machine; fault-tolerant control; induction motor; one phase open circuit fault (1-Ph); adjacent two-phase open circuit fault (A2-Ph); volt-per-hertz control (scalar control)

---

1. Introduction

In electric drives and machines, a three-phase machine is the default implementation in industrial applications. Emphasis should be placed on possibilities with more than a three-phase machine which is difficult to achieve with conventional three-phase machines. The simple expansion of three-phase drives to multiphase drives is not sufficient. It is highly important to investigate inventive employment of the extra degrees of flexibility. Incorporation of more than three phases is advised to improve performance. The advantages that can be achieved with the utilization of multiphase systems are investigated in [1]. Numerous endeavors concluded that multiphase machines have some inherent advantages such as higher reliability, higher frequency of torque pulsation with lower amplitude, lower rotor harmonic current, reduction in current per phase without expanding the voltage per phase, and less current ripple in the DC link [1–5].

Multiphase system reliability is most important in safety-critical applications, such as, electric ships, compressors, pumps, electric aircraft, hybrid vehicles and marine applications. In recent high power industrial applications, a multi-leg voltage source inverter (VSI) was used for multi-phase induction motors for variable-voltage, variable-frequency control.

In many industrial applications, if open-circuit fault exists in any phase of three-phase machine, it leads to considerably large torque oscillations. These oscillations are double the electrical line frequency, which may affect the shaft of the machine. For fault-tolerant control of three-phase machine requires separate current control for remaining healthy phases by enabling the connection of motor star point to the DC link midpoint [6]. Broad investigations have been accounted for open-circuit fault-tolerant mode of operation for three-phase AC machines [7–14]. Increasing the number of phases provides better sinusoidal Magneto Motive Force (MMF) distribution, which decreases torque ripples and harmonic currents compared to three-phase machines [15,16]. A five-phase machine is superior to a three-phase machine for fault-tolerant operation modes. When single-phase (1-Ph) or adjacent double-phase (A2-Ph) open circuit faults occur, the machines can remain in operation using other healthy phases without additional hardware and control [17–20].

A five-phase machine with the star-connected stator winding with no neutral connection can work as a four-phase machine when a single-phase open circuit fault (1-Ph) occurs. Similarly, it works as a three-phase machine when adjacent double phase open circuit faults (A2-Ph) occur. These faulty conditions generate torque oscillations due to unbalanced rotating MMF present in the air gap [21]. Connecting a load neutral point to the DC link midpoint reduces the negative sequence MMF component in the air gap and the oscillation without any additional control strategy.

Phase sequences are highly important when considering AC motors, as the production of the torque via the sequential "rotation" of the applied five-phase power is responsible for the mechanical rotation of the rotor. The frequency of positive-sequence is used to drive the rotor in the required direction, whereas the frequency of negative-sequence operates motor in the opposite direction of the rotation of the rotor. However, the frequency of the zero-sequence neither adds to nor detracts from the torque of the rotor. Because of the distortion in the current, an excessive number of harmonics of negative-sequence (5th, 11th, 17th and/or 23rd) is observed in the power, and if this power is applied to a five-phase AC machine, it will result in deterioration of the performance as well as possible overheating.

Many investigations have been accounted for the open-phase fault-tolerant operation of multiphase induction machines [6,20,21], developed fault-tolerant control algorithm including non-linearities of machine and converter in the modeling of open-phase fault drive system. The speed control of five-phase induction motor by using finite-control set model-based predictive control for fault-tolerant condition is introduced in [22]. The fundamental and third-harmonic component of current is used as a fault-tolerant control technique for the excitation of healthy stator phases has been proposed in [23]. The aim of this work is to represent reconfiguration of motor phase currents under one-phase and two-phase open fault condition. This paper presents the implementation of a remedial strategy to neutralize ripple in the torque and analyzed the motor-performance in four-phase and three-phase modes of operation.

The contribution of this work is

i.   Insight into the asymmetrical post-fault mode of operation and the remedial strategy compensates the unbalanced rotating MMF present in the air gap of the machine by a modified reference signal.
ii.   The control strategy is emphasized on the reduction in torque oscillations and verified with a reduction in unbalanced line current.
iii. By using volt-per-hertz (V/f) scalar control, a voltage compensation control algorithm is developed in the dsPIC33EP256MU810 Digital Signal Controller.
iv.   Pre-fault and post-fault mode of operation with fault remedial technique is experimentally verified and discussed.
v.   The method presented here enhances the continuity of the star-connected five-phase induction motor in case of one-phase and two-phase open faults.

It is assumed that the stator winding is opened in a five-phase induction motor because of gate failure of the inverter, i.e., an open switch condition.

2. Fault-Tolerant Remedial Strategy of a Five-Phase System

The schematic arrangement for a five-leg inverter with an induction motor is represented in Figure 1. The arrangement is composed of a five-phase voltage source inverter (VSI) with dc-link. Based on the industrial application dc-link voltage (VDC) can be supplied through a DC source. Five-phase motor drive system consists of phase shift of 72o symmetrical connection of the stator windings and separate neutral connection (n) [24]. The switch S1 denotes a gate drive open fault, and the switch S2 denotes the short-circuit device fault. It is not recommended to run the drive under foresaid faulty conditions even though another device in the same leg of the inverter is in healthy condition. The switch S3 is included in phase "a" to isolate the faulty leg in order to analyze the continuous operation of drive in four-phase mode of operation under a healthy and open-phase fault condition. Similarly, a gate drive open fault or switch short circuit fault in two adjacent or alternate legs of the inverter, may cause a two-phase open fault. The two faulty legs should be isolated and the drive runs in three-phase mode operation [22]. The modeling and performance of the five-phase voltage source inverter with the five-phase induction machine in pre-fault and post-fault conditions is briefly explained in section A, B, and C, respectively.

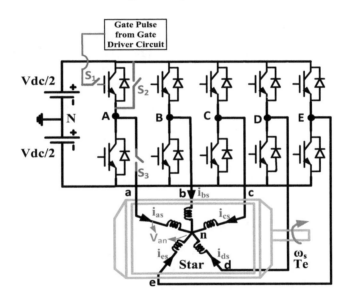

Figure 1. Five-phase system with 1-Ph fault.

2.1. Voltage Source Inverter

A five-phase drive for a machine can be obtained by developing a five-leg voltage source inverter (VSI). The phase voltages of the motor with this inverter are signified in (1).

Lowercase alphabetical letters (a–e) represent the phase voltages and the inverter leg voltages are represented by capital letters (A, B, C, D, E). Each switch conducts for 180°, giving a ten-step mode of operation.

The phase difference between two conducting switches in any sequential two phases is 720 [3,20]. For star-connected load phase-to-neutral voltages are obtained by determining the difference between the voltage of the neutral point 'n' of the load and the negative point of the dc-bus 'N'.

$$V_i = V_j + V_{nN} \tag{1}$$

where, i is {A, B, C, D, E} and j is {a, b, c, d, e}.

Since in a star-connected load the aggregate of phase voltages equals to zero and the sum of the equations yields.

$$V_{nN} = \frac{1}{5} \times (V_A + V_B + V_C + V_D + V_E) \tag{2}$$

Replacing (2) into (1), the loads with phase-to-neutral voltages are as follows:

$$
\left.\begin{aligned}
V_{nN} &= \tfrac{1}{5} \times (V_A + V_B + V_C + V_D + V_E) \\
V_b &= \tfrac{4}{5}V_B - \tfrac{1}{5}(V_A + V_C + V_D + V_E) \\
V_c &= \tfrac{4}{5}V_C - \tfrac{1}{5}(V_A + V_B + V_D + V_E) \\
V_d &= \tfrac{4}{5}V_D - \tfrac{1}{5}(V_A + V_B + V_C + V_E) \\
V_e &= \tfrac{4}{5}V_E - \tfrac{1}{5}(V_E + V_B + V_C + V_D)
\end{aligned}\right\}
\tag{3}
$$

The phase values of inverter-leg voltages are $\pm 0.5 VDC$. For a fixed modulation index M and dc-link voltage VDC, the fundamental inverter leg voltages analogous to a star-connected winding can be given as [3,20].

$$
V_{Ph\ Star} = M\frac{V_{DC}}{2} \times \sin(\omega_s t)
\tag{4}
$$

2.2. Modeling of a Five-Phase Induction Motor

The five-phase induction motor is provided with an IGBT-based five-phase voltage source converter (VSCs) drive system. A DC-link voltage is provided from a diode bridge rectifier, which exclusively permits unidirectional power flow. The Clarke matrix for this particular case is:

$$
T(\theta) = \frac{2}{5}\begin{bmatrix}
Cos\,\theta & Cos\left(\theta - \tfrac{2\pi}{5}\right) & Cos\left(\theta - \tfrac{4\pi}{5}\right) & Cos\left(\theta + \tfrac{4\pi}{5}\right) & Cos\left(\theta + \tfrac{2\pi}{5}\right) \\
Sin\,\theta & Sin\left(\theta - \tfrac{2\pi}{5}\right) & Sin\left(\theta - \tfrac{4\pi}{5}\right) & Sin\left(\theta + \tfrac{4\pi}{5}\right) & Sin\left(\theta + \tfrac{2\pi}{5}\right) \\
Cos\,\theta & Cos\left(\theta + \tfrac{4\pi}{5}\right) & Cos\left(\theta - \tfrac{2\pi}{5}\right) & Cos\left(\theta + \tfrac{2\pi}{5}\right) & Cos\left(\theta - \tfrac{4\pi}{5}\right) \\
Sin\,\theta & Sin\left(\theta + \tfrac{4\pi}{5}\right) & Sin\left(\theta - \tfrac{2\pi}{5}\right) & Sin\left(\theta + \tfrac{2\pi}{5}\right) & Cos\left(\theta - \tfrac{4\pi}{5}\right) \\
\tfrac{1}{2} & \tfrac{1}{2} & \tfrac{1}{2} & \tfrac{1}{2} & \tfrac{1}{2}
\end{bmatrix}
\tag{5}
$$

$$
\begin{bmatrix} i_{ds} & i_{qs} & i_{xs} & i_{ys} & i_{os}\end{bmatrix}^t = T(\theta)\begin{bmatrix} i_{as} & i_{bs} & i_{cs} & i_{ds} & i_{es}\end{bmatrix}^t
\tag{6}
$$

Clark transformation is used to disintegrate the phase a, b, c, d, e variable into two subspaces, the d-q, x-y and the zero variable components. The d-q subspaces are orthogonal to each other and provide basic torque and flux production. In healthy operation, they can be independently controlled; the x-y subspace is not coupled with the d-q subspace. In case of 1-Ph and A2-Ph open circuit faults, the d-q and x-y components are coupled with each other. The mapping of the various harmonics with the subspaces are as follows: order of the harmonics $10n \pm 1$ (where n = 1, 2, 3, 4 ...) are mapped with the q-d subspace including the fundamental component while the order of the harmonics $5n \pm 1$ (n = 1, 3, 5, 7 ...) are mapped with the x-y subspace.

The mathematical modelling equations of the machine assuming sinusoidal distributed symmetrical windings and linear flux path are represented below. Using vector space decomposition phase voltage equations of the stator winding in a stationary reference frame [1] are,

$$
\left.\begin{aligned}
V_{qs} &= r_s i_{qs} + \frac{d\{L_{ls}i_{qs} + L_m(i_{qs} + i_{qr})\}}{dt} + \omega\{L_{ls}i_{ds} + L_m(i_{ds} + i_{dr})\} \\
V_{ds} &= r_s i_{ds} + \frac{d\{L_{ls}i_{ds} + L_m(i_{ds} + i_{dr})\}}{dt} - \omega\{L_{ls}i_{qs} + L_m(i_{qs} + i_{qr})\} \\
V_{xs} &= r_s i_{xs} + \frac{d(L_{ls}i_{xs})}{dt} \\
V_{ys} &= r_s i_{ys} + \frac{d(L_{ls}i_{ys})}{dt} \\
V_{os} &= r_s i_{os} + \frac{d(L_{ls}i_{os})}{dt}
\end{aligned}\right\}
\tag{7}
$$

Since the neutral of the five-phase winding disconnected, zero-sequence currents i_0 cannot flow and are precluded from the investigation. The x-y currents are not connected with the rotor side, leaving only circulating currents which flow in the stator winding and create stator copper loss. The production of torque is associated with the d-q subspace as in the case of three-phase. By using Equation (7), the electromagnetic torque and machine's rotor speed can be determined as follows.

$$T_e = \frac{5}{2}\frac{p}{2}\frac{L_m}{(L_{lr} + L_m)}(\lambda_{dr}i_{qs} - \lambda_{qr}i_{ds})$$

$$\omega_r = \int \frac{p}{2J}(T_e - T_L) \tag{8}$$

where,

$$\lambda_{qr} = L_{ls}i_{qr} + L_m(i_{qs} + i_{qr}) \text{ and } \lambda_{dr} = L_{ls}i_{dr} + L_m(i_{ds} + i_{dr})$$

The five-phase machine drive works on normal/healthy operation with zero x-y currents. The conventional variable-voltage, variable-frequency (V/f) controller is used to control the d-q currents rotating in the positive reference frame apart from the faulty five-phase drive system required to incorporate a controller to control the circulating x-y currents. The control scheme of V/f and the equations of the d-q plane remain same as in three-phase machines.

2.3. Four-Phase and Three-Phase Modes of Operation

In the incident of an open-gate drive circuit and switch short circuit it is compulsory to diagnose and isolate the faulty leg before the control strategy is reconstructed [24]. For the concept of fault-tolerant control technique, suppose the induction motor carries regulated balanced five-phase sinusoidal currents, which gives positive sequence rotating MMF.

$$i_a = I_m \cos(\omega_e t)$$

$$i_b = I_m \cos\left(\omega_e t - \frac{2\pi}{5}\right)$$

$$i_c = I_m \cos\left(\omega_e t - \frac{4\pi}{5}\right) \tag{9}$$

$$i_d = I_m \cos\left(\omega_e t + \frac{4\pi}{5}\right)$$

$$i_e = I_m \cos\left(\omega_e t + \frac{2\pi}{5}\right)$$

By considering stator winding sinusoidal distribution, the stator current generates rotating MMF, hence effective resultant rotating MMF is the summation of the MMFs generated by each of the five phases. Under normal healthy operation, five-phase stator currents give balanced healthy positively rotating MMF. The resultant MMF is specified by,

$$F_s = \frac{5}{2}NI_m \cos(\omega_e t - \phi) = i_a + ai_b + a^2i_c + a^3i_d + a^4i_e \tag{10}$$

where, $a = e^{j2\pi/5}$ and N is the active stator turns per phase with spatial angle denoted by Ø. For "disturbance-free" operation during 1-Ph, A2-Ph, or A3-Ph open circuit faults, the winding of themachine carries harmonic distributed currents. This current produces MMF that should be the same as that in the healthy condition. For example, if phase "a" is isolated due to an open gate drive fault or device fault or machine windings fault, a rotating positive forward field is feasible by setting ia, equal to zero, Equation (10) becomes,

$$\frac{5}{2}NI_m \cos(\omega_e t - \varnothing) = ai'_b + a^2i'_c + a^3i'_d + a^4i'_e \tag{11}$$

By separating real and imaginary terms of the Equation

$$5\frac{I_m}{2}\cos(\omega t) = \cos\left(\frac{2\pi}{5}\right)(i'_b + i'_e) + \cos\left(\frac{4\pi}{5}\right)(i'_c + i'_d)$$

$$5\frac{I_m}{2}\sin(\omega t) = \sin\left(\frac{2\pi}{5}\right)(i'_b - i'_e) + \sin\left(\frac{4\pi}{5}\right)(i'_c - i'_d) \tag{11a}$$

To find a solution by assuming that each winding has the same current magnitude, so that

$$i'_b = -i'_d \text{ and } i'_c = -i'_e \tag{11b}$$

which gives the currents in the remaining phases are

$$\left.\begin{array}{l}
i'_b = \frac{5i_m}{4(\sin\frac{2\pi}{5})^2}\cos\left(\omega t - \frac{\pi}{5}\right) = 1.382 I_m \cos\left(\omega t - \frac{\pi}{5}\right) \\[2mm]
i'_c = \frac{5i_m}{4(\sin\frac{2\pi}{5})^2}\cos\left(\omega t - \frac{4\pi}{5}\right) = 1.382 I_m \cos\left(\omega t - \frac{4\pi}{5}\right) \\[2mm]
i'_d = \frac{5i_m}{4(\sin\frac{2\pi}{5})^2}\cos\left(\omega t + \frac{4\pi}{5}\right) = 1.382 I_m \cos\left(\omega t + \frac{4\pi}{5}\right) \\[2mm]
i'_e = \frac{5i_m}{4(\sin\frac{2\pi}{5})^2}\cos\left(\omega t + \frac{\pi}{5}\right) = 1.382 I_m \cos\left(\omega t + \frac{\pi}{5}\right)
\end{array}\right\} \tag{12}$$

With the modified stator currents, d-q currents get modified which gives the electromagnetic torque,

$$T'_e = \frac{5}{2}\frac{p}{2}\frac{L_m}{(L_{lr} + L_m)}\left(\lambda'_{dr}i'_{qs} - \lambda'_{qr}i'_{ds}\right) \tag{12a}$$

To clarify this, by assuming phase "a" is isolated, the resultant rotating MMF produced by the stator winding currents will be composed of a negative-sequence component and a positive-sequence component. The remaining phase currents are expressed in such a way that there is only forward rotating MMF [13]. Hence, if any phase is open-circuited, "disturbance-free" control is possible with the modification of adjacent phases. If phase "a" is an open phase "b" advanced by 360 and phase "c" is retarded by 360. Figure 2b shows the phasor relationships before and after phase "a" is suddenly open-circuited.

However, due to the open circuit fault coupled d-q and x-y current components, the x-y currents cannot be zero. Because the d-q currents remain unchanged, it is necessary to maintain a forward rotating MMF and smooth post-fault operation (Figure 2b) [3]. By using the Clark transformation, "x" current is equal to the "d" current with the negative value: $i_x = -i_d$ and it is possible to remove the "x" component by using (12). If two adjacent phases open i.e., "a" and "b" as shown in Figure 2c,d, then the equation with real and imaginary terms of,

$$\frac{5}{2}NI_m\cos(\omega_e t - \varnothing) = a^2 i''_c + a^3 i''_d + a^4 i''_e \tag{13}$$

With the assumption is that no neutral connection is required

$$i''_c + i''_d + i''_e = 0 \tag{14}$$

Solving Equations (13) and (14)

$$i'_e = \frac{5i_m}{4(\sin\frac{2\pi}{5})^2}\cos\left(\omega t + \frac{\pi}{5}\right) = 1.382 I_m \cos\left(\omega t + \frac{\pi}{5}\right)$$

$$i''_d = \frac{5I_m \cos\left(\frac{\pi}{5}\right)^2}{\left(\sin\frac{2\pi}{5}\right)^2} \cos\left(\omega t + \frac{4\pi}{5}\right) = 3.618I_m \cos\left(\omega t + \frac{4\pi}{5}\right)$$

$$i''_e = \frac{5I_m \cos\left(\frac{\pi}{5}\right)}{2\left(\sin\frac{2\pi}{5}\right)^2} \cos(\omega t) = 2.236I_m \cos(\omega t) \tag{15}$$

Similarly with modified stator currents under two adjacent open-phase fault, d-q currents get modified which gives the electromagnetic torque,

$$T''_e = \frac{5}{2}\frac{p}{2} \frac{L_m}{(L_{lr} + L_m)}\left(\lambda''_{dr} i''_{qs} - \lambda''_{qr} i''_{ds}\right) \tag{15a}$$

If three phases are open circuited i.e., "a", "b" and "c", for remedial strategy and disturbance-free operation the motor neutral must be connected to the dc mid-point so that remaining two phase currents can be individually controlled.

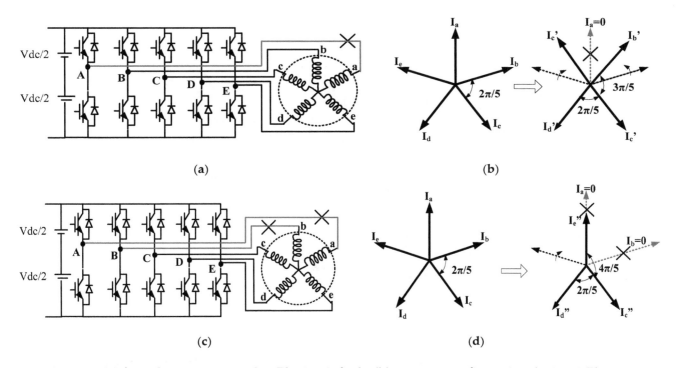

(a)   (b)

(c)   (d)

Figure 2. (**a**) five-phase system with 1-Ph circuit fault; (**b**) vector reconfiguration during 1-Ph open fault; (**c**) five-phase system with A2-Ph open circuit fault; (**d**) vector reconfiguration during A2-Ph open circuit fault.

3. Description of Fault-Tolerant Control Strategy

The general block diagram of a volt/Hz controlled power circuit of a five-phase induction motor drive is shown in Figure 3, which is in the fault mode condition. In most drives, conventional scalar control is used. Hence, the conventional scalar control method for fault-tolerant control strategy is represented. In conventional scalar control, reference signals are given to the pulse-width modulator which operates based on the common speed reference. Generation of these reference signals is as follows. V/f block is multiplied by the reference frequency (ω).

Generally, this is a fixed value depending on the rating of frequency and voltage of the machine. Five sinusoidal reference signals, vref a, vref b, vref c, vref d and vref e are the outputs of this block. These reference signals have an appropriate magnitude and operating frequency with a phase shift of 72° (Refer Figure 2b). These reference signals are fed to the modulator, which depends on the operating speed of the machine. In a modulator using a comparator, the reference signal and a saw-tooth signal are compared. The frequency of the saw-tooth signal is equal to the required switching frequency. The reference signal changes either at the healthy condition or at a faulty four-phase or three-phase modes of operation with the reference operating speed.

The cause of a negative-sequence component in the distorted stator current is due to the disconnection of a faulty phase in the motor winding. Accordingly, the negative-sequence current appears in the x-y component, which gives MMF in the negative sequence reference frame [13]. The proposed control strategy will try to reduce this x-y component to zero or a minimum. The output signal of the control strategy is the modified reference signal (refer Equations (12) and (13)), which is generated by using a standard constant V/f (scalar) control system. The resultant modified signal is then given to the modulator to get the proper switching pattern. In the healthy condition the modified reference signal making the negative-sequence component zero results in balanced five-phase line currents which are equal in peak values with a phase shift of 72°. If a 1-Ph open fault occurs, the remaining four active phase currents are rearranged by the controller so that they are equal in peak and have a phase shift of 72°, 108°, 72° and 108°, which cause only clockwise (i.e., positive-sequence) rotating MMF (refer to Equation (12)).

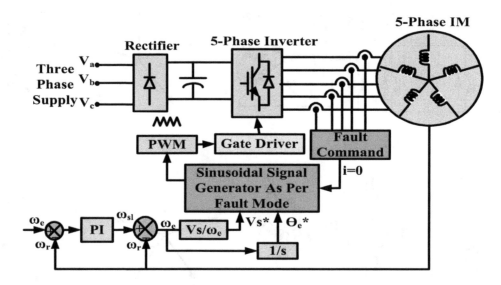

Figure 3. Block diagram of fault-tolerant control strategy.

This can be described as a virtual four-phase connected winding, with phase "a" isolated (refer to Figure 2a). The equivalent active four-phase currents of phase-b, c, d, e in the stator winding are equal in peak value (magnitude) with each other and a phase shift equal to 72°, 108°, 72° and 108° between them (refer to Figure 2b). Hence by keeping the position of vector "c" and "d" as it is and moving phase "b" vector in advance by 360 and phase "c" is retarded by 360 modified the switching pattern. In a similar manner with two-phase "a" and "b" isolated, the other two active currents of phases "c" and "e" in the stator winding are the same in magnitude. The magnitude of phase "d" current is 1.62 time of magnitude of other two currents. By keeping the position of vector "c" and "d" as it is and moving phase "e" vector at the position of "a" modified the switching pattern for two-phase open. (Refer to Figure 2c,d).

In a practical case, if a gate driver open circuit fault or switched short-circuit fault is occurring then it is recommended to disconnect the power lines of the inverter. Isolation switches are inserted in between inverter power lines and motor for laboratory experimentation. The current sensor is used to measure the line current. If the summation of all measured current is zero, then the induction motor drive is working in healthy condition. The opening of particular phase can be done using isolation switch for single phase or two phase open fault. Hence, phase current in specific phase becomes zero. Now summation of other phase currents are no longer being zero identifies the faulty condition. Also, it can be seen that due to an open-circuit fault, torque control is lost, i.e., torque is oscillating which oscillates the speed. Hence in V/f control, the speed control loop gets weak and oscillation in torque.

By checking each current with zero detects particular phase open fault. Proposed control logic provides the switching pattern to the adjacent phases of the specific open phase. Similarly, by checking adjacent current with zero gives an idea about adjacent phase open fault. For this, the proposed control logic is used to get the switching pattern for remaining three phases of the inverter. Accordingly, the proposed new fault-tolerant control strategy decide required switching pattern for four-phase and three-phase modes of operation in V/f control technique. The complete control logic is shown in the flowchart of Figure 4.

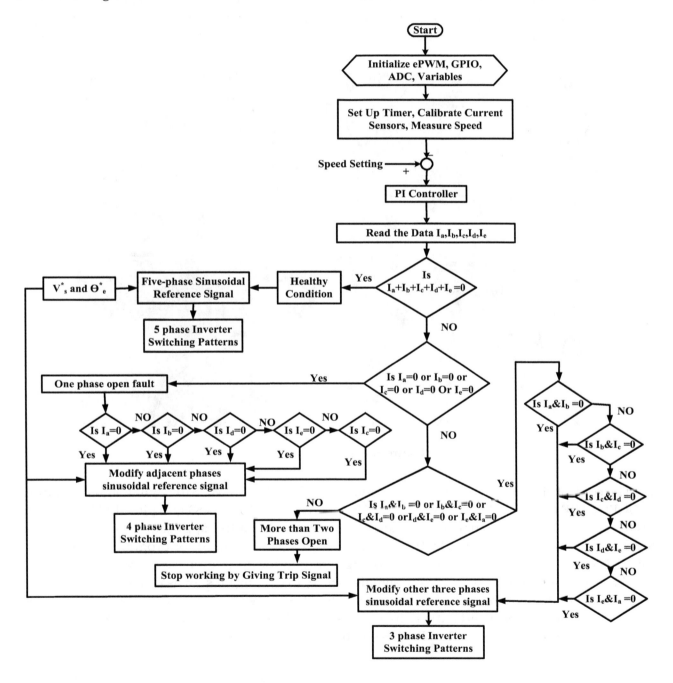

Figure 4. Flow-chart of fault-tolerant control strategy.

4. Experimental Results

4.1. Test Experimental Setup

To investigate the performance of the proposed fault-tolerant control technique, the drive comprises a 1 Hp five-phaseinduction squirrel cage symmetrically distributed induction motor. It has rs = 0.499 Ω, Lls = 2.7 mH, rr = 0.926 Ω, Llr =2.7 mH, Lm = 223mH, P = 2 and rotor inertia (J) = 0.047 kg-m^2.

The motor was designed so that it can be configured either as a star, pentagon or pentacle-connected stator [19]. The motor is composed of 40 stator slots with closed rotor bars. The five-phase induction machine is provided by an IGBT-based two-level five-phase inverter (Fairchild, Sunnyvale, CA, USA). The machine is driven in normal/healthy operation with modified reference signal controlled by a variable-voltage, variable-frequency (V/f) controller Vdc = 300 V (refer to Figure 5).

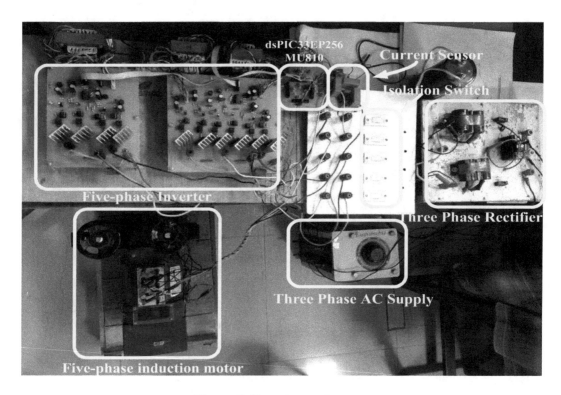

Figure 5. Experimental setup.

The control circuit of the drive is performed by dsPIC33EP256MU810 Digital Signal Controller, of MICROCHIP (MICROCHIP, Chandler, AZ, USA). This controller has 83 I/O pins, 12 PWM outputs, 2 ADC modules with 32 channels which are useful for motor-control applications. Code Composer Studio software (Version 7, Texas Instruments, Dallas, TX, USA) is used for programming of control unit. It is capable of simultaneously controlling two two-level three-phase inverters. A two-level, five-phase inverter requires only 10 gate signals; hence, PWM output signals can be directly given to the gate driver circuit of the five-phase inverter (Micrel, San Jose, CA, USA). Five hall-effect current sensors (LEM, Geneva, Switzerland) are used to measure the line current. The main processor has the fault-tolerant control technique. The PWM switching frequency was set to 10 kHz.

4.2. Experimental Results

The fault-tolerant strategy described in Section 3 is experimentally performed in the laboratory and results are shown in Figures 5–7. The load condition was at one fourth, i.e., 2 Nm, 50 Hz. This is because while doing experimentation the motor will be under loaded during the fault conditions, i.e., four-phase and three-phase modes of operation.

The motor-drive performance with line current, torque and circulating d-q current is indicated in these figures. For the healthy and faulty conditions, Figures 6–8 with a,b and d show the results of 1-Ph and A2-Ph open circuit faults without a control strategy. Figures 6–8 with c, e show when a control strategy was introduced. The experimental test results of line current for the 1-Ph open circuit fault are illustrated in Figure 6b,c) without and with a control strategy, respectively. Similarly, the line current waveform for the A2-Ph open circuit fault is represented in Figure 6d,e without and with a control strategy, respectively.

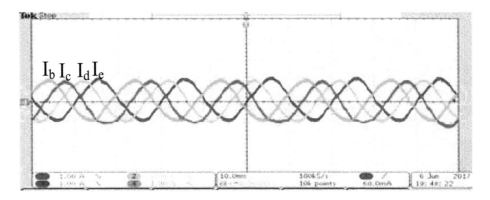

(**a**) Line current in amp (Scale: 1 A/div, 10 ms/div)

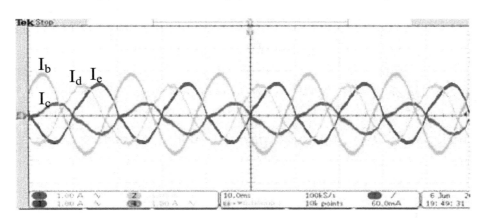

(**b**) Line current in amp (Scale: 1 A/div, 10 ms/div)

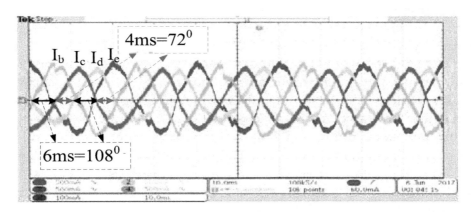

(**c**) Line current in amp (Scale: 500 mA/div, 10 ms/div)

Figure 6. *Cont.*

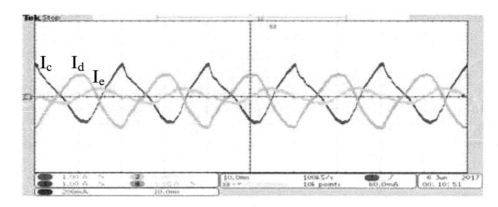

(**d**) Line current in amp (Scale: 1 A/div, 10 ms/div)

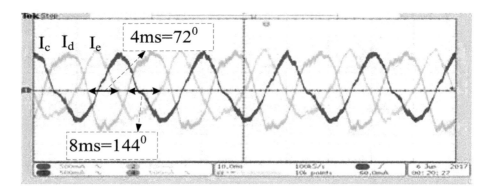

(**e**) Line current in amp (Scale: 500 mA/div, 10 ms/div)

Figure 6. Line current at (**a**) Healthy operation; (**b**) 1-Ph open circuit fault without control strategy; (**c**) 1-Ph open circuit fault with control strategy; (**d**) A2-Ph open circuit fault without control strategy; (**e**) A2-Ph open circuit fault with control strategy.

Unequal increase in the peak value of the line currents at any instant is due to one of the phases being disconnected. This increases the negative sequence component of the current. This control strategy maintained the equal magnitude of the active current and phase displacement which ensured that the torque pulsation was reduced (As shown in Figure 7c,e). Additionally, the current reduces the negative sequence component. The capability of this control strategy is verified by analyzing the unbalanced line current with balanced line current for both the cases of open phase fault (as shown in Figure 6). The line current obtained from the experiment for the four-phase and three-phase modes of operation with a deactivated control strategy is represented in Figure 6b,d. For the case of the activated control strategy, the line current waveforms are represented in Figure 6c,e. The effectiveness of this controller shows in the reduced magnitude of the line current with properly balanced current, which remarkably enhances the quality of the output torque of the five-phase machine.

Induction motor torque profile during the transition of a five-phase healthy mode to the four-phase and three-phase faulty modes of operation is represented in Figure 7. These figures show the effectiveness of this fault-tolerant method and corresponding quality of the fault control strategy, which maintains the quality of the motor's torque under faulty conditions. The motor output torque waveforms with control strategy deactivated for 1-Ph and A2-Ph faults is shown in Figure 7b,d. The motor output torque waveforms with control strategy activated is shown in Figure 7c,e. The torque pulsation is of approximately 3 N-m, while the developed torque is 2 N-m when this control strategy was not used at the steady-state condition. The torque pulsation decreased to less than 2.5 N-m when the control strategy was introduced, as shown in Figure 7c,e. Since for smooth post-fault operation, the MMF remains unchanged, the d-q currents describe nearly a circle as in healthy operation has shown Figure 8a. In contrast, currents cannot be circular, as shown in Figure 8b,c. The use of the control strategy current makes a near circular current as in the case of Figure 8d.

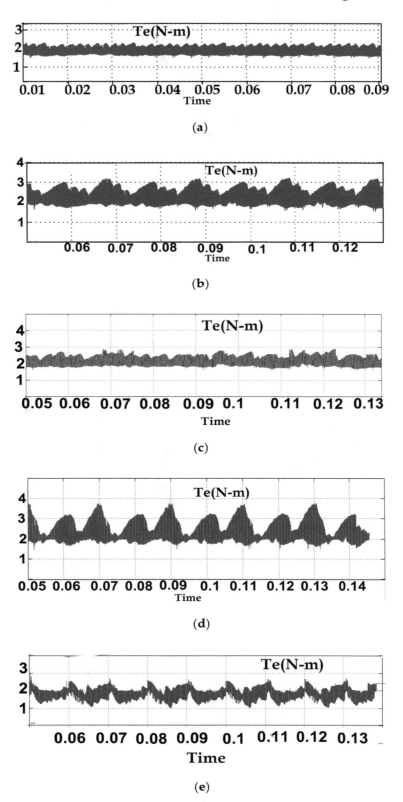

Figure 7. Torque (N-m) under (**a**) Healthy operation; (**b**) 1-Ph open circuit fault without control strategy; (**c**) 1-Ph open circuit fault with control strategy; (**d**) A2-Ph open circuit fault without control strategy; (**e**) A2-Ph open circuit fault with control strategy.

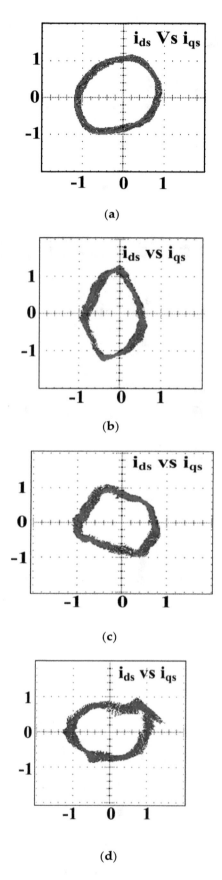

(a)

(b)

(c)

(d)

Figure 8. *Cont.*

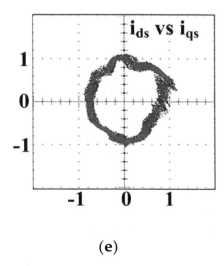

(e)

Figure 8. d-q current in amp (both the scale is in amperes) under (**a**) Healthy operation; (**b**) 1-Ph open circuit fault without control strategy; (**c**) A2-Ph open circuit fault without control strategy; (**d**) 1-Ph open circuit fault with control strategy; (**e**) A2-Ph open circuit fault with control strategy.

5. Conclusions

The theoretical and conceptual background of a new control technique validated with experimental results is presented here. The control technique enables the four-phase and three-phase operation modes of a star-connected induction motor. The experimental results show that a five-phase induction motor drive supplied by a faulty five-phase voltage source inverter can be successfully operated in the four-phase operation mode when 1-Ph open circuit fault occurs, and can be operated in the three-phase operation mode when an A2-Ph open circuit fault occurs. Smoothly controlling the speed of the machine using V/f control, improves the reliability control. The machine performance in the four-phase and three-phase operations was thoroughly analyzed. It shows that the current in the remaining active phases is independently controlled and can produce a positive-sequence rotating MMF component. The proposed controller is able to remarkably reduce the torque pulsations. The available torque in the four-phase and three-phase operation modes is considerably smaller than that of the five-phase drive in the healthy condition. This control strategy is suitable for a drive in steady state operation (or slow acceleration/deceleration, such as transportation drives) with minimum losses in the stator winding. Furthermore, the torque capacity can be enhanced by over-designing the power circuit and control strategy.

Author Contributions: S.C.R. and H.M.S. developed the concept; H.M.S. and S.C.R. performed the experiment; S.C.R. and H.M.S. wrote the paper; M.R. analyzed the data. These authors contributedequally to this work.

Acknowledgments: Authors acknowledges the VNIT, Nagpur for providing infrastructure support and RCOEM, Nagpur for all institutional facilities.

References

1.   Levi, E.; Bojoi, R.; Profumo, F.; Toliyat, H.A.; Williamson, S. Multiphase induction motor drives—A technology status review. *IET Electr. Power Appl.* **2007**, *1*, 489–516. [CrossRef]
2.   Levi, E.; Barrero, F.; Duran, M.J. Multiphase machines and drives-revisited. *IEEE Trans. Ind. Electron.* **2016**, *63*, 429–432. [CrossRef]
3.   Barrero, F.; Duran, M.J. Recent advances in the design, modeling, and control of multiphase machines—Part I. *IEEE Trans. Ind. Electron.* **2016**, *63*, 449–458. [CrossRef]

4.   Duran, M.J.; Barrero, F. Recent advances in the design, modeling, and control of multiphase machines—Part II. *IEEE Trans. Ind. Electron.* **2016**, *63*, 459–468. [CrossRef]

5.   Levi, E. Advances in converter control and innovative exploitation of additional degrees of freedom for multiphase machines. *IEEE Trans. Ind. Electron.* **2016**, *63*, 433–448. [CrossRef]

6.   Liu, T.H.; Fu, J.R.; Lipo, T.A. A strategy for improving reliability of field-oriented controlled induction motor drives. *IEEE Trans. Ind. Appl.* **1993**, *29*, 910–918.

7.   Wallmark, O.; Harnefors, L.; Carlson, O. Control algorithms for a fault-tolerant PMSM drive. *IEEE Trans. Ind. Electron.* **2007**, *54*, 1973–1980. [CrossRef]

8.   Zhao, W.; Cheng, M.; Hua, W.; Jia, H.; Cao, R. Back-EMF harmonic analysis and fault-tolerant control of flux-switching permanent-magnet machine with redundancy. *IEEE Trans. Ind. Electron.* **2011**, *58*, 1926–1935. [CrossRef]

9.   De Lillo, L.; Empringham, L.; Wheeler, P.W.; Khwan-On, S.; Gerada, C.; Othman, M.N.; Huang, X. Multiphase power converter drive for fault-tolerant machine development in aerospace applications. *IEEE Trans. Ind. Electron.* **2010**, *57*, 575–583. [CrossRef]

10. Bianchi, N.; Bolognani, S.; Zigliotto, M.; Zordan, M.A.Z.M. Innovative remedial strategies for inverter faults in IPM synchronous motor drives. *IEEE Trans. Energy Convers.* **2003**, *18*, 306–314. [CrossRef]

11. Errabelli, R.R.; Mutschler, P. Fault-tolerant voltage source inverter for permanent magnet drives. *IEEE Trans. Power Electron.* **2012**, *27*, 500–508. [CrossRef]

12. Aghili, F. Fault-tolerant torque control of BLDC motors. *IEEE Trans. Power Electron.* **2011**, *26*, 355–363. [CrossRef]

13. Sayed-Ahmed, A.; Mirafzal, B.; Demerdash, N.A. Fault-tolerant technique for Δ-connected AC-motor drives. *IEEE Trans. Energy Convers.* **2011**, *26*, 646–653. [CrossRef]

14. Mendes, A.M.; Cardoso, A.M. Fault-tolerant operating strategies applied to three-phase induction-motor drives. *IEEE Trans. Ind. Electron.* **2006**, *53*, 1807–1817. [CrossRef]

15. Abdel-Khalik, A.S.; Ahmed, S.; Elserougi, A.A.; Massoud, A.M. Effect of stator winding connection of five-phase induction machines on torque ripples under open line condition. *IEEE/ASME Trans. Mechatron.* **2015**, *20*, 580–593. [CrossRef]

16. Yepes, A.G.; Riveros, J.A.; Doval-Gandoy, J.; Barrero, F.; López, O.; Bogado, B.; Jones, M.; Levi, E. Parameter identification of multiphase induction machines with distributed windings—Part 1: Sinusoidal excitation methods. *IEEE Trans. Energy Convers.* **2012**, *27*, 1056–1066. [CrossRef]

17. Mecrow, B.C.; Jack, A.G.; Haylock, J.A.; Coles, J. Fault-tolerant permanent magnet machine drives. *IEE Proc.-Electr. Power. Appl.* **1996**, *143*, 437–442. [CrossRef]

18. Parsa, L. On advantages of multi-phase machines. In Proceedings of the 31st Annual Conference of IEEE Industrial Electronics Society, Raleigh, NC, USA, 6–10 November 2005; pp. 1574–1579.

19. Parsa, L.; Toliyat, H.A. Fault-tolerant interior-permanent-magnet machines for hybrid electric vehicle applications. *IEEE Trans. Veh. Technol.* **2007**, *56*, 1546–1552. [CrossRef]

20. Mohammadpour, A.; Sadeghi, S.; Parsa, L. A generalized fault-tolerant control strategy for five-phase PM motor drives considering star, pentagon, and pentacle connections of stator windings. *IEEE Trans. Ind. Electron.* **2014**, *61*, 63–75. [CrossRef]

21. Jasim, O.; Sumner, M.; Gerada, C.; Arellano-Padilla, J. Development of a new fault-tolerant induction motor control strategy using an enhanced equivalent circuit model. *IET Electr. Power Appl.* **2011**, *5*, 618–627. [CrossRef]

22. Guzman, H.; Duran, M.J.; Barrero, F.; Bogado, B.; Sergio, L.; Marín, T. Speed control of five-phase induction motors with integrated open-phase fault operation using model-based predictive current control techniques. *IEEE Trans. Ind. Electron.* **2014**, *61*, 4474–4484. [CrossRef]

23. Dwari, S.; Parsa, L. Fault-tolerant control of five-phase permanent-magnet motors with trapezoidal back EMF. *IEEE Trans. Ind. Electron.* **2011**, *58*, 476–485. [CrossRef]

24. Kastha, D.; Bose, B.K. Fault mode single-phase operation of a variable frequency induction motor drive and improvement of pulsating torque characteristics. *IEEE Trans. Ind. Electron.* **1994**, *41*, 426–433. [CrossRef]

SHIL and DHIL Simulations of Nonlinear Control Methods Applied for Power Converters using Embedded Systems

Arthur H. R. Rosa *, Matheus B. E. Silva, Marcos F. C. Campos, Renato A. S. Santana, Welbert A. Rodrigues, Lenin M. F. Morais and Seleme I. Seleme Jr.

Graduate Program in Electrical Engineering, Universidade Federal de Minas Gerais, Av. Antônio Carlos 6627, Belo Horizonte 31270-901, MG, Brazil; matbeiras@gmail.com (M.B.E.S.); engemarcoscampos@gmail.com (M.F.C.C.); rass.eletrica@gmail.com (R.A.S.S.); welbertalves@gmail.com (W.A.R.); lenin@cpdee.ufmg.br (L.M.F.M.); seleme@cpdee.ufmg.br (S.I.S.J.)
* Correspondence: arthurcpdee@gmail.com

Abstract: In this work, a new real-time Simulation method is designed for nonlinear control techniques applied to power converters. We propose two different implementations: in the first one (Single Hardware in The Loop: SHIL), both model and control laws are inserted in the same Digital Signal Processor (DSP), and in the second approach (Double Hardware in The Loop: DHIL), the equations are loaded in different embedded systems. With this methodology, linear and nonlinear control techniques can be designed and compared in a quick and cheap real-time realization of the proposed systems, ideal for both students and engineers who are interested in learning and validating converters performance. The methodology can be applied to buck, boost, buck-boost, flyback, SEPIC and 3-phase AC-DC boost converters showing that the new and high performance embedded systems can evaluate distinct nonlinear controllers. The approach is done using matlab-simulink over commodity Texas Instruments Digital Signal Processors (TI-DSPs). The main purpose is to demonstrate the feasibility of proposed real-time implementations without using expensive HIL systems such as Opal-RT and Typhoon-HL.

Keywords: real-time simulation; power converters; nonlinear control; embedded systems; high level programing; SHIL; DHIL

---

1. Introduction

The rapid advance of digital and embedded systems has enabled the use of such systems in different applications [1]. Although still little explored, one of these utilities includes Hardware in The Loop (HIL) Simulations, in which both software and hardware are tested.

Real-time simulation (RTS) methods can be a feasible way to verify controllers performance and stability of dynamic systems. Commercial platforms, such as OPAL-RT Technologies Inc. (Montreal, QC, Canada), that implemented sophisticate and expensive test bench, are widely available [2]. Examples of Digital real-time simulator (DRTS) with applications attaining high accuracy results are: TYPHON HIL [2], OPAL-RT [3], dSPACE [4] and RTDS [5].

On the other hand, a real-time simulation platform with less complexity than those previously mentioned may be desirable. On these terms, the employment of powerful computational devices does not justify increased costs. Along these lines, a Processor in the Loop (PIL) applying the SimCoder platform of PSIM (Power System Simulator) is designed in [6], where a F28335 Texas Instruments micro-controller is employed to embed a PFC (Power Factor Correction) and motor drive circuits via software simulation. Also, Ref. [7] presents a simple and interesting real-time implementation.

In view of these concerns, an RTS based method is proposed in order to verify the power converters dynamics and validate the stability of their implemented control equations. The approach is made in a way that justifies the required computational power needed to simulate the elementary converters in real-time, with lower cost and spent time. In the proposed Single Hardware in The Loop (SHIL), both control and state equations implemented on Matlab/Simulink development environment are directly embedded in a C2000 F28377 Texas Instruments device, through Simulink Coder and Embedded Coder packages. In Double Hardware in The Loop (DHIL), the equations are embedded in distinctive DSPs, as illustrated in Figure 1. In the first DSP, the converter model is embedded (usually described by state-space equations or switched model). In the second DSP, the equation of the duty cycle d is calculated for the input control of the switch.

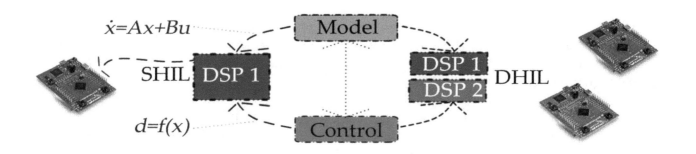

Figure 1. SHIL and DHIL.

As illustrated in Figure 1, the methodology of this work differs from the concepts found in the literature about Software in The Loop (SIL), Processor in The Loop (PIL) [7] and HIL [8] implementation. That is the reason we call it "Single Hardware In The Loop" (SHIL), since it contains hybrid characteristics of these methodologies. Given this, we can easily test the control law without needing a desktop computer and a real plant. In addition, the data transfer occurs directly and more quickly when both model and control are inserted in the same DSP or different DSPs.

The main objective of this work is to validate nonlinear control laws in embedded systems using the proposed real-time simulation methods. When dealing with unconventional control equations the following question appears: are these new methods feasible? To achieve this goal, it is not necessary to use complex models, since such models will be replaced by the real prototype. In fact, it is worth highlighting here that the controllers are usually performed by embedded systems in power electronics applications. This is the preponderant trend.

So, this work presents two different RTS approaches, where model and control equations are executed on DSP processors. The models and control equations are demonstrated in Section 2 and Appendix A. The proposed SHIL and DHIL simulation methods and their experimental results are explained in Section 3. Furthermore, the additional contribution of this work is the comparison of nonlinear control techniques (SFL, PBC and IDAPBC) applied to static power converters. As a whole, three converters models (Table 1) and nine control equations (Table 2) are validated using the proposed methods. Finally, results and conclusions are presented in Sections 4 and 5.

Table 1. Converters models.

	Boost	Buck	Buck-Boost
SS	$\dot{x}_1 = -(1-d)\frac{1}{L}x_2 + \frac{E}{L}$ $\dot{x}_2 = (1-d)\frac{1}{C}x_1 - \frac{G}{C}x_2$	$\dot{x}_1 = -\frac{1}{L}x_2 + d\frac{E}{L}$ $\dot{x}_2 = \frac{1}{C}x_1 - \frac{G}{C}x_2$	$\dot{x}_1 = (1-d)\frac{1}{L}x_2 + d\frac{E}{L}$ $\dot{x}_2 = -(1-d)\frac{1}{C}x_1 - \frac{G}{C}x_2$

EL:

$$x = \begin{bmatrix} \dot{x}_1 \\ \dot{x}_2 \end{bmatrix}; D_B = \begin{bmatrix} L & 0 \\ 0 & C \end{bmatrix}; R_B = \begin{bmatrix} 0 & 0 \\ 0 & G \end{bmatrix}; F = \begin{bmatrix} E \\ 0 \end{bmatrix}$$

Boost	Buck	Buck-Boost
$D_B\dot{x} + (1-d)J_B x + R_B x = F$ $J_B = \begin{bmatrix} 0 & 1 \\ -1 & 0 \end{bmatrix};$	$D_B\dot{x} + (J_B + R_B)x = dF$ $J_B = \begin{bmatrix} 0 & 1 \\ -1 & 0 \end{bmatrix}$	$D_B\dot{x} + (1-d)J_B x + R_B x = dF$ $J_B = \begin{bmatrix} 0 & -1 \\ 1 & 0 \end{bmatrix};$

PCH:

$$x = \begin{bmatrix} x_1 \\ x_2 \end{bmatrix}; \dot{x} = [J_H(d) - R_H]\frac{\partial H}{\partial x}(x) + g_H E; H(x) = \frac{1}{2}Lx_1^2 + \frac{1}{2}Cx_2^2$$

Boost	Buck	Buck-Boost
$J_H = \begin{bmatrix} 0 & -\frac{1-d}{LC} \\ \frac{1-d}{LC} & 0 \end{bmatrix};$	$J_H = \begin{bmatrix} 0 & -\frac{1}{LC} \\ \frac{1}{LC} & 0 \end{bmatrix};$	$J_H = \begin{bmatrix} 0 & \frac{1-d}{LC} \\ -\frac{1-d}{LC} & 0 \end{bmatrix};$
$R_H = \begin{bmatrix} 0 & 0 \\ 0 & \frac{1}{RC^2} \end{bmatrix};$	$R_H = \begin{bmatrix} 0 & 0 \\ 0 & \frac{1}{RC^2} \end{bmatrix};$	$R_H = \begin{bmatrix} 0 & 0 \\ 0 & -\frac{1}{RC^2} \end{bmatrix};$
$g_H = \begin{bmatrix} \frac{1}{L} \\ 0 \end{bmatrix}$	$g_H = \begin{bmatrix} \frac{d}{L} \\ 0 \end{bmatrix};$	$g_H = \begin{bmatrix} \frac{d}{L} \\ 0 \end{bmatrix};$

Table 2. Control equations.

	SFL	PBC	IDA-PBC
Boost	$d = 1 - \frac{[E + Lk_1(x_1 - x_{1d}) - L\dot{x}_{1d}]}{x_2}$	$d = 1 - \frac{[E + R_{1damp}(x_1 - x_{1d}) - L\dot{x}_{1d}]}{x_{2d}}$ $\dot{x}_{2d} = \frac{(1-d)x_{1d} - Gx_{2d} + R_{2damp}(x_2 - x_{2d})}{C}$	$\bar{d}_1 = 1 - \frac{E}{V_d}$ $d = 1 - (1 - \bar{d}_1)\left(\frac{x_2}{V_d}\right)^\alpha$
Buck	$d = \frac{L\dot{x}_{1d} - Lk_1(x_1 - x_{1d}) + x_2}{E}$	$d = \frac{L\dot{x}_{1d} - R_{1damp}(x_1 - x_{1d}) + x_{2d}}{E}$ $\dot{x}_{2d} = \frac{x_{1d} - Gx_{2d}}{C}$	$\bar{d}_1 = 1 - \frac{E - V_d}{E}$ $d = 1 - (1 - \bar{d}_1)\left(\frac{x_2}{V_d}\right)^\alpha$
Buck-Boost	$d = \frac{-L\dot{x}_{1d} + Lk_1(x_1 - x_{1d}) + x_2}{x_2 - E}$	$d = \frac{-L\dot{x}_{1d} + R_{1damp}(x_1 - x_{1d}) + x_{2d}}{x_{2d} - E}$ $\dot{x}_{2d} = \frac{-(1-d)x_{1d} - Gx_{2d}}{C}$	$\bar{d}_1 = 1 - \frac{E}{E - V_d}$ $d = 1 - (1 - \bar{d}_1)\left(\frac{x_2}{V_d}\right)^\alpha$

2. Modeling and Control Equations

The basic power converters, such as boost, buck and buck-boost (shown in Figure 2), are typical switching-mode nonlinear systems, which customarily adopt conventional linear control method. These classic linear controllers, as mentioned in [9], exhibit some natural inconsistencies (for example, the intrinsic non-minimum phase characteristic related in [10]) and cannot satisfy the meaningful prerequisites of high performance control. The boost has inductor positioned in the input to reducing spikes in grid voltage, so is recommended to power factor (PFC) systems. Buck-boost inverts the polarity of the output voltage signal relative to the input signal and allows up-down output voltages. Note that the state variables are those related to energy store elements, i.e., capacitors and inductors.

In this context, there is a growing demand for new controllers to deal with this problem. Some nonlinear methods, such as SFL [11,12], PBC [13,14], IDA-PBC [15–17], fuzzy logic control [18], backstepping approach [19], predictive control [20], piecewise affine (PWA) [21] and repetitive control [22] have been designed and implemented in power converters.

This section presents the relevant models and control equations used in this work, collected in a literature review [10–17]. Notice in Figure 3, the Euler Lagrange (EL) is the base model to find the others. With the EL model, the PBC control equations are obtained. But the SFL control uses the model description in state space (SS). In turn, the IDA-PBC control requires the Port-controlled Hamiltonian model (PCH) system. Note that each model is associated with a control technique. Despite having specific mathematical and physical interpretations, the Euler-Lagrange and Hamiltonian models are mathematically similar to the models described in state space. It should be noted that the controllers are designed for continuous mode operation [23].

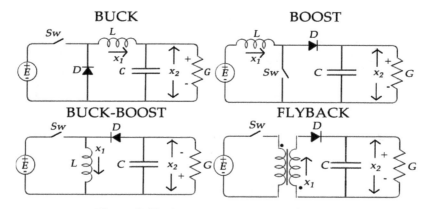

Figure 2. Basic power converters circuits.

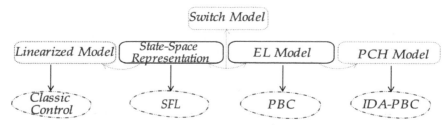

Figure 3. Models associated with non-linear control techniques.

As case studies, the control methods used in this work are SFL, PBC and IDA-PBC. A study and comparison of these methods are presented in [14]. The SFL control uses state space equations. PBC and IDAPBC include passivity properties, applying Lagrangian and Hamiltonian approaches, respectively. It should be noted that nonlinear control methods are currently widely discussed in the literature. However, another important trend in the design of these new controllers is the practical implementation.

To make the paper self-contained, we recall non linear concepts, including and intercalating different control equations associated with specific converter (see Appendix A). Further details and concepts of the applied methodology can be seen in the Section 3.

2.1. Buck-Boost and Flyback Examples

In the following paragraphs, the control equations (summarized in Table 3) and models of the Buck-Boost and Flyback converters will be described. Figure 2 presents the converters topologies. The readers that are familiar with nonlinear control can go directly to Section 3.

Table 3. Control equations.

	SFL	PBC	IDA-PBC
Buck-Boost	$d = \frac{-L\dot{x}_{1d}+Lk_1(x_1-x_{1d})+x_2}{x_2-E}$	$d = \frac{-L\dot{x}_{1d}+R_{1damp}(x_1-x_{1d})+x_{2d}}{x_{2d}-E}$ $\dot{x}_{2d} = \frac{-(1-d)x_{1d}-Gx_{2d}}{C}$	$d = 1 - \left(\frac{E}{E-V_d}\right)\left(\frac{x_2}{V_d}\right)^{\alpha}$.
Flyback	$d = \frac{-L\dot{x}_{1d}+L_{eq}k_1(x_1-x_{1d})+x_2}{x_2+E_{eq}}$	$d = \frac{-L_{eq}\dot{x}_{1d}+R_{1damp}(x_1-x_{1d})+x_{2d}}{x_{2d}+E_{eq}}$ $\dot{x}_{2d} = \frac{(1-d)x_{1d}-Gx_{2d}}{C}$	$d = 1 - \left(\frac{E_{eq}}{E_{eq}+V_d}\right)\left(\frac{x_2}{V_d}\right)^{\alpha}$.

According to [24] the average Buck-Boost converter circuit can be written by equivalent state space equations:

$$\dot{x}_1 = (1-d)\frac{1}{L}x_2 + d\frac{E}{L}, \tag{1}$$

$$\dot{x}_2 = -(1-d)\frac{1}{C}x_1 - \frac{G}{C}x_2. \tag{2}$$

where, d is the converter duty cycle, $0 \leq d < 1$. As it can be seen from (1) and (2), there are two state variables, x_1 and x_2 and an input (control) variable, the duty cycle d.

2.2. SFL Control

Summarily, the procedure to obtain the state feedback linearization [25] includes the steps:

1. Select the state variable to be controlled. Two possibilities: indirect control (current x_1) or direct control (voltage x_2);
2. Derivation of the output (n) times until an explicit relation between output (y) and the input (E) is achieved;
3. Determine $d = d(v, x)$ in order to perform the feedback linearization;
4. Investigate the stability of internal dynamics.

Defining L_f as the derivative of Lie [25] and consider that:

$$x_1 = h(x), \quad y = x_1, \quad x_2 = L_f h,$$
$$\dot{y} = \dot{x}_1, \quad \dot{x}_1 = (1-d)\frac{1}{L}x_2 + d\frac{E}{L}. \tag{3}$$

Since we have to accomplish one derivation to obtain a relation between the input and output, the relative degree is unitary ($n = 1$). On these terms, the general expression for the duty cycle equation is:

$$d_{SFL} = \frac{L\left[\dot{x}_1 d - k_1\left(x_1 - x_{1d}\right)\right] - x_2}{E - x_2}. \tag{4}$$

A literature review of the main stability analysis methods applied to power converters is presented in [26]. PBC and IDAPBC control techniques, reported in [27,28], are demonstrated in details on appendices.

2.3. Flyback Modelling and Control Equations

Derived from a mathematical formulation, the flyback converter can be interpreted as an isolated buck-boost converter. As shown by [29], the average state-space model of the circuit illustrated in Figure 2 are given by:

$$\dot{x}_1 = \frac{L_1}{L_1 L_2 - L_M^2}(1-d)x_2 - \frac{L_M}{L_1 L_2 - L_M^2}dE. \tag{5}$$

$$\dot{x}_2 = (1-d)\frac{1}{C}x_1 - \frac{G}{C}x_2. \tag{6}$$

where L_1 and L_2 are the primary and secondary inductances, respectively, and L_M is the mutual inductance. After replacing:

$$L_{eq} = \frac{L_1 L_2 - L_M^2}{L_1}, \quad E_{eq} = \frac{L_M}{L_1}E. \tag{7}$$

the flyback converter Equations (8) and (9) become similar to the buck-boost Equation:

$$\dot{x}_1 = \frac{1}{L_{eq}}(1-d)x_2 - d\frac{E_{eq}}{L_{eq}}. \tag{8}$$

$$\dot{x}_2 = (1-d)\frac{1}{C}x_1 - \frac{G}{C}x_2. \tag{9}$$

Thus, the adapted mathematical models from buck-boost are evaluated to represent and withdraw the flyback control equations. All control equations (detailed in [30]) are summarized in Table 3.

By collecting and manipulating the terms, it is possible to obtain the general expression for the duty cycle(by considering $\dot{x}_{1d} = 0$), defined by:

$$d_{SFL} = \frac{x_2 + R_{1damp}(x_1 - x_{1d})}{E_{eq} + x_2}. \tag{10}$$

One of the lessons learned from previous researches [31] is that the nonlinear controllers need an integral action to achieve voltage regulation. Therefore, in order to improve stead-state performance and assure the convergence of error between the output voltage and desired value V_d, a proportional integrative term is recommended, given by:

$$G_{int} = -k_{int} \int_0^t [x_2(s) - V_d]ds. \tag{11}$$

3. SHIL and DHIL Proposed Methods

In order to validate a simulation, a modeling or a controller design it is necessary to obtain experimental results through hardware implementation. In the context of Power Electronics, as systems complexity increases [3,32]:

- Costs with semiconductor devices and power components rise significantly.
- Implementation of controller conditioning and communication systems complexity increase.
- Time spent for concluding the hardware implementation may become a problem.

In this session, we present a procedure for high-level programming of a DSP (Digital Signal Processor) using SHIL and DHIL Simulations. The HIL based method simulation is a technique that mixes both virtual and real elements. Currently, this technique is often used to test embedded control systems, where both the hardware and system software are tested. Also, we can verify the control and the system operations without the need of a physical circuit.

Besides the independence of the physical prototype, the proposed methodology has other advantages:

- There is no need for costly real-time Simulators (RTS) systems, such as those offered by OPAL RT, Typhon HIL, dSPACE and RTDS. In the same way, it is possible to use the method remotely, in residences, in the laboratory, using desktop pc, laptop, without being conditioned to a complex system—which involves both hardware and software—previously installed;
- It is possible to emulate only the converter model and perform several tests, regardless of control;
- The control of the system is embedded and its proper functionality can be evaluated in DSP; therefore, the determination of the processing time of each step of the algorithm can also be achieved. It is possible to monitor, make initial parameter updates, controller gains, input and load disturbances, etc., through the friendly interface offered by Matlab/Simulink.
- In addition, there is the possibility of the DSP to emulate the model or the control independently of the pc/laptop. In other words, its possible to upload the codes into the flash memory of the embedded system (tests are limited to DSP input/output capabilities, for example, DAC and other digital/analog ports).
- Simple and complex converters can be evaluated;

The major drawbacks are also listed:

- It needs Matlab installation.
- The approach depends on the mathematical model of the converter.

3.1. SHIL

Figure 4 shows the overview of SHIL proposed methodology. The first step is obtaining the plant and the control equation models. After that, such models can be simulated using commonly softwares as Matlab or PSIM. As an example, Figure 4 illustrates the Flyback's converter model and the SFL control equations, designed with Matlab/Simulink tools. Figure 5 shows the simulation in PSIM software. After the implementation, it is necessary to run the simulation and verify if the control and state variables are converging to the desired state.

Once this is achieved, the next step is to embed the simulated system in the DSP. By using the *external mode* of Simulink and a compiler for the DSP, the model will be converted in code and then embedded to the target. Finally, the target will run the code, emulating the converter and control models. It is possible to verify desired signals in an oscilloscope by programming the DSP pins through DAC (Digital Analog Converter) blocks available in the C2000 *Texas Instruments* package. An important detail is the need of a scale adjustment for voltage compatibility between the simulation and target.

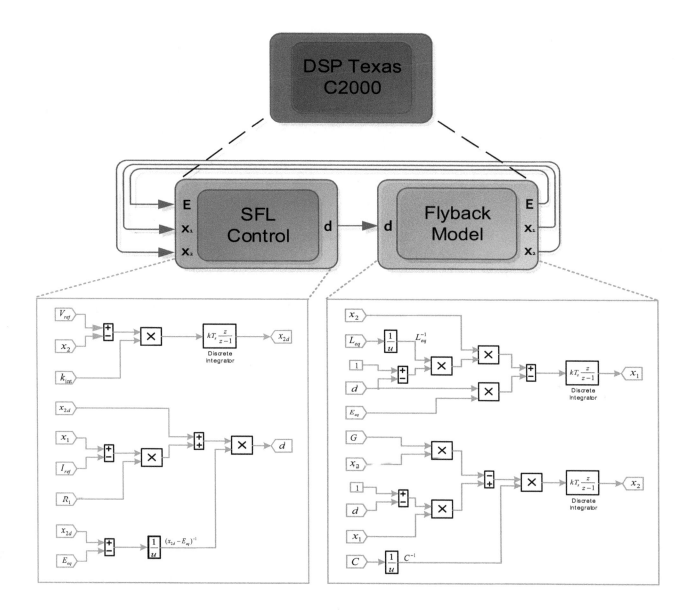

Figure 4. Flyback model, Equations (8) and (9), and SFL control Equation (10) in block diagrams (high level programing) that is embedded in DSP C2000.

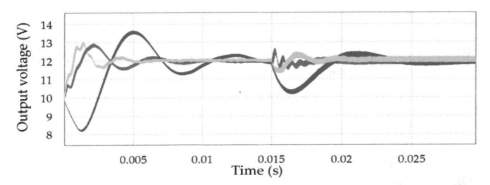

Figure 5. Simulation of flyback switched model using PSIM software. Normalized output voltage x_2 for load perturbation with a fixed time step 100 times smaller than the switching period. SFL (red), PBC (green) and IDAPBC (blue).

3.2. DHIL

With the proposed SHIL and DHIL methodologies, we can easily test the control law without needing a desktop computer or a real plant. In SHIL, the data transfer occurs directly and faster when both model and control are inserted in the same Digital Signal Processor. However, DHIL is best suited for synchronization, measurement and data communication tests between different systems.

The supporting package for C2000 microcontrollers available in Matlab/Simulink can be found in the Simulink library. Basic information about block functions, simulation configurations for real-time simulation or external mode and examples of systems implementation using the fundamental blocks such as PWM, ADC, DAC and interruptions can be found in Mathworks [33] website, or in Matlab "Help" area. Once selected, a list of C2000 DSP family will be displayed. By choosing the corresponding DSP, available blocks for the microcontroller are displayed. it is possible to build block systems with other Simulink blocks, by simply dragging them to the model window.

In DHIL, the use of a PWM (Pulse Width Modulation) block is necessary in order to control the converter switching sequence and also to synchronize 3 ADCs (Analog Digital Converter) available for measuring the inductor current (x_1), capacitor voltage (x_2) and the input voltage E. As seen in Figure 6, the control laws and the converter model are embedded in different microcontrollers. For computing the control laws, the inputs of the model are (x_1), (x_2) and E. Since those inputs are originated from the converter model computation—configured as analog signals type—it is necessary to convert those signals to a digital one, through analog-digital converter (ADC). After conversion, the output of the control law is the duty cycle. Since this control variable is digital, it will be converted to analog (DAC), for reading in the ADC of the DSP embedded with the converter model. For closing the loop, the variables (x_1), (x_2) and E are calculated and consequently converted from digital to analog type. It is important to notice that the conversions are based on the PWM sample rate, then requires the synchronization between both DSPs for a correct computation of control laws and converter model.

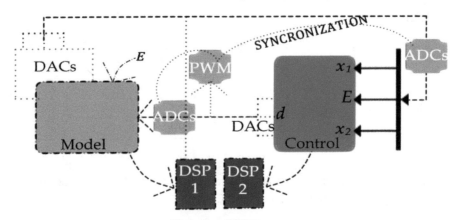

Figure 6. DHIL.

3.3. Details of the Implementation

Also, the approach used for programming the microcontroller diverges from the conventional one, being unnecessary the development of code lines. By using code generation tools and software libraries it is possible to resort the implementation of converter models and controllers through Matlab/Simulink blocks. The evident advantages of this approach are the clear visualization of the programming process and the time spared for development. Figure 7 illustrates the proposed steps for this methodology.

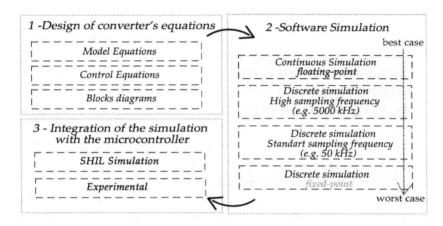

Figure 7. Methodology steps for SHIL simulation. The optional steps are highlighted in gray.

The first stage consists in developing an equation model for the system, demonstrating the relation between the state and control variables. It is up to the user to consider or not the nonlinearities of the system. Next, it is necessary to choose a control technique for actuating in the variable of interest. The nature of the control technique is wide, and can include since classical techniques, as PID controllers, to nonlinear control approaches. This stage ends with the implementation of the model and control equations in the simulation software. Figure 8 presents the implemented model and control equations of a buck-boost converter, as an example. Figure 9 shows the general SHIL simulation scheme for any converter. The main control goal is to calculate the duty cycle d (used in mathematical model), then the corresponding PWM signal is generated as an input control to command the switch of the physical converter. The state variables (inductor current x_1 and voltage capacitor x_2) and input voltage E are the required measures. An integral action is added to better regulate the output voltage.

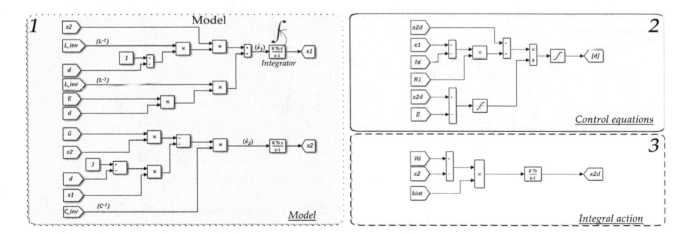

Figure 8. Buck-boost model and SFL control equations in block diagrams. Model (1), control Equation (2) and integral action (3).

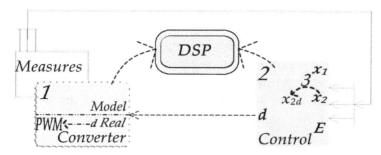

Figure 9. General SHIL Simulation Scheme.

The second phase consists of simulating the implemented system. Aiming at minimizing the errors, it is advised to simulate the system in continuous time and afterly in discrete time. For the continuous system it is important to work with float data type, by programming the operations of multiplication, division and constant blocks to "single/double" types. Although the increasing in processing time, this data type conversion configures one less source of error in the continuous time simulation.

After that, it is necessary to discretize the system, by changing the controllers structure and continuous operators to the corresponding discrete blocks (defining a sample time in which the blocks will be sampled). In a way to approach the results of the first discretized system to the continuous simulation it is proposed the use of a high sample frequency (or a small sample time).

The next task is to simulate the system, considering a standard sample frequency (e.g., nominal 50 kHz), in such way to obtain a discrete model that is less approached by the continuous model. The selection of this frequency must be cautiously chosen, since the system can converge to instability. Another common problem associated to a bad choice of the sample frequency is the signal aliasing (Nyquist rule). The final objective of this stage consists in transforming the floating point data in fixed point data. This conversion can be done by programming the operations of multiplication, division and constant blocks to fixed-point type, or by using operational blocks offered in specific libraries for microcontrollers (libraries that are offered by Mathworks for users of C2000's microcontrollers family, by Texas Instruments, for example), as shown in Figure 10. Some recomendations: avoid operations with floating data and divisions that increase the processing time. Always when possible, use multiplication operations instead of division operations (ex: when the denominator is a constant). Declare the variables as fixed point, preventing calculus with float and optmizing the code's execution. Discrete models can be embedded for HIL simulation, since the user compiler can convert the blocks in code. However, a discrete model that converges, when working with fixed point data, makes the compilation and the processing time of the microcontroller smaller (and also the memory used smaller). In this way, the last step brings a discrete model more appropriated for a HIL procedure.

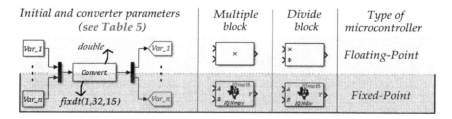

Figure 10. Fixed-point and Floating-point different implementations.

The final stage is the SHIL simulation itself. Once the compiler has generated the code and the system is embedded to the microcontroller (also called target) the communication between the target and the computer, in which Matlab/Simulink is running, begins. Usually microcontrollers of C2000 family communicates to the computer through USB or ethernet cable. In this application it is proposed the use of a USB cable. For running the system as a real-time simulation it is proposed to set the simulation as "external mode" simulation on Simulink. Also, it is necessary to set on the simulation configurations which target the connection must occur.

Since we are interested in plotting or viewing the gathered data, it is necessary to use specific blocks in the simulation for real-time plotting or setting DSP pins as analog outputs [33]. By doing that the user can see the data generated in the target in a Simulink "scope" or in an oscilloscope (by analog output reading). Each case can be achieved by using the RTDX (real-time Data Exchanged) or the DAC blocks, as seen in Figure 11. Since the R2016b version of Matlab there is a DAC block for DSP28377S of C2000 family (Texas Instruments). This block configures 3 digital inputs as 3 analog outputs (also called channels A, B and C). In this way it is possible to use 3 channels of an oscilloscope and view the curves in real-time.

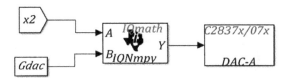

Figure 11. DAC Block: Used for showing state and internal variables in a scale between 0 V–3.3 V (12-bit resolution).

More details of the control algorithms implementation in block diagrams can be seen in the Figures 12 and 13. Since the didactical background available in the literaute lacks of information, details of the functional blocks are shown, providing a development base for future works. Although the approach of this chapterdeals with a specific study of case, the available content makes it simple to adapt the program to other applications. General files, containing the control methods for Buck-Boost, Boost and Buck converters are shown in Figures 8, 12 and 13, respectively.

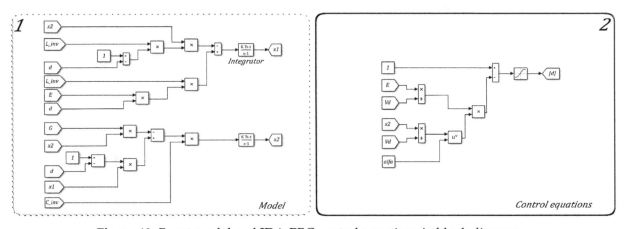

Figure 12. Boost model and IDA-PBC control equations in block diagrams.

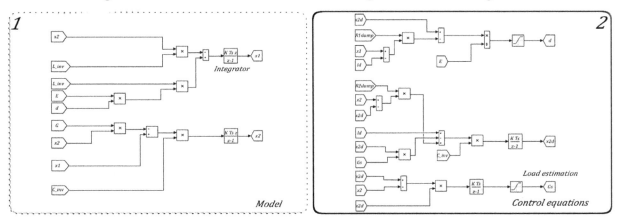

Figure 13. Buck model and PBC control equations in block diagrams.

4. SHIL and DHIL Results for Buck-Boost and Flyback

This section shows the digital simulations using Matlab (Model in The Loop: MIL) and SHIL/DHIL experimental results. The converters are designed according to specifications listed on Table 4 and the three control laws studied in this work. The experimental setup is sketched in Figure 14. The plots displayed in Figure 15a,c present the capacitor voltage response when an input voltage and a load variations, respectively, are included in the Buck-Boost converter simulated in software. The same effect is reported for the Flyback converter in Figure 15b,d. In both input and load variation, consecutive steps of 70% to 100% are applied in the simulated systems. An open-loop control test is also presented in Figure 16. Figures 17 and 18 show the output capacitor voltage of Flyback and Buck-Boost converters, respectively, for the SHIL and DHIL applications. A load variation (70–100%) is applied for evaluate the three control laws.

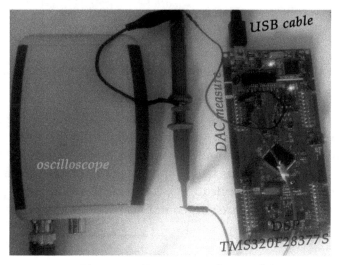

Figure 14. Experimental setup.

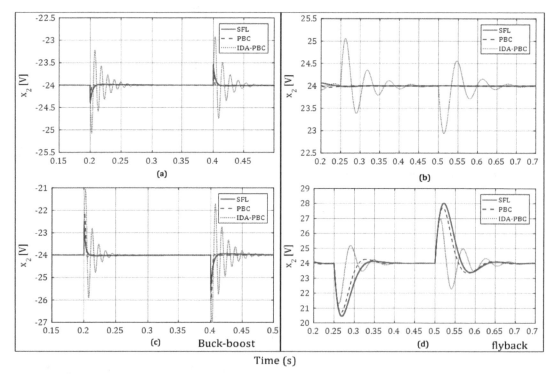

Figure 15. Software simulation result for Buck-Boost (**a,c**) and Flyback (**b,d**) (using control techniques SFL (red), PBC (blue) and IDA-PBC (green). Output voltage x_2 for input voltage (**a,b**) and for load variation (**c,d**).

Table 4. Initial and converters parameters.

Parameters	Buck-Boost	Flyback
V_d	-24 V	24 V
E	50 V	50 V
R	10 Ω	11.5 Ω
L	0.6 mH	146 µH, 35 µH
C	470 µF	470 µF
f_{sw}	50 kHz	100 kHz
$R_{1damp} = Lk_1$	100	15
k_{int}	200	-5000

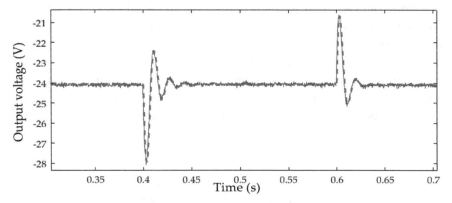

Figure 16. Open loop control comparing SHIL (blue line) with Software Simulation (red line) results for Buck-Boost converter. Normalized output voltage x_2 for load perturbation (70–100%) using fixed $d = 0.325$.

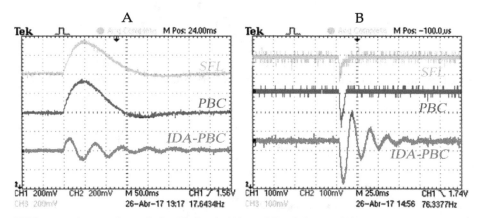

Figure 17. SHIL experimental result for Flyback (**A**) and Buck-boost (**B**) converters. Normalized output voltage x_2 for load perturbation (70–100%).

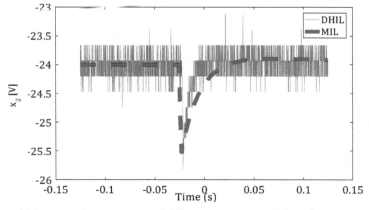

Figure 18. DHIL experimental compared with Software Simulation (MIL) result for Buck-Boost converter (SFL control). Normalized output voltage x_2 for load perturbation (70–100%).

It is possible to appreciate in the figures how the waveforms in both software and real-time simulations are compatibles. Although consecutive perturbations can be applied, the convergence of the variables to steady-state values is assured. In addition, similar transient dynamics can be observed, none of the implemented systems in software or in HIL present instability. Therefore, the embedded models and control equations are validated. The processing time for the DSP to compute the control law and converter state, and therefore to run a real-time simulation, is 1.2 µs.

5. SHIL Results for Second Order Power Converters and Comparison

This section presents the digital simulation results using Matlab and SHIL method for buck, boost and buck-boost. The converters are implemented according to design specifications of Table 5 and the three control laws studied in this work. Figure 19 shows the capacitor voltage response to a load voltage variation (70% G to 100% G), respectively, in the converters simulated in software. Figure 20 shows the capacitor voltage of the converters for the SHIL application. A load variation (70–100%) is applied to evaluate the three control laws. Figure 21 shows the PWM signal generated in HIL simulation for the buck-boost converter in steady state operation.

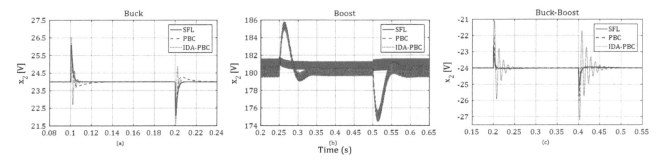

Figure 19. Software simulation result using control techniques SFL (red), PBC (blue) and IDA-PBC (green). Output voltage x_2 for buck (**a**) and boost PFC (**b**); buck-boost (**c**) for load variation in 0.25 s and 0.75 s.

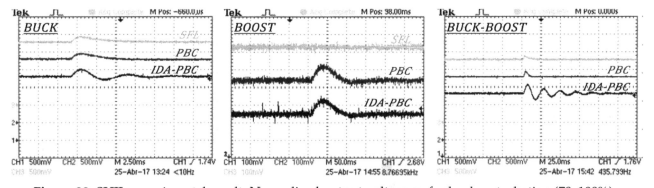

Figure 20. SHIL experimental result. Normalized output voltage x_2 for load perturbation (70–100%).

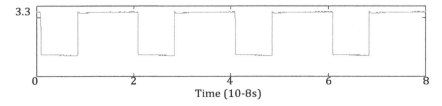

Figure 21. PWM signal generated in SHIL Simulation for buck-boost permanent condition ($d = 0.37$).

Table 5. Initial and converters parameters.

Parameters	Boost	Buck	Buck-Boost
I_d	$\frac{G}{E}V_d^2$	GV_d	$GV_d(\frac{V_d}{E}-1)$
R	52.5 Ω	10 Ω	10 Ω
L	0.6 mH	0.6 mH	0.6 mH
C	2800 μF	470 μF	470 μF
E	100 V	50 V	50 V
V_d	180 V	24 V	−24 V
$Pout$	630 W	57.6 W	57.6 W
f	50 kHz	50 kHz	50 kHz
R_{1damp}	33	500	100
R_{2damp}	50	0	0
k_g	0.0356	2.5	14
k_{int}	−150	2000	200
α	0.8	−10	0.8

As seen in the figures, both software simulations and SHIL results converged to the steady state value after the consecutive step applications. It should be noticed that the dynamical response of the systems is compatible, since the same transient dynamics is seen, even for the IDA-PBC's oscillatory dynamics. None of the implemented systems in software or in SHIL presented instability. Therefore, it means that the embedded models and control equations are capable in controlling the systems, and therefore, validating the control techniques.

In general, the SFL, PBC and IDA-PBC control laws show satisfactory results for the three types of converters studied: boost, buck and buck-boost. It was observed that the inductor current and output voltage in the capacitor, the main variables, follow the reference set points, reaching the control objectives. Since one of the main objectives of this work is the comparison of control methods, Table 6 shows, from an implementation point of view, the advantages and disadvantages of each method.

It is seen that SFL and PBC present similar results. Reminding that, for both control techniques the indirect control is the only possibility [10]. On one hand the SFL is a didactic and easier solution than the PBC. On the other hand, the control complexity of PBC is justified for load estimation and better voltage output regulation. However, there is a trade-off, since the overshoot and undershoot increase with the adaptative control law and the integral gain used.

SFL and PBC control present small error in steady state for the output voltage of the capacitor, without the integral action. On the other hand, the IDA-PBC control presented larger overshoot/ undershoot and accommodation times, notably for input voltage variations. This is because this technique is even more dependent on the exact knowledge model parameters. Thus, to improve the results it is necessary to include other non-modeled effects like parasitic resistances, diode and switches voltage drops [34].

Once IDA-PBC control is a direct control, it does not require the measurement of the current x_1, which is an advantage in terms of implementation. In general terms, we can verify:

- SFL: simpler, didactic, effective, dispenses load estimation for DC-DC systems when using integrative gain.
- PBC: has medium complexity, needs more measurements and control parameters to better estimate the load and regulate the output capacitor voltage. It is the most recommended technique for PFC systems, since it offers lower THD levels [13]. It has the same processing time as the SFL control because of the same amount of division operations, which effectively determine the total processing time (sums and multiplications offer irrelevant contributions).
- IDA-PBC: non-trivial control solution, allows direct control, which exempts current measurement and eventual problems. By the same direct nature of the control, does not work correctly for AC-DC systems, since the objective is to impose the current in phase with grid input voltage.

Table 6. Comparison of non linear control methods.

Comparison	SFL Advantage	SFL Disadvantage	PBC Advantage	PBC Disadvantage	IDA-PBC Advantage	IDA-PBC Disadvantage
Control law	simple			medium	open loop	
Solution	easy			medium complexity		Nontrivial
Parameters	2			4	1	
Measurements		3		3	2	
AC-DC System	low THD levels and high PF		low THD levels and high PF			incorrectly, high THD rates
Control Type		direct control is only for the buck		direct control is only for the buck	Direct control is possible for all	
Integral term	Soft integration		Soft integration			very sensitive
Dependence on realistic model		medium		medium		very dependent on realistic mode
Parametric dependence control	Does not depend on C			It depends on the G, L and C.	Control law does not depend on G, L or C	
Processing Time (TMS320F2812 fixed-point DSP)	Fast (2.4 µs)		Fast (3.2 µs)		Fast (6 µs)	
Processing Time (TMS320F28377S floating-point DSP)	Fast (1.2 µs)		Fast (1.2 µs)		Fast (1.2 µs)	
sum	4			8	3	
div	1			1	0	
mult	3			6	3	
other operation						1 (exp)
Load Estimation		Cannot estimate by the output voltage error.	Can estimate the load by voltage error		It is not necessary estimate the load	

6. Conclusions

This work showed the buck, boost, buck-boost, flyback, SEPIC and 3-phase AC/DC boost converters modeling as well as the development of nonlinear control techniques using SHIL and DHIL implementations. In addition, the control and converter models were implemented in a DSP, resulting in a quick and cheap HIL realization of the proposed systems, ideal for students and engineers interested in learning and validating converters performance. Using a switching frequency of 50 kHz (20 μs), the processing time of the model/control equations (1.2 μs) demands 6% of the bandwidth (for buck, boost, buck-boost, flyback and SEPIC). According to the data in Table 6, there is a clear preference for new embedded systems with floating-point operation. As illustrated in Figures 16 and 18 and despite measurement noises, the SHIL/DHIL results remained close to the model-simulated one. The advantages provided by the proposed method are: security, saving development time, facilitating the understanding of the programming process, standardization, concurrent simulation, rapid prototyping and, mainly, an easy and cheap way to validate linear and nonlinear controllers.

Author Contributions: Conceptualization, A.H.R.R., L.M.F.M. and S.I.S.J.; Data curation, M.F.C.C., R.A.S.S. and W.A.R.; Formal analysis, A.H.R.R., W.A.R. and L.M.F.M.; Investigation, A.H.R.R., M.B.E.S. and L.M.F.M.; Methodology, A.H.R.R.; Project administration, S.I.S.J.; Resources, A.H.R.R., L.M.F.M., W.A.R.; Software, A.H.R.R., M.B.E.S., M.F.C.C. and R.A.S.S.; Supervision, A.H.R.R. and L.M.F.M.; Validation, A.H.R.R., M.B.E.S., M.F.C.C., R.A.S.S. and W.A.R.; Visualization, A.H.R.R. and S.I.S.J.; Writing—original draft, A.H.R.R. and M.B.E.S.; Writing—review & editing, A.H.R.R., L.M.F.M. and S.I.S.J.

Nomenclature

E	Input voltage.
d	Duty cycle.
x_1	Inductor current.
x_{1d}	Desired inductor current.
I_d	Constant desired inductor current.
x_2	Capacitor voltage.
x_{2d}	Desired capacitor voltage.
V_d	Constant desired capacitor voltage.
L	Converters inductance.
L_1	Flyback primary-side inductance.
L_2	Flyback secondary-side inductance.
L_M	Flyback mutual inductance.
C	Converters capacitance.
G	Load conductance.
R_{damp}	Nonlinear PBC gain.
k_g	Load estimation gain.
k_{int}	Integral gain.
α	IDAPBC control gain.
k	SFL control gain.

Appendices

These appendices are optionals for those who are familiar with non-linear control applied to converters. So, we exemplify how to obtain SFL control for the buck, PBC for the buck-boost and IDA-PBC equations for the boost. Notice that we mix the three control laws and the three converters distinctly, to cover the maximum information in a smaller space. In time, the boost converter will be analyzed with power factor correction (PFC), since it is the most suitable for this specific application. The other systems are analyzed as voltage regulators (DC-DC) in which the input voltage comes up to a constant value. We also included SEPIC (Appendix B) and 3-phase AC-DC boost converter (Appendix C).

Appendix A. Nonlinear Controllers

The state feedback linearization control is used in this work mainly because it represents a didactic and effective procedure. This method facilitates the understanding of the system being useful for an initial contact with nonlinear control techniques and leads to a change Of coordinates that shows an interesting structure and mathematical properties. Moreover, it allows dynamic change of a nonlinear system into a linear dynamics through a nonlinear feedback of the output state conveniently chosen. For this purpose, it is necessary to perform a change of state variable input and an auxiliary input variable. Then, it is possible to use familiar linear techniques to effect control of the proposed system.

Appendix A.1. SFL Control Equations of the Buck Converter

According to [35,36] the average Buck converter circuit can be written as:

$$\dot{x}_1 = -\frac{1}{L}x_2 + d\frac{E}{L}, \tag{A1}$$

$$\dot{x}_2 = \frac{1}{C}x_1 - \frac{G}{C}x_2, \tag{A2}$$

where, d is the converter duty cycle, $0 \le d < 1$. As it can be seen from (A1) and (A2), there two state variables, x_1 and x_2 and an input (control) variable, the duty cycle d. Defining L_f as the derivative of Lie [25] and choosing:

$$x_1 = h(x), y = x_1, x_2 = L_f h, \dot{y} = 1\dot{x}_1,$$
$$\dot{x}_1 = -\frac{1}{L}x_2 + d\frac{E}{L}. \tag{A3}$$

Since we have to derive $g_r = 1$ times to obtain a relation between the input and output, the relative degree is $g_r = 1$. In this way, the new coordinate system is:

$$[\dot{z}_1] = [z_2] = [v],$$
$$\begin{bmatrix} z_1 \\ z_2 \end{bmatrix} = \begin{bmatrix} x_1 \\ v \end{bmatrix}. \tag{A4}$$

Using the control law $v = r^{(g_r)} - k^T e$, with k and e given by:

$$k = [k_1],$$
$$e = [e_1] = [x_1 - r] \tag{A5}$$

obtains:

$$v = \dot{r} - k_1 e, \tag{A6}$$

$$-\frac{1}{L}x_2 + d\frac{E}{L} = \dot{r} - k_1(x_1 - r). \tag{A7}$$

Isolating d and considering the reference $r = x_{1d}$, the general expression for the duty cyclic equation is:

$$d_2 = d_{SFL},$$
$$d_{SFL} = \frac{L[\dot{r} - k_1(x_1 - r)] + x_2}{E}. \tag{A8}$$

We can observe that as the system relative degree is one (there is only one switch to control two variables), we need to perform only one branch, which is already inferred directly from (A1). Thus, we need only control equation given by (A8).

Appendix A.2.   Passivity-Based Control (PBC)

The goal of the passivity-based control is to modify the dissipative structure since the inputs and store elements are constant. The basic premise is to keep the energy stored in the capacitors and inductors less than injected by the source. This effect is achieved by the addition of "virtual" resistors in parallel or in series with the load. Such resistances are emulated by the controller through the duty cycle signal conditioning.

Other definitions about passivity, as well as the equations necessary to control in view of this method, can be visualized in [10,13].

Appendix A.3. PBC Control of Buck-Boost Converter

According to [35,37] the average Buck-Boost converter circuit can be written by Euler-Lagrange equations, as:

$$D_B \dot{x} + (1 - d)J_B x + R_B x = dF, \tag{A9}$$

with

$$x = \begin{bmatrix} x_1 \\ x_2 \end{bmatrix}, D_B = \begin{bmatrix} L & 0 \\ 0 & C \end{bmatrix},$$

$$R_B = \begin{bmatrix} 0 & 0 \\ 0 & G \end{bmatrix}, F = \begin{bmatrix} E \\ 0 \end{bmatrix}, J_B = \begin{bmatrix} 0 & -1 \\ 1 & 0 \end{bmatrix}. \tag{A10}$$

The equivalent state space equations are:

$$\dot{x}_1 = (1 - d)\frac{1}{L}x_2 + d\frac{E}{L}, \tag{A11}$$

$$\dot{x}_2 = -(1 - d)\frac{1}{C}x_1 - \frac{G}{C}x_2, \tag{A12}$$

For PBC control, let us consider the state error in function of desired vector x_d:

$$\tilde{x} = e,$$
$$\tilde{x} = x - x_d. \tag{A13}$$

The error equation formulated as in (A11) and (A12) becomes:

$$D_B \dot{\tilde{x}} + (1 - d)J_B \tilde{x} + R_B \tilde{x} + R_{damp}\tilde{x} = \psi,$$
$$\psi = F - [D_B \dot{x}_d + (1 - d)J_B x_d + R_B x_d] + R_{damp}\tilde{x} \tag{A14}$$

In order to guarantee the error vector to converge to zero, one has to impose $\Psi = 0$, which can be written as:

$$L\dot{x}_{1d} - (1 - d)\, x_{2d} - R_{1damp}\tilde{x}_1 = dE,$$
$$C\dot{x}_{2d} + (1 - d)\, x_{1d} + Gx_{2d} = 0. \tag{A15}$$

where R_{damp} is the damping matrix defined as:

$$R_{damp} = \begin{bmatrix} R_{1damp} & 0 \\ 0 & R_{2damp} \end{bmatrix}. \tag{A16}$$

R_{damp} is the damping added to the system which shapes its energy. Some fundamental definitions regarding passivity, and the derivation of the control equations in view of this method, can be found in [10,38]. Aiming at rendering the system passive, via the condition established by (A15), one has:

$$d_3 = d_{PBC},$$

$$\dot{x}_{2d} = \frac{-(1-d)x_{1d} - Gx_{2d}}{C}, \tag{A17}$$

$$d_{PBC} = \frac{R_{1damp}(x_1 - x_{1d}) + x_{2d}}{x_{2d} - E} \tag{A18}$$

The load estimation is given by (A19):

$$\dot{G}_s = -k_g x_{2d}(x_2 - x_{2d}). \tag{A19}$$

Equation (A19) can be used for all four converters.

Appendix A.4. IDA-PBC Control

The IDA-PBC control methodology provides a clear separation between elements of the system in terms of their energy functions, enabling the controllers design with a clear physical interpretation [17]. Based on the Hamiltonian model, in which the term $H(x)$ is represented explicitly, describes how the energy flows within the system and between the subsystems interconnections, represented by the H matrix, and energy dissipation elements, represented by R_H matrix. The IDA-PBC controller design is to find the solution that leads to the stabilization of the system in closed loop, by modifying the matrix interconnection and system damping. Thus, it is necessary to solve partial differential equations from the interconnected subsystems, to enter the desired damping energy function.

Based on [15,17,36,39] the IDA-PBC control equations are obtained for the boost, buck and buck-boost converters.

Appendix A.5. IDA-PBC Control for Boost Converter

The average boost converter circuit can be written by equivalent state space equations, as:

$$\dot{x}_1 = -(1-d)\frac{1}{L}x_2 + \frac{E}{L}, \tag{A20}$$

$$\dot{x}_2 = (1-d)\frac{1}{C}x_1 - \frac{G}{C}x_2, \tag{A21}$$

The modeling and IDA-PBC control of boost converter is presented in [39]. Consecutively, PCH can be obtained by EL model:

$$x = \begin{bmatrix} x_1 \\ x_2 \end{bmatrix}, H(x) = \frac{1}{2}Lx_1^2 + \frac{1}{2}Cx_2^2,$$

$$J_H = \begin{bmatrix} 0 & \frac{-1-d}{LC} \\ \frac{1-d}{LC} & 0 \end{bmatrix}, R_H = \begin{bmatrix} 0 & 0 \\ 0 & \frac{1}{RC^2} \end{bmatrix}, g_H = \begin{bmatrix} \frac{1}{L} \\ 0 \end{bmatrix},$$

$$\dot{x} = [J_H(d) - R_H]\frac{\partial H}{\partial z}(z) + g_H E \tag{A22}$$

The equilibrium points of the boost converter system obtained when $\dot{x}_1 = 0$ and $\dot{x}_2 = 0$ on Equations (A20) and (A21) are:

$$\bar{x}_1 = \frac{EG}{(1-d)^2}, \bar{x}_2 = \frac{E}{(1-d)} \tag{A23}$$

Considering the desired output capacitor voltage as $x_{2d} = \bar{x}_2 = V_d$, the equilibrium point to stabilize $\bar{x}$ and the constant input control $\bar{d}$ given by:

$$\bar{d} = 1 - \frac{E}{V_d}, \bar{x} = [\bar{x}_1, \bar{x}_2]^T = \left[GV_d\left(\frac{V_d}{E}\right), V_d\right]^T. \tag{A24}$$

The main objective of IDA-PCB control is to find a static function through space state feedback, $d = v(x)$. In this way the closed loop dynamics becomes a Port Controlled Hamiltonian, given by:

$$\dot{x} = [J_d(x,d) - R_d] \frac{\partial H_d}{\partial x}(x) \tag{A25}$$

Given this, the IDA-PBC control equation:

$$d = 1 - (1 - \bar{d}) \left(\frac{x_2}{V_d}\right)^{\alpha}. \tag{A26}$$

Substituting (A24) in (A26) derives:

$$d = 1 - \left(\frac{E}{V_d}\right) \left(\frac{x_2}{V_d}\right)^{\alpha}. \tag{A27}$$

Appendix A.6. Boost PFC

For Boost PFC converter, considering a rectified sinusoidal input voltage, the inductor desired current, x_{1d}, must be sinusoidal and in phase with the input voltage E. So, if:

$$E = E_{max} |sin(wt + \text{\OE})|, I_d = \frac{2V_d^2 G}{E_{max}}, \tag{A28}$$

then

$$x_{1d} = I_d |sin(wt + \text{\OE})|. \tag{A29}$$

Note that for a system with input E constant, we obtain some simplifications

$$\dot{x}_{1d} = 0, I_d = \frac{G}{E} V_d^2, x_{1d} = I_d. \tag{A30}$$

By deriving from a direct control, IDA-PBC control Equation (A27) does not correct the power factor. Thus, we make the following adaptation based on PBC control law:

$$d_3 = d_{PBC},$$
$$\dot{x}_{2d} = \frac{(1-d)x_{1d} - Gx_{2d} + R_{2damp}(x_2 - x_{2d})}{C},$$
$$\bar{d}_3 = 1 - \frac{\left[E + R_{1damp}(x_1 - x_{1d}) - L\dot{x}_{1d}\right]}{x_{2d}}, \tag{A31}$$
$$d = 1 - (1 - \bar{d}_3) \left(\frac{x_2}{V_d}\right)^{\alpha}$$

Appendix A.7. Integral Action

In order to minimize errors in steady state of the output voltage at a desired value V_d, it is useful to add a proportional integrative term in the control law, given by:

$$G_{Int} = -k_{int} \int_0^t [x_2(s) - V_d]ds. \tag{A32}$$

Equation (A32) can be used for all converters and SFL, PBC and IDA-PBC control laws. For SFL control and boost converter:

$$\mu = 1 - \frac{[E + Lk_1(x_1 - x_{1d}) - L\dot{x}_{1d}]}{x_{2d}}, \tag{A33}$$
$$x_{2d} = -k_{int} \int_0^t [x_2(s) - V_d]ds$$

Appendix B. SHIL of SEPIC Converter

The state equation describing the behaviour of the CCM SEPIC converter [40], shown in Figure A1, is given by:

$$\begin{cases} L_1 \frac{dx_1}{dt} = E - (1-d) \cdot (x_4 + x_2) \\ C_o \frac{dx_2}{dt} = (1-d) \cdot (x_1 + x_3) - Gx_2 \\ L_2 \frac{dx_3}{dt} = d \cdot x_4 - (1-d).x_2 \\ C_1 \frac{dx_4}{dt} = (1-d) \cdot x_1 - d \cdot x_3 \end{cases} \tag{A34}$$

where d is the duty cycle of the semiconductor switch.

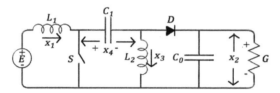

Figure A1. SEPIC converter.

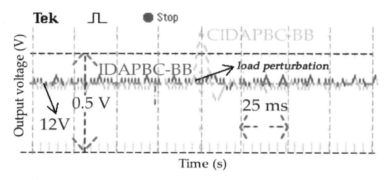

Figure A2. SHIL result for SEPIC: IDAPBC-BB (magenta) and CIDAPBC-BB (cyan). Normalized output voltage x_2 for load perturbation (30–100%).

Now, let us illustrate applications of SHIL using nonlinear equations, previously and recently found in the literature. Firstly, we consider two control laws based on IDA-PBC. In [39], Classic IDAPBC, which will refer as CIDAPBC, is applied to boost converters attaining a simplified control equation described by:
CIDAPBC-BB (CIDAPBC—Based on Boost converter):

$$\bar{d} = 1 - \frac{E}{V_d}, \quad d = 1 - (1 - \bar{d}) \left(\frac{x_2}{V_d} \right)^{k_\alpha}. \tag{A35}$$

Yet, Ref. [41] accomplish an evolution of (A35) given by:
IDAPBC-BB (IDAPBC—Based on Boost converter):

$$d = 1 - \frac{k_z E}{2Ex_2 + (k_z - 2E)x_{2d}} \tag{A36}$$

For more details, refer to [42]. In Figure A2 is sketched the output voltage x_2 in view of load change and nonlinear controllers IDAPBC-BB and CIDAPBC-BB. The following nominal conditions are: R = 10 Ω, L_1 = 146 μH, L_2 = 35 μH, C = 470 μF, E = 50 V, k_α = −0.77, k_z = −150, V_d = 12 V.

Appendix C. 3-Phase AC/DC Boost

Based on the work of [43,44], the average model(a-b-c) of the three-phase boost converter, shown in Figure A3, are given by:

$$\begin{cases} L\frac{di_a}{dt} = E_a - R_p i_a - \frac{1}{2} d_a v_o, \\ L\frac{di_b}{dt} = E_b - R_p i_b - \frac{1}{2} d_b v_o, \\ L\frac{di_c}{dt} = E_c - R_p i_c - \frac{1}{2} d_c v_o, \\ C\frac{dv_o}{dt} = \frac{1}{2}(d_a i_a + d_b i_b + d_c i_c) - G v_o. \end{cases} \tag{A37}$$

where E_a, E_b and E_c represent the line input voltages, L, C and R_p denote the inductance, the filter capacitance and the line resistance, respectively. The bipolar functions that control the semiconductor switches are d_a, d_b and d_c.

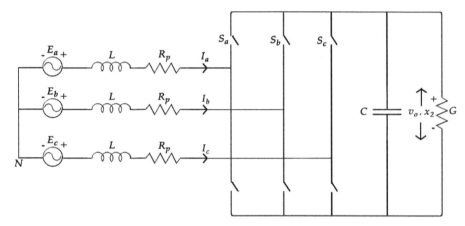

Figure A3. 3-phase AC-DC boost converter circuit.

The space-state variables of (A38) are already represented on the dq axis:

$$\begin{cases} L\dot{x}_1 = E - R_p x_1 - \frac{1}{2} x_2 \mu_d - wL x_3, \\ \frac{2}{3} C\dot{x}_2 = \frac{1}{2} x_1 \mu_d + \frac{1}{2} x_3 \mu_q - \frac{2}{3} G x_2, \\ L\dot{x}_3 = wL x_1 - -\frac{1}{2} x_2 \mu_q - R_p x_3 \end{cases} \tag{A38}$$

where w is the angular frequency of the sinusoidal voltage, x_1 is the mean current in the d-axis, x_2 is the output voltage in the capacitor, x_3 is the average current in the q-axis, μ_d and μ_q are the duty cycles and E_m is the amplitude of input voltage:

Thereby, the control goal is always to find the equations for μ_d and μ_q. With dq/abc transformations, the signals of the duty cycle $d_{a,b,c}$ are synthesized and then the corresponding PWM signal is produced for input to the converter. The converter state variables (currents on the dq axis of the inductor x_1 and x_3 and the voltage on the capacitor x_2) and the references (desired values of the output voltage V_d and the currents on the inductor $x_{1d} - I_d$ and $x_{3d} = 0$) feedback the nonlinear controller, given by:

$$\mu_d = \frac{2}{V_d}\left[-R_p x_{1d} + k_1(x_1 - x_{1d}) + E_m\right], \tag{A39}$$

$$\mu_q = \frac{2}{V_d}\left[wL x_{1d} + k_2 x_3\right], \tag{A40}$$

$$x_{1d} = \frac{1}{2}\left[\frac{E_m}{R_p} - \sqrt{\frac{E_m^2}{R_p^2} - \frac{4V_d^2}{RR_p}}\right]. \tag{A41}$$

For demonstrating the evaluation of the method applied to 3-phase boost converter, Figure A4 shows the SHIL experimental results in view of the nominal conditions: R = 40 Ω, R_p = 0.1 Ω, L= 5 mH, C = 2200 μF, E_m = 80 V, k_1 = 50, k_2 = 20, V_d = 200 V. For further information, refer to [43].

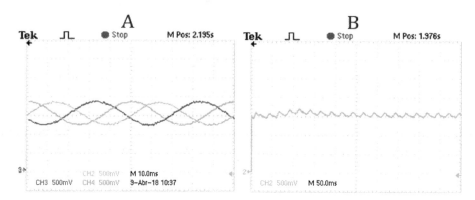

Figure A4. SHIL experimental result of 3-phase boost converter). Input voltage ((**A**): E_a—blue, E_a—magenta, E_c—green) and normalized output voltage x_2 (1.5 V DAC output voltage correspond to $V_d = 200$ V) using nonlinear control (A39-19)—(**B**).

Appendix D. Other Applications

The proposed methodology is used to evaluate another study cases: a PFC (Power Factor Correction) Boost converter [45] and the SST (Solid-State Transformer) [46]. In the last case, 32 control loops are embedded in the same DSP, which also justifies the use of the approach in more complex applications.

References

1. Shoushtari, M.; Dutt, N. SAM: Software-Assisted Memory Hierarchy for Scalable Manycore Embedded Systems. *IEEE Embed. Syst. Lett.* **2017**, *9*, 109–112. [CrossRef]
2. Noureen, S.; Shamim, N.; Roy, V.; Bayne, S. real-time Digital Simulators: A Comprehensive Study on System Overview, Application, and Importance. *Int. J. Res. Eng.* **2017**, *4*, 266–277. [CrossRef]
3. Grégoire, L.; Cousineau, M.; Seleme, S., Jr.; Ladoux, P. real-time Simulation of Interleaved Converters with Decentralized Control. In Proceedings of the International Conference on Renewable Energies and Power Quality (ICREPQ 2016), Madrid, Spain, 4–6 May 2016; pp. 1–6.
4. Li, Y.; Xu, X.; Sun, X.; Xue, H.; Jiang, H.; Qu, Y. Theoretical and experimental analytical study of powertrain system by hardware-in-the-loop test bench for electric vehicles. *Int. J. Veh. Syst. Model. Test.* **2017**, *12*, 44–71. [CrossRef]
5. Fernández-Álvarez, A.; Portela-García, M.; García-Valderas, M.; López, J.; Sanz, M. HW/SW Co-Simulation System for Enhancing Hardware-in-the-Loop of Power Converter Digital Controllers. *IEEE J. Emerg. Sel. Top. Power Electron.* **2017**, *5*, 1779–1786. [CrossRef]
6. Vardhan, H.; Akin, B.; Jin, H. A Low-Cost, High-Fidelity Processor-in-the Loop Platform: For Rapid Prototyping of Power Electronics Circuits and Motor Drives. *IEEE Power Electron. Mag.* **2016**, *3*, 18–28. [CrossRef]
7. Motahhir, S.; El Ghzizal, A.; Sebti, S.; Derouich, A. MIL and SIL and PIL tests for MPPT algorithm. *Cogent Eng.* **2017**, *4*, 1378475. [CrossRef]
8. Bélanger, J.; Venne, P.; Paquin, J. The What, Where and Why of Real-Time Simulation. Available online: https://www.opal-rt.com/wp-content/themes/enfold-opal/pdf/L00161_0436.pdf (accessed on 1 June 2018).
9. Erickson, R.W.; Maksimovic, D. *Fundamentals of Power Electronics*; Springer Science & Business Media: New York, NY, USA, 2007.
10. Sira-Ramirez, H.; Perez-Moreno, R.A.; Ortega, R.; Garcia-Esteban, M. Passivity-based controllers for the stabilization of DC-to-DC power converters. *Automatica* **1997**, *33*, 499–513. [CrossRef]
11. Sanders, S.R. Nonlinear Control of Switching Power Converters. Ph.D. Thesis, Massachusetts Institute of Technology, Cambridge, MA, USA, 1989.
12. Rosa, A.H.R. Estudo e Comparação de Técnicas de Controle não Lineares Aplicadas a Conversores Estáticos de Potência. Ph.D. Thesis, UFMG, Belo Horizonte, Brazil, 2015.

13. Seleme, S.I.; Rosa, A.H.R.; Morais, L.M.F.; Donoso-Garcia, P.F.; Cortizo, P.C. Evaluation of adaptive passivity-based controller for power factor correction using a boost converter. *IET Control Theory Appl.* **2012**, 6, 2168–2178. [CrossRef]

14. Rosa, A.; Morais, L.M.; Seleme, I. A study and comparison of nonlinear control techniques apply to second order power converters using HIL simulation. In Proceedings of the 2016 12th IEEE International Conference on Industry Applications (INDUSCON), Curitiba, Brazil, 20–23 November 2016; pp. 1–8.

15. Wang, Y.; Yu, H.; Yu, J. The modeling and control of buck-boost converter based on energy-shaping theory. In Proceedings of the IEEE International Conference on Industrial Technology (ICIT 2008), Chengdu, China, 21–24 April 2008; pp. 1–6.

16. Ortega, R.; Garcia-Canseco, E. Interconnection and damping assignment passivity-based control: A survey. *Eur. J. Control* **2004**, 10, 432–450. [CrossRef]

17. Ortega, R.; Van Der Schaft, A.; Maschke, B.; Escobar, G. Interconnection and damping assignment passivity-based control of port-controlled Hamiltonian systems. *Automatica* **2002**, 38, 585–596. [CrossRef]

18. Raviraj, V.; Sen, P.C. Comparative study of proportional-integral, sliding mode, and fuzzy logic controllers for power converters. *IEEE Trans. Ind. Appl.* **1997**, 33, 518–524. [CrossRef]

19. Fu, J.; Jin, Y.; Zhao, J. Nonlinear control of power converters: A new adaptive backstepping approach. *Asian J. Control* **2009**, 11, 653–656. [CrossRef]

20. Beccuti, A.G.; Papafotiou, G.; Morari, M. Explicit model predictive control of the boost dc-dc converter. *IFAC Proc. Vol.* **2006**, 39, 315–320. [CrossRef]

21. Almér, S.; Mariéthoz, S.; Morari, M. Piecewise affine modeling and control of a step-up DC-DC converter. In Proceedings of the American Control Conference (ACC), Baltimore, MD, USA, 30 June–2 July 2010; pp. 3299–3304.

22. Morais, L.M.F.; Santos Filho, R.M.; Cortizo, P.C.; Seleme, S.I.; Garcia, P.F.D.; Seixas, P.F. Pll-based repetitive control applied to the single-phase power factor correction using boost converter. In Proceedings of the 35th Annual Conference of IEEE Industrial Electronics (IECON'09), Porto, Portugal, 3–5 November 2009; pp. 737–742.

23. Kancherla, S.; Tripathi, R. Nonlinear average current mode control for a DC-DC buck converter in continuous and discontinuous conduction modes. In Proceedings of the 2008 IEEE Region 10 Conference (TENCON 2008), Hyderabad, India, 19–21 November 2008; pp. 1–6.

24. He, W.; Ortega, R.; Machado, J.E.; Li, S. An Adaptive Passivity-Based Controller of a Buck-Boost Converter with a Constant Power Load. *arXiv* **2017**, arXiv:1712.07792.

25. Khalil, H.K. *Noninear Systems*; Prentice-Hall: Upper Saddle River, NJ, USA, 1996.

26. El Aroudi, A.; Giaouris, D.; Iu, H.H.; Hiskens, I. A review on stability analysis methods for switching mode power converters. *IEEE J. Emerg. Sel. Top. Circuits Syst.* **2015**, 5, 302–315. [CrossRef]

27. Sira-Ramírez, H.; Ortega, R.; García-Esteban, M. Adaptive passivity-based control of average dc-to-dc power converter models. *Int. J. Adapt. Control Signal Proc.* **1998**, 12, 63–80. [CrossRef]

28. Hosseinzadeh, M.; Yazdanpanah, M.J. Robust adaptive passivity-based control of open-loop unstable affine non-linear systems subject to actuator saturation. *IET Control Theory Appl.* **2017**, 11, 2731–2742. [CrossRef]

29. Seker, M.; Zergeroglu, E. Nonlinear control of flyback type DC to DC converters: An indirect backstepping approach. In Proceedings of the 2011 IEEE International Conference on Control Applications (CCA), Denver, CO, USA, 28–30 September 2011; pp. 65–69.

30. Rosa, A.; Silva, M.; Campos, M.; Santana, R.; Cortizo, P.; Mendes, M.; Morais, L.; Seleme, I. Hil simulation of non linear control methods applied for buck-boost and flyback converters. In Proceedings of the 2017 Brazilian Power Electronics Conference (COBEP), Juiz de Fora, Brazil, 19–22 November 2017; pp. 1–6.

31. Zhang, M.; Borja, P.; Ortega, R.; Liu, Z.; Su, H. PID Passivity-Based Control of Port-Hamiltonian Systems. *IEEE Trans. Autom. Control* **2018**, 63, 1032–1044. [CrossRef]

32. Balluchi, A.; Benvenuti, L.; Engell, S.; Geyer, T.; Johansson, K.H.; Lamnabhi-Lagarrigue, F.; Lygeros, J.; Morari, M.; Papafotiou, G.; Sangiovanni-Vincentelli, A.L.; et al. Hybrid control of networked embedded systems. *Eur. J. Control* **2005**, 11, 478–508. [CrossRef]

33. Parameter Tuning and Signal Logging with Serial External Mode. Document Description. Available online: https://www.mathworks.com/help/supportpkg/texasinstrumentsc2000/examples/parameter-tuning-and-signal-logging-with-serial-external-mode.html (accessed on 30 May 2017).

34. Márquez-Contreras, R.; Rodríguez-Cortés, H.; Spinetti-Rivera, M. Revisiting IDA-PBC, Open-Loop Control, and Modeling for the Boost DC-DC Power Converter. In Proceedings of the Latin American Congress of Automatic Control, Rio de Janeiro, Brazil, 5–6 November 2008.

35. Sira-Ramirez, H.; deNieto, M.D. A Lagrangian approach to average modeling of pulsewidth-modulation controlled DC-to-DC power converters. *IEEE Trans. Circuits Syst. I Fundam. Theory Appl.* **1996**, *43*, 427. [CrossRef]

36. Stadlmayr, R.; Schlacher, K. An energy-based control strategy for DC/DC power converters. In Proceedings of the 2009 European Control Conference (ECC), Budapest, Hungary, 23–26 August 2009; pp. 3967–3972.

37. Yildiz, H.A.; Goren-Sumer, L. Lagrangian modeling of DC-DC buck-boost and flyback converters. In Proceedings of the European Conference on Circuit Theory and Design (ECCTD 2009), Antalya, Turkey, 23–27 August 2009; pp. 245–248.

38. Seleme, S.I., Jr.; Morais, L.M.F.; Rosa, A.H.R.; Torres, L.A.B. Stability in passivity-based boost converter controller for power factor correction. *Eur. J. Control* **2013**, *19*, 56–64. [CrossRef]

39. Rodriguez, H.; Ortega, R.; Escobar, G. A new family of energy-based non-linear controllers for switched power converters. In Proceedings of the IEEE International Symposium on Industrial Electronics (ISIE 2001), Pusan, Korea, 12–16 June 2001; Volume 2, pp. 723–727.

40. Ma, H.; Li, Y.; Lai, J.; Zheng, C.; Xu, J. An Improved Bridgeless SEPIC Converter without Circulating Losses and Input Voltage Sensing. *IEEE J. Emerg. Sel. Top. Power Electron.* **2018**, *6*, 1447–1455. [CrossRef]

41. Zhang, M.; Ortega, R.; Liu, Z.; Su, H. A new family of interconnection and damping assignment passivity-based controllers. *Int. J. Robust Nonlinear Control* **2017**, *27*, 50–65. [CrossRef]

42. Rosa, A.; de Souza, T.; Morais, L.; Seleme, S., Jr. Adaptive and Nonlinear Control Techniques Applied to SEPIC Converter in DC-DC, PFC, CCM and DCM Modes Using HIL Simulation. *Energies* **2018**, *11*, 602. [CrossRef]

43. Lee, T.S. Lagrangian modeling and passivity-based control of three-phase AC/DC voltage-source converters. *IEEE Trans. Ind. Electron.* **2004**, *51*, 892–902. [CrossRef]

44. Rodriguez, L.; Jones, V.; Oliva, A.R.; Escobar-Mejía, A.; Balda, J.C. A new SST topology comprising boost three-level AC/DC converters for applications in electric power distribution systems. *IEEE J. Emerg. Sel. Top. Power Electron.* **2017**, *5*, 735–746. [CrossRef]

45. Rosa, A.; Morais, L.; Seleme, S., Jr. Practical hybrid solutions based on nonlinear controllers applied to PFC boost converter. *Przeglkad Elektrotech.* **2018**, *1*, 10–16. [CrossRef]

46. Rodrigues, W.; Oliveira, T.; Morais, L.; Rosa, A. Voltage and Power Balance Strategy without Communication for a Modular Solid State Transformer Based on Adaptive Droop Control. *Energies* **2018**, *11*, 1802. [CrossRef]

Permissions

All chapters in this book were first published in MDPI; hereby published with permission under the Creative Commons Attribution License or equivalent. Every chapter published in this book has been scrutinized by our experts. Their significance has been extensively debated. The topics covered herein carry significant findings which will fuel the growth of the discipline. They may even be implemented as practical applications or may be referred to as a beginning point for another development.

The contributors of this book come from diverse backgrounds, making this book a truly international effort. This book will bring forth new frontiers with its revolutionizing research information and detailed analysis of the nascent developments around the world.

We would like to thank all the contributing authors for lending their expertise to make the book truly unique. They have played a crucial role in the development of this book. Without their invaluable contributions this book wouldn't have been possible. They have made vital efforts to compile up to date information on the varied aspects of this subject to make this book a valuable addition to the collection of many professionals and students.

This book was conceptualized with the vision of imparting up-to-date information and advanced data in this field. To ensure the same, a matchless editorial board was set up. Every individual on the board went through rigorous rounds of assessment to prove their worth. After which they invested a large part of their time researching and compiling the most relevant data for our readers.

The editorial board has been involved in producing this book since its inception. They have spent rigorous hours researching and exploring the diverse topics which have resulted in the successful publishing of this book. They have passed on their knowledge of decades through this book. To expedite this challenging task, the publisher supported the team at every step. A small team of assistant editors was also appointed to further simplify the editing procedure and attain best results for the readers.

Apart from the editorial board, the designing team has also invested a significant amount of their time in understanding the subject and creating the most relevant covers. They scrutinized every image to scout for the most suitable representation of the subject and create an appropriate cover for the book.

The publishing team has been an ardent support to the editorial, designing and production team. Their endless efforts to recruit the best for this project, has resulted in the accomplishment of this book. They are a veteran in the field of academics and their pool of knowledge is as vast as their experience in printing. Their expertise and guidance has proved useful at every step. Their uncompromising quality standards have made this book an exceptional effort. Their encouragement from time to time has been an inspiration for everyone.

The publisher and the editorial board hope that this book will prove to be a valuable piece of knowledge for researchers, students, practitioners and scholars across the globe.

List of Contributors

Jin-Wook Kang
Hanwha Defense, Seongnam 13488, Korea

Seung-Wook Hyun and Yong Kan
LG Electronics, Seoul 07336, Korea

Hoon Lee
Department of Electrical and Computer Engineering, Sungkyunkwan University, Suwon 16419, Korea

Jung-Hyo Lee
Department of Electrical Engineering, Kunsan National University, Gunsan, Jeollabuk-do 54150, Korea

Ahmad Alzahrani
Electrical Engineering Department, Faculty of Engineering, Najran University, Najran 66446, Saudi Arabia

R. Balasubramanian, K. Parkavikathirvelu, R. Sankaran and Rengarajan Amirtharajan
School of Electrical & Electronics Engineering (SEEE) S, SASTRA Deemed University, Thirumalaisamudram, Thanjavur 613401, India

Rui Zhang
College of Safety Engineering, Chongqing University of Science and Technology, Chongqing 401331, China

Wei Ma, Min Hu, Hongjun Zhou and Yihui Zhang
College of Electrical Engineering, Chongqing University of Science and Technology, Chongqing 401331, China

Lei Wang
The School of Electric Power Engineering, South China University of Technology, Guangzhou 510641, China

Longhan Cao
Department of Electrical Egineering, Chongqing Institute of Communication, Chongqing 400035, China

Tounsi Kamel and Djahbar Abdelkader
Department of Electrical Engineering, LGEER laboratory, U.H.B.B-Chlef University, Chlef 02000, Algeria

Barkat Said
Laboratoire de Génie Électrique, Faculté de Technologie, Université de M'Sila, M'Sila 28000, Algeria

Sanjeevikumar Padmanaban
Department of Energy Technology, Aalborg University, 6700 Esberg, Denmark

Atif Iqbal
Department of Electrical Engineering Qatar University, Doha, Qatar

Serkan Öztürk and Işık Çadırcı
Department of Electrical and Electronics Engineering, Hacettepe University, Beytepe, Ankara 06800, Turkey

Mehmet Canver and Muammer Ermiş
Department of Electrical and Electronics Engineering, Middle East Technical University, Ankara 06800, Turkey

Hong Cheng, Wenbo Chen, Cong Wang and Jiaqing Deng
School of Mechanical Electronic & Information Engineering, China University of Mining and Technology Beijing, Ding No. 11 Xueyuan Road, Haidian District, Beijing 100083, China

Ming Lu and Xiaodong Li
Faculty of Information Technology, Macau University of Science and Technology, Taipa, Macau 999078, China

Tianqing Yuan and Dazhi Wang
School of Information Science and Engineering, Northeastern University, Shenyang 110819, China

Muhammad Ali, Muhammad Mansoor Khan, Muhammad Talib Faiz, Yaqoob Ali, Khurram Hashmi and Houjun Tang
School of Electronics, Information & Electrical Engineering (SEIEE), Smart Grid Research & Development Centre, Shanghai Jiao Tong University, Shanghai 200240, China

Jianming Xu
Changzhou Power Supply Company, Changzhou 213176, Jiangsu, China

Sonali Chetan Rangari and Mohan Renge
Department of Electrical Engineering, Shri Ramdeobaba College of Engineering and Management, Nagpur 440013, India

Hiralal Murlidhar Suryawanshi
Department of Electrical Engineering, Visvesvaraya National Institute of Technology, Nagpur 44013, India

Arthur H. R. Rosa, Matheus B. E. Silva, Marcos F. C. Campos, Renato A. S. Santana, Welbert A. Rodrigues, Lenin M. F. Morais and Seleme I. Seleme Jr.
Graduate Program in Electrical Engineering, Universidade Federal de Minas Gerais, Av. Antônio Carlos 6627, Belo Horizonte 31270-901, MG, Brazil

Index

A

Active Power Filter, 43-45, 56, 61-62, 123, 169, 187-188

B

Bifurcation, 63-64, 69-71, 73

Boost Converter, 33, 63-64, 73, 95, 120, 122, 209, 213, 216-218, 221, 223-224, 226-230

Buck-boost, 73, 205, 209, 213, 215-219, 221, 223-224, 229-230

C

Capacitors, 1, 20, 56, 64, 72, 99, 102, 106, 110, 123-124, 141, 169-170, 175, 180-181, 223

Control Circuitry, 98-99, 101

Control Equations, 206, 209, 213, 215, 218-219, 221-224

Control Loop, 43-44, 48, 63-64

Control Strategies, 26, 61, 92, 136, 152-153, 164-167

Control Techniques, 44, 59-60, 150, 204-206, 209, 216, 218-219, 221-222, 229-230

Coordinate System, 222

Current Sensors, 56, 198

D

Decoupled Control, 75, 89

Direct Torque Control, 91, 152-153, 166-167

E

Effect Factors, 152, 162-163, 166

Electromagnetic Torque, 89, 193

Embedded Systems, 205-206, 221, 228-229

Equivalent Circuits, 100, 141-145

Error Compensations, 153, 155, 157-158, 161

Extended Kalman Filter, 75, 84, 89-90, 92

F

Fault Diagnosis, 124-125, 128-129, 132-133, 135-138

Fault-tolerant Control, 92, 137, 189-190, 193, 196, 198, 204

Five-phase Machine, 75, 189-190, 193, 200

Flux Density, 109-110

Flux Linkage, 152-158, 160-166

H

Harmonic Content, 54, 103, 115-116

Harmonic Filters, 44-45

Harmonics, 14, 25, 43-44, 48, 50, 56, 58, 60-61, 91, 95-96, 98, 114-115, 120, 123, 169-170, 172-173, 177, 180-181, 187, 190, 192

Healthy Operation, 192-193, 198, 200-201, 203

Hybrid Converter, 168, 185-186

I

Induction Motor, 92, 137, 166, 189-193, 196, 198, 200, 203-204

Inductor Current, 37, 63, 70-72, 143-144, 146, 148, 172, 212-213, 219, 221

Inductors, 56, 102, 124-126, 223

Input Side, 128

Input Voltage, 14, 63, 70-72, 127-128, 151, 212-213, 216, 219, 221, 227-228, 230

L

Line Frequency Instability, 63, 70-71

Load Side, 14, 126, 128-129

Load Torques, 75, 84, 86, 89-90

M

Modulation Index, 1, 101, 170, 172-173, 177-178, 184-185, 192

Multi-level Converters, 98, 169

Multi-machine Drives, 76, 84

N

Nanocrystalline Core, 94, 96, 99, 109-110

Neutral Point, 1-3, 5, 25-26, 95, 124-125, 166, 169, 187, 190-191

Non-linear Phenomena, 63, 69-70

Nonlinear Control, 55, 59-61, 76, 92, 205-206, 213, 221-222, 228-230

Nonlinear Load, 49-51, 53, 55, 57-59

O

One-cycle-controlled, 63-64

Operation Mode, 145, 148, 172, 176, 203

Oscillations, 103, 189-190

Output Voltage, 1, 3, 6, 8, 27, 37, 70-72, 76, 92-93, 124, 126, 128, 139-140, 169-170, 172-178, 180-181, 184-186, 210, 212-213, 216-219, 226-228

P

Parameter Variations, 60, 76

Permanent Magnet, 14, 75-76, 91-92, 152, 155-156, 166-167, 204

Power Compensator, 44-45, 49, 55, 58

Power Converters, 26, 73-74, 95, 100, 170, 178, 205-206, 209, 218, 228-230

Power Factor Correction, 63, 73, 168, 206, 221, 228-230

Power Quality, 43, 56, 60-62, 98, 136, 169, 228

Power Range, 139-140, 150

Power Ratings, 98, 118, 168, 181, 186

Prototype System, 44, 94, 96

R

Real-time Simulation, 205-206, 212, 214, 218, 228

Reference Frame, 43-45, 60-62, 94-96, 103-104, 155, 192-193, 196

S

Semiconductor Devices, 74, 121, 140, 146, 168, 170-172, 184, 186, 210

Series Active Filter, 168, 170, 177-178, 180-181

Sliding Mode Control, 75-76, 89-92

State Operation, 100, 117, 203, 218

Stator Currents, 75, 90, 161, 193

Stator Winding, 91, 190, 192-193, 196, 203-204

Switching Behaviour, 139-140, 145, 148

Synchronous Motor, 14, 91-92, 152, 155, 166-167, 204

T

Terminal Voltage, 100

Total Harmonic Distortion, 2, 43, 114, 169, 186

U

Utility Grid, 96, 98, 116, 120

V

Variation Ranges, 94, 96, 101, 104

Voltage Control, 14, 43, 49, 63-64, 169

Voltage Ripple, 71, 140, 180

Voltage Source, 1, 25, 50, 53, 58-59, 61, 75-76, 90, 92-96, 98, 112, 114, 122-123, 150, 153, 169-170, 189, 191-192, 203-204

W

Waveforms, 20, 23-24, 50, 52, 55, 58, 60, 70-72, 104, 108, 111-117, 125, 127-128, 141, 143, 148-150, 164, 200, 218